Neutrino Physics

AF768342

Neutrino Physics

Proceedings of an International Workshop
Held in Heidelberg, October 20-22, 1987

Editors: H.V. Klapdor and B. Povh

With 167 Figures

Springer-Verlag Berlin Heidelberg New York
London Paris Tokyo

Professor Dr. Hans Volker Klapdor
Professor Dr. Bogdan Povh
Max-Planck-Institut für Kernphysik, Postfach 103980,
D-6900 Heidelberg, Fed. Rep. of Germany

ISBN-13: 978-3-642-73681-0 e-ISBN-13: 978-3-642-73679-7
DOI: 10.1007/ 978-3-642-73679-7

This work is subject to copyright. All rights are reserved, whether the whole or part of the material is concerned, specifically the rights of translation, reprinting, reuse of illustrations, recitation, broadcasting, reproduction on microfilms or in other ways, and storage in data banks. Duplication of this publication or parts thereof is only permitted under the provisions of the German Copyright Law of September 9, 1965, in its version of June 24, 1985, and a copyright fee must always be paid. Violations fall under the prosecution act of the German Copyright Law.

© Springer-Verlag Berlin Heidelberg 1988
Softcover reprint of the hardcover 1st edition 1988

The use of registered names, trademarks, etc. in the publication does not imply, even in the absence of a specific statement, that such names are exempt from the relevant protective laws and regulations and therefore free for general use.

2156/3150-543210

Preface

Neutrinos play an intriguing role in modern physics, linking central questions of particle physics, cosmology and astrophysics.

Since the time of Fermi and Pauli our knowledge of the weak interaction has been intimately connected with the understanding of the nature of the neutrino. Today they play a corresponding role within the framework of theories of grand unification (GUTs, SUSYs, superstrings). A nonvanishing neutrino mass would yield an important boundary condition for models of grand unification. Neutrinos with a Majorana mass (which would be a unique property among the elementary fermions) would correspond to the breaking of baryon number-lepton number (B-L) conservation. Since neutrinos are the best candidates for dark matter in the universe, their mass could determine its large scale structure and evolution. They are also the only direct probes of processes in the cores of collapsing stars, and their properties might yield a solution of the solar neutrino puzzle.

On the other hand, today, 30 years after their direct observation, it is still not known whether neutrinos have a mass or not. The discovery of a nonzero neutrino mass would be a simple, yet extremely profound, example of "new physics". This justifies the many efforts to detect it, ranging from neutrino oscillation experiments using reactors, accelerators or the sun as neutrino sources, to tritium decay experiments and the search for neutrinoless double beta decay. While most of these efforts are non-accelerator experiments, they are on the other hand beyond-accelerator experiments since the mass scales of GUTs probed by a neutrino mass in the eV range are far beyond the direct reach of present accelerators.

This workshop was the tenth in a series of meetings on problems of particles and nuclei, started by one of us three years ago. We thank the (about 150) participants for creating such a highly stimulating atmosphere during the workshop, reflecting the enthusiasm with which the boundaries of this field are being pushed forward.

We would also like to thank our young colleagues Evi Bender, A. Piepke, J. Schneider, C. Stamopoulos and H. Strecker, who have contributed so efficiently to the organization and to the success of the workshop. The concern and excellent cooperation of Professor W. Beiglböck and Mrs. C. Pendl are gratefully acknowledged.

Heidelberg, March 1988 *H.V. Klapdor*
B. Povh

Contents

Part I

Neutrinos in Gauge Theories and Cosmology

Massive Neutrinos

*P. Langacker**

Deutsches Elektronen Synchrotron, DESY, D-2000 Hamburg,
Fed. Rep. of Germany
*Permanent address: Department of Physics, University of Pennsylvania, Philadelphia, PA 19104, USA

Most extensions of the standard $SU_2 \times U_1$ electroweak model predict nonzero neutrino masses. In this talk I summarize the theoretical expectations for neutrino mass in a variety of possible extensions and discuss their implications for various types of experiments.

1 Introduction

The question of whether the neutrino has a nonzero mass is one of the most important issues in both particle physics and astrophysics. In the minimal $SU_2 \times U_1$ model the neutrinos are predicted to be massless. However, extensions of the standard model involving new SU_2-singlet neutral fermions (the right-handed neutrino partners needed for Dirac mass terms) or new Higgs representations (to generate Majorana masses) allow non-zero masses. [1] In fact, most extensions of the standard model (e.g. most grand unified theories other than SU_5) involve one or both of these mechanisms.

Neutrino mass is also of great importance for astrophysics and cosmology. Masses in the 10 eV range could account for the dark matter of the universe, while masses $\leq 10^{-2} eV$ could resolve the Solar neutrino problem.

In this talk I will describe the complicated subject neutrino mass: the principle theoretical models and their implications will be described, and the experimental situation will be briefly summarized.

2 Weyl, Majorana, and Dirac Neutrinos.

For the weak interactions it is convenient to deal with Weyl two-component spinors ψ_L or ψ_R, each of which represents two physical degrees of freedom. The field ψ_L can annihilate a left-handed (L) particle or create a right-handed (R) antiparticle, while $\psi_L^\dagger$ annihilates a L-particle or creates an R-antiparticle. For a ψ_R field the roles of L and R are reversed. An ordinary four-component Dirac field ψ can be written as the sum $\psi = \psi_L + \psi_R$ of two Weyl fields, where ψ_L and ψ_R are just the chiral projections $\psi_{L,R} = P_{L,R}\psi$, with $P_{L,R} = (1 \pm \gamma_5)/2$.

Alternatively, one can consider Weyl fermions that do not have distinct partners of the opposite chirality. We will see below that such spinors correspond to particles that are either massless or carry no conserved quantum numbers.

In the free field limit a Weyl field ψ_L can be written as

$$\psi_L(x) = \sum_{\vec{p}} \left[b_L(\vec{p}) u_L(\vec{p}) e^{-ip\cdot x} + d_R^\dagger(\vec{p}) v_R(\vec{p}) e^{+ip\cdot x} \right], \tag{1}$$

where $\sum_{\vec{p}}$ represents $\int d^3\vec{p}/\sqrt{(2\pi)^3 2E}$. In(1), b_L and d_R are annihilation operators for L particles and R-antiparticles, respectively, and u_L and v_R are the corresponding (4-component) spinors satisfying $P_L u_L = u_L$, $P_L v_R = v_R$, $P_R u_L = P_R v_R = 0$. For a ψ_R spinor one simply interchanges L and R. Equation (1) differs from an ordinary (Dirac) free field in that there is no sum over spin.

It is apparent from (1) that each left-handed (right-handed) particle is necessarily associated with a right-handed (left-handed) antiparticle. The right-handed antiparticle [2] field ψ_R^c is not independent of ψ_L, but is closely related to $\psi_L^\dagger$. One has $\psi_R^c = C\bar{\psi}_L^T$, where C is the charge conjugation matrix, defined by $C\gamma_\mu C^{-1} = -\gamma_\mu^T$. Similarly, for a R-Weyl spinor, $\psi_L^c = C\bar{\psi}_R^T$. In the special case that ψ_L is the chiral projection $P_L\psi$ of a Dirac field ψ, ψ_R^c is just the R-projection $P_R\psi^c$ of the antiparticle field $\psi^c = C\bar{\psi}^T$.

If ψ_R and ψ_R^c both exist, they have the opposite values for all additive quantum numbers. Since the quarks and charged leptons carry conserved quantum numbers (e.g. color and electric charge), they must be Dirac fields - i.e. ψ_R and ψ_R^c must be distinct. The only quantum number associated with the neutrinos is lepton number, however, and it is conceivable that that is violated in nature. As we will see, that will allow for two very different possibilities for neutrino mass.

The known neutrinos of the first family are the left-handed electron neutrino ν_{eL} and its CP partner, the right-handed "antineutrino" $\nu_{eR}^c = C\bar{\nu}_{eL}^T$. These are associated with the e_L^- and e_R^+, respectively, in ordinary charged current weak interactions.

Mass terms always take left- and right-handed fields into each other. If one introduces a new field N_R (distinct from ν_R^c) and its CP conjugate $N_L^c = C\bar{N}_R^T$ into the theory, then one can write a Dirac (lepton number conserving) mass term

$$-L_{Dirac} = m_D\bar{\nu}_L N_R + h.c., \tag{2}$$

which connects N_R and ν_L. In this case ν_L, N_R, N_L^c and ν_R^c form a four component Dirac particle - i.e. one can define $\nu \equiv \nu_L + N_R$, $\nu^c = N_L^c + \nu_R^c = C\bar{\nu}^T$, so that $-L_{Dirac} = m_D\bar{\nu}\nu$. Clearly lepton number is conserved in this case, because there is no transition between ν and ν^c. In the free field limit the Dirac neutrino field ν has the canonical expression

$$\nu_{Dirac}(x) = \sum_{\vec{p}} \sum_{S=L,R} \left[b_S(\vec{p})u_S(\vec{p})e^{-ip\cdot x} + d_S^\dagger(\vec{p})v_S(\vec{p})e^{+ip\cdot x}\right], \tag{3}$$

Usually, the N_R is an $SU_2 \times U_1$ singlet, with m_D generated by an ordinary Higgs doublet, and $L = L_e + L_\mu + L_\tau$ is conserved in the three family generalization. This possibility is most similar to the way in which masses are generated for the other fermions (e^-, u, d, etc.) in the standard model, but it is difficult to understand why m_{ν_e} is so small in this case.

Another possibility [3] is that N_R is a known doublet neutrino, such as $\nu_{\tau R}^c$. This is a variation on the Konopinski-Mahmoud model. [4] Then ν_{eL}, $\nu_{\tau R}^c$, $\nu_{\tau L}$ and ν_{eR}^c can be combined to form a Dirac neutrino with $L_e - L_\tau$ conserved.

For the generalization of (2) to F fermion families one has

$$-L_{Dirac} = \bar{n}_L^0 m_D N_R^0 + h.c., \tag{4}$$

where m_D is an arbitrary [5] $F \times F$ mass matrix, and n_L^0 and N_R^0 are F-component vectors; thus $n_L^0 = (n_{1L}^0\ n_{2L}^0 \ldots n_{FL}^0)^T$, where n_{iL}^0 are the "weak eigenstate" neutrinos - i.e. n_{iL}^0 is associated with e_{iL}^- in weak transitions. The weak eigenstate neutrinos are related to the neutrinos n_{iL}, N_{iR} of definite mass by unitary transformations $n_L^0 = V_L n_L$, $N_R^0 = V_R N_R$, where V_L and V_R are $F \times F$ unitary matrices, determined by

$$V_L^\dagger m_D V_R = m_d = \mathrm{diag}(m_1\ m_2 \cdots m_F) \tag{5}$$

and m_d is the diagonal matrix of physical neutrino masses. V_L and V_R can be determined by

$$V_L^\dagger m_D m_D^\dagger V_L = V_R^\dagger m_D^\dagger m_D V_R = m_d^2 \tag{6}$$

($m_D m_D^\dagger$ and $m_D^\dagger m_D$ are Hermitian). In general V_L and V_R are unrelated. If there are no degeneracies then V_L and V_R are determined uniquely by (6) up to diagonal phase matrices; i.e. if $V_{L,R}$ satisfy (6) then so do $V_{L,R}K_{L,R}$, where $K_{L,R}$ are diagonal phase matrices associated with the unobservable phases of the n_{iL} and N_{jR} fields. Usually one chooses K_L to put V_L into a simple conventional form. Then K_R is determined by the requirement that m_d be real.

In the presence of neutrino mass the leptonic weak charged current becomes

$$J_W^{\mu\dagger} = (\bar{\nu}_e \; \bar{\nu}_\mu \; \bar{\nu}_\tau) \; V_L^\dagger \gamma^\mu (1+\gamma^5) \begin{pmatrix} e^- \\ \mu^- \\ \tau^- \end{pmatrix} \tag{7}$$

so that $V_L^\dagger$ is just the analogue of the CKM quark mixing matrix. It describes the relative strengths [7] of the weak transition between the various charged leptons and neutrinos of definite mass.

In a Majorana (lepton number violating) mass term one avoids the need for a new fermion field by coupling the ν_L to its CP conjugate ν_R^c:

$$-L_M = \tfrac{1}{2} m \bar{\nu}_L \nu_R^c + h.c. = \tfrac{1}{2} m \bar{\nu}_L C \bar{\nu}_L^T + h.c. \tag{8}$$

L_M can be thought of as creating or annihilating two neutrinos, and violates lepton number by $\Delta L = \pm 2$. ν_L and ν_R^c can be combined to form a two component Majorana neutrino $\nu = \nu_L + \nu_R^c$, so that $-L_M = \frac{1}{2} m \bar{\nu} \nu$. One has $\nu = C\bar{\nu}^T$, i.e. a Majorana neutrino is its own antiparticle. In the free field limit ν is just

$$\nu(x) = \sum_{\vec{p}} \sum_{S=L,R} \left[b_S(\vec{p}) u_S(\vec{p}) e^{-ip\cdot x} + b_S^\dagger(\vec{p}) v_S(\vec{p}) e^{+ip\cdot x} \right], \tag{9}$$

i.e. it has the same form as for a free Dirac field (cf (3)) except that there is no distinction between b and d annihilation operators.

The Majorana mass m in (8) can be generated by the vacuum expectation value (VEV) of a new Higgs triplet [8] or as a higher order effective operator. Majorana masses are popular amongst theorists because they are so different from quark and lepton masses, and there is therefore the possibility of explaining why m_{ν_e} is so small (if it is non-zero).

For F fermion families, the Majorana mass term is

$$-L_M = \tfrac{1}{2} \bar{n}_L^0 M n_R^{0c} + h.c. \tag{10}$$

where M is an $F \times F$ Majorana mass matrix and n_L^0 and n_R^{0c} are F component vectors: i.e. $n_L^0 = (n_{1L}^0 \ldots n_{FL}^0)^T$, $n_R^{0c} = (n_{1R}^{0c} \ldots n_{FR}^{0c})^T$, where n_{iL}^0 and n_{iR}^{0c} are weak eigenstate neutrinos and "antineutrinos", related by

$$n_{iR}^{0c} = C \bar{n}_{iL}^{0T} \tag{11}$$

From (11) one can prove the identity $\bar{n}_{iL}^0 n_{jR}^{0c} = \bar{n}_{jL}^0 n_{iR}^{0c}$, from which it follows that the Majorana mass matrix M must be symmetric: $M = M^T$. Proceeding in analogy to the Dirac case, one can relate the n_{iL}^0 and n_{jR}^{0c} to mass eigenstate neutrino fields by $n_L^0 = U_L n_L$, $n_R^{0c} = U_R n_R^c$, where U_L and U_R are $F \times F$ unitary matrices chosen so that

$$U_L^\dagger M U_R = M_d = \mathrm{diag}(m_1 \; m_2 \; \ldots m_F), \tag{12}$$

where M_d is a diagonal matrix of Majorana mass eigenvalues. Unlike the Dirac case (for which m_D was an arbitrary matrix and V_L and V_R unrelated), the symmetry of M implies a relation between U_L and U_R, viz

$$U_L = U_R^* K^\dagger, \tag{13}$$

where K is unitary and symmetric. That is, just as in the Dirac case, U_L is determined from $U_L^\dagger M M^\dagger U_L = M_d^2$ to be of the form $U_L = \hat{U}_L K_L$, where K_L is a matrix of phases that can be chosen for convenience. U_R is then determined from (13), where K is chosen so that M_d is real and positive. If there are no degeneracies then K is just a matrix of phases. [9] One can always pick K_L such that $K = I$, but it is not always convenient to do so.

In terms of the mass eigenstates, (10) reduces to

$$-L_M = \tfrac{1}{2} \sum_{i=1}^{F} m_i \bar{n}_{iL} n_{iR}^c + h.c. = \tfrac{1}{2} \sum_{i=1}^{F} m_i \bar{n}_i n_i, \tag{14}$$

where $n_i = n_{iL} + n_{iR}^c$ is the ith Majorana mass eigenstate. [10] Written in terms of the n_{iL}, the weak charged current assumes a form analogous to (7), with $U_L^\dagger$ replacing $V_L^\dagger$ to describe the leptonic mixing. [11]

There are several physical distinctions between Dirac and Majorana neutrinos. If the ν_e is Majorana, for example, one could have the sequence $\pi^+ \to e^+\nu_e$ followed by $\nu_e p \to e^+ n$. The combined process violates lepton number by two units and is allowed for Majorana but not Dirac neutrinos. Similarly, a hypothetical heavy neutrino N would undergo the decays $N \to e^+ q_1 \bar{q}_2$ and $N \to e^- \bar{q}_1 q_2$ with equal rates if it is Majorana, while for a Dirac particle one would have $N \to e^- \bar{q}_1 q_2$, $N^c \to e^+ q_1 \bar{q}_2$ only [12]. There are differences due to Fermi statistics in the production of $\nu\nu$ (Majorana) or $\nu\nu^c$ (Dirac) pairs near threshold [13], and finally Majorana neutrinos cannot have electromagnetic form factors, such as magnetic moments [14].

It is important to keep in mind, however, that these distinctions must all disappear in the limit that the neutrino mass can be neglected. For $m_\nu \to 0$ the ν_R component of a Dirac neutrino decouples, and both Majorana and Dirac neutrinos reduce to Weyl two-component neutrinos - there is no difference between them. [15] In particular, lepton number conservation is reestablished smoothly as $m_\nu \to 0$ for a Majorana neutrino, because in that limit helicity - which is conserved up to corrections of order m_ν/E_ν - plays the role of an approximate lepton number. For example, the ν_e produced in $\pi^+ \to e^+\nu_e$ has $h_\nu = -1$ up to corrections of order $(m_\nu/E_\nu)^2$ (in rate), while the reaction $\nu_e p \to e^+ n$ has a cross section that is suppressed by $(m_\nu/E_\nu)^2$ for the wrong (negative) helicity.

In many models Dirac and Majorana mass terms are both present. For one doublet neutrino ν_L^0 (with $\nu_R^{0c} = C\bar{\nu}_L^{0T}$) and one new singlet N_R^0 (with $N_L^{0c} = C\bar{N}_R^{0T}$), for example, one could have the general mass term

$$-L = \tfrac{1}{2}\begin{pmatrix}\bar{\nu}_L^0 & \bar{N}_L^{0c}\end{pmatrix}\begin{pmatrix} m_t & m_D \\ m_D^T & m_S\end{pmatrix}\begin{pmatrix}\nu_R^{0c} \\ N_R^0\end{pmatrix} + h.c., \tag{15}$$

where $m_D = m_D^T$ is a Dirac mass generated by a Higgs doublet (analogous to (2)), m_t is a Majorana mass for ν_L^0 generated by a Higgs triplet or effective interaction (cf. (8)), and m_S is a Majorana mass for N_R^0, generated by a Higgs singlet or bare mass. Similarly, for F families (15) still holds provided one interprets ν_L^0 , N_L^{0c} , ν_R^{0c} , and N_R^0 as F component vectors, and m_t, m_D, and m_S as $F \times F$ matrices (with $m_t = m_t^T$, $m_S = m_S^T$). Then, (15) becomes simply

$$-L = \tfrac{1}{2}\bar{n}_L^0 M n_R^{oc} + h.c., \tag{16}$$

where $n_L^0 \equiv (\nu_L^0,\ N_L^{0c})^T$ and $n_R^{0c} \equiv (\nu_R^{0c},\ N_R^0)^T$ are $2F$ component vectors and M is the symmetric $2F \times 2F$ Majorana mass matrix in (15). Equation (16) can be diagonalized in exact analogy with (10-14), yielding finally

$$-L = \tfrac{1}{2}\sum_{i=1}^{2F} m_i \bar{n}_{iL} n_{iR}^c + h.c. \tag{17}$$

i.e. there are in general $2F$ Majorana neutrinos, related to n_L^0, n_R^{0c} by unitary transformations. Unlike the pure Majorana case, however, there is now mixing between particles with different weak interaction properties (e.g. $n_{iL} = (U_L^\dagger)_{ij} n_{jL}^0$ is a mixture of SU_2 doublets and singlets), which can have important consequences for neutrino oscillations [16] and decays.

It is instructive to see how the Dirac case ($m_t = m_S = 0$) emerges as a limiting case of (15). For a single family one has $M = m_D\begin{pmatrix}0 & 1\\ 1 & 0\end{pmatrix}$. Since M is Hermitian (for m_D real) one can diagonalize it by a unitary transformation U_L. One finds

$$U_L^\dagger M U_L = m_D\begin{pmatrix}1 & 0\\ 0 & -1\end{pmatrix}, \tag{18}$$

with $U_L = \frac{1}{\sqrt{2}}\text{diag}(1\ \ -1)$; i.e. the mass eigenstates are

$$
\begin{aligned}
n_{1L} &= \frac{1}{\sqrt{2}}(\nu_L^0 + N_L^{0c}) \\
n_{2L} &= \frac{1}{\sqrt{2}}(\nu_L^0 - N_L^{0c}) \\
n'_{1R} &= \frac{1}{\sqrt{2}}(\nu_R^{0c} + N_R) \\
n'_{2R} &= \frac{1}{\sqrt{2}}(\nu_R^{0c} - N_R).
\end{aligned} \tag{19}
$$

The negative mass eigenvalue in (18) can be removed by redefining [17] the right-handed fields $n_{1R} = n'_{1R}$, $n_{2R} = -n'_{2R}$. This is nothing more than taking $U_L^\dagger M U_R = m_d = m_D$ diag(1 1), where U_R is given by (13) with K = diag(1 − 1). Finally, the two Majorana states $n_1 = n_{1L} + n_{1R}^c$ and $n_2 = n_{2L} + n_{2R}^c$ are degenerate. We can therefore reexpress L in the new basis

$$
\begin{aligned}
\nu &\equiv \frac{1}{\sqrt{2}}(n_1 + n_2) = \nu_L^0 + N_R^0 \\
\nu^c &\equiv \frac{1}{\sqrt{2}}(n_1 - n_2) = N_L^{0c} + \nu_R^{0c},
\end{aligned} \tag{20}
$$

yielding

$$
\begin{aligned}
-L &= \tfrac{1}{2} m_D(\bar{n}_{1L} n_{1R}^c + \bar{n}_{2L} n_{2R}^c) + h.c. \\
&= m_D \bar{\nu}_L^0 N_R^0 + h.c. \;=\; m_D \bar{\nu}\nu.
\end{aligned} \tag{21}
$$

This is just a standard Dirac mass term, with a conserved lepton number (i.e. no transition between ν and ν^c). A Dirac neutrino is therefore nothing but a pair of degenerate two-component Majorana neutrinos (n_1 and n_2), combined to form a 4-component neutrino with a conserved lepton number.

One sometimes refers to a pseudo-Dirac neutrino, which is just a Dirac neutrino to which is added as small lepton number-violating perturbation. For example, one could modify the Dirac mass to

$$
M = \begin{pmatrix} \epsilon & m_D \\ m_D & 0 \end{pmatrix}, \tag{22}
$$

with $\epsilon << m_D$. One then finds two Majorana mass eigenstates $n_\pm$, with

$$
\begin{aligned}
n_{+L} &= n_{1L} + \frac{\epsilon}{4} n_{2L} \\
n_{-L} &= -\frac{\epsilon}{4} n_{1L} + n_{2L},
\end{aligned} \tag{23}
$$

(n_{1L} and N_{2L} are defined in (19)), with masses $m_D \pm \frac{\epsilon}{2}$.

Other important special cases of (15) are considered below.

3 Models of Neutrino Mass

There are many models for neutrino mass [1], all of which have good and bad features. The major classes of models are listed in Table 1, along with the most natural scales for the neutrino masses and for $\langle m_{\nu_e} \rangle$, an effective mass relevant to neutrinoless double β decay.

Dirac neutrinos are exactly like other fermions. They involve a conserved total lepton number (though the individual L_e, L_μ, and L_τ lepton numbers are violated by mixing in general) and therefore do not lead to neutrinoless double beta decay. The problem with Dirac neutrinos is that it is hard to understand why the neutrinos are so much lighter than the other fermions. In the standard model Dirac mass are generated by the vacuum expectation value (VEV) $v = \sqrt{2}\langle \varphi^0 \rangle \simeq$

Table 1: Models of neutrino mass, along with their most natural scales for the light neutrino masses.

Model	m_{ν_e}	$\langle m_{\nu_e}\rangle$	m_{ν_μ}	m_{ν_τ}
Dirac	$1-10\ MeV$	0	$100\ MeV - 1\ GeV$	$1-100\ GeV$
pure Majorana [8] (Higgs triplet)	arbitrary	m_{ν_e}	arbitrary	arbitrary
GUT seesaw [18,19] ($M \sim 10^{14}\ GeV$)	$10^{-11}\ eV$	m_{ν_e}	$10^{-6}\ eV$	$10^{-3}\ eV$
intermediate seesaw [20] ($M \sim 10^{9}\ GeV$)	$10^{-7}\ eV$	m_{ν_e}	$10^{-2}\ eV$	$10\ eV$
$SU_{2L}\times SU_{2R}\times U_1$ seesaw [21] ($M \sim 1\ TeV$)	$10^{-1}\ eV$	m_{ν_e}	$10\ KeV$	$1\ MeV$
light seesaw [22] ($M \ll 1\ GeV$)	$1-10\ MeV$	$\ll m_{\nu_e}$	-	-
charged Higgs [23]	$< 1\ eV$	$\ll m_{\nu_e}$	-	-

$(\sqrt{2}G_F)^{-1/2} \simeq 246\ GeV$ of the neutral component of a doublet of Higgs scalar fields. One has $m_D = h_\nu v$, where h_ν is the Yukawa coupling

$$L = -\sqrt{2}h_\nu(\bar{\nu}_L\ \bar{e}_L)\begin{pmatrix}\varphi^0\\ \varphi^-\end{pmatrix}N_R + h.c. \tag{24}$$

of the neutrino to φ^0.

A ν_e mass in the 20 eV range would require an anomalously small Yukawa coupling $h_{\nu_e} \leq 10^{-10}$. Moreover, h_{ν_e} would have to be smaller by $m_{\nu_e}/m_e \leq 10^{-4}$ than the analogous Yukawa coupling for the electron. Of course, we do not understand the masses of the other fermions either (or why they range over at least five orders of magnitude), so it is hard to totally exclude the possibility that h_{ν_e} is simply small. Nevertheless, the possibility seems sufficiently ugly that it is hard to take seriously unless some mechanism (other than fine-tuning) for the smallness is proposed.

One possibility is that h_ν is actually zero to lowest order (tree level) due to some new symmetry, and that h_ν is only generated as a higher order correction (i.e. so that m_ν/m_e is some power of α.) This is a very attractive possibility, but no particularly compelling models to implement it have emerged. The idea has recently been resurrected in some superstring inspired models [24], which have difficulty incorporating the seesaw type ideas described below.

Majorana mass terms for the ordinary SU_2-doublet neutrinos involve a transition from ν_R^c ($t_3 = -\frac{1}{2}$) into ν_L ($t_3 = +\frac{1}{2}$), and therefore must be generated by an operator transforming as a triplet under weak SU_2.

The simplest possibility is the Gelmini-Roncadelli model [8], in which one introduces a triplet of Higgs fields $\vec{\varphi}_t = (\varphi_t^0,\ \varphi_t^-,\ \varphi_t^{--})$ into the theory. The Yukawa coupling

$$\begin{aligned} L &= \tfrac{1}{2}h_t(\bar{\nu}_L\ \bar{e}_L)\vec{\tau}\cdot\vec{\varphi}_t\begin{pmatrix}e_R^c\\ -\nu_R^c\end{pmatrix}\\ &= \tfrac{1}{2}h_t(\bar{\nu}_L\ \bar{e}_L)\begin{pmatrix}\varphi_t^- & \sqrt{2}\varphi_t^0\\ \sqrt{2}\varphi_t^{--} & -\varphi_t^-\end{pmatrix}\begin{pmatrix}e_R^c\\ -\nu_R^c\end{pmatrix} \end{aligned} \tag{25}$$

then generates a Majorana mass $m_t = h_t v_t$ for the ν, where $v_t = \sqrt{2}\langle\varphi_t^0\rangle$ is the VEV of the Higgs triplet. Since both h_t and v_t are unknown the neutrino mass is unrelated to the other fermions and can in principle be arbitrarily small, at least at tree level.

However, small m_{ν_e} is not explained in such models - it is merely parametrized and in fact is almost as problematic as a Dirac mass. The weak neutral current (and W and Z masses) require [25] $v_t \leq 0.08v \sim 20\ GeV$. For v_t close to this limit one requires $h_t \leq 10^{-9}$, i.e. almost as bad a fine-tuning as the Dirac case. For $v_t \ll v$ one can tolerate more reasonable values for h_t, but then it is difficult to understand the large hierarchy in vacuum expectation values. One generally expects all non-zero VEV's to be comparable in magnitude unless fine-tunings are performed on the parameters in the Higgs potential. Even if one does this, higher order corrections are likely to upset the hierarchy. [26]

The VEV $\langle\varphi_t^0\rangle \neq 0$ necessarily violates lepton number conservation by two units (the Yukawa coupling in (25) does not by itself violate L because φ_t can be regarded as carrying two units of L). If the rest of the Lagrangian conserves L then lepton-number is spontaneously broken, and there will be an associated massless Goldstone boson, the triplet-Majoron. (This is the version of the model that is usually considered [8].) In this case limits based on stellar energy loss (carried off by Majorons) require [27] $v_t \leq 2 - 10\ KeV$. Implications of the Majoron for neutrino decay and annihilation, cosmology, and neutrinoless double beta decay will be mentioned below.

It is also possible introduce other couplings into the Higgs triplet model which explicitly break lepton number conservation, such as a cubic interaction between $\vec{\varphi_t}$ and two Higgs doublets. (This violates L since $\vec{\varphi_t}$ was assigned $L = 2$ to make (25) invariant). In that case all of the new scalar particles associated with φ_t become massive - i.e. there is no Majoron.

Another mechanism for introducing a Majorana mass is to consider the induced interaction (Fig. 1).

$$L_{eff} = \frac{1}{2}\frac{C}{M}(\bar{\nu}_L\ \bar{e}_L)\vec{\tau}\begin{pmatrix} e_R^c \\ -\nu_R^c \end{pmatrix}\cdot(\varphi^- \ -\varphi^0)\vec{\tau}\begin{pmatrix} \varphi^0 \\ \varphi^- \end{pmatrix} \tag{26}$$

between two leptons and two Higgs doublets. The Higgs fields in (26) are arranged to transform as an SU_2 triplet, so L_{eff} is $SU_2 \times U_1$ invariant; however, L_{eff} is non-renormalizable, as is evidenced by the dimensional coupling C/M, where M is a mass. L_{eff} cannot therefore be an elementary coupling, but it could be an effective four-particle interaction induced [28] by new physics at some large mass scale M (just as the four-fermion weak interaction is a nonrenormalizable effective interaction that is really generated by W and Z exchange). When φ^0 is replaced by its vacuum

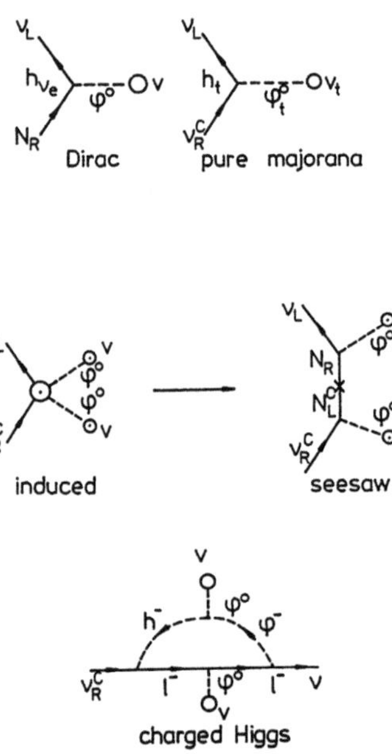

Figure 1: Dirac, pure Majorana, induced, and charged Higgs generated neutrino masses.

expectation value, (26) yields an effective Majorana mass $m \sim Cv^2/M$, which is naturally small for $M \gg v$. For example, if (26) were somehow induced by quantum gravity one would expect $M \sim 10^{19}$ GeV (the Planck scale). Then for $C \sim 1$ one would have $m_\nu \sim 10^{-5}$ eV.

The most popular realisation of this idea is the seesaw model, [18] in which the underlying physics is the exchange of a very heavy SU_2-singlet Majorana neutrino N_R^0, as indicated in Fig. 1. The seesaw model for one family is a special case of the general mass matrix in (15), in which m_D is a typical Dirac mass (typically assumed to be comparable to m_u or m_e for the first family) connecting ν_L^0 to a new SU_2-singlet N_R^0 and $m_S \gg m_D$ is a Majorana mass for N_R^0, presumably comparable to some new (large) physics scale. One typically assumes that $m_t = 0$ in the seesaw model, i.e. that there is not a Higgs triplet as well. [29] In that case, (15) yields two Majorana mass eigenstates n_1 and n_2 with

$$\begin{aligned} \nu_L^0 &= n_{1L}\cos\theta + n_{2L}\sin\theta \\ N_L^{0c} &= -n_{1L}\sin\theta + n_{2L}\cos\theta \\ \nu_R^{0c} &= -(n_{1R}^c\cos\theta + n_{2R}^c\sin\theta) \\ N_R^0 &= -n_{1R}^c\sin\theta + n_{2R}^c\cos\theta. \end{aligned} \tag{27}$$

The physical masses [30] are

$$\begin{aligned} m_1 &\simeq \frac{m_D^2}{m_S} \ll m_D \\ m_2 &\simeq m_S \end{aligned} \tag{28}$$

and the mixing angle is

$$\tan\theta = \left(\frac{m_1}{m_2}\right)^{1/2} \simeq \frac{m_D}{m_S} \ll 1. \tag{29}$$

Hence, one naturally obtains one very light neutrino, which is mainly the ordinary SU_2 doublet (ν_L^0, ν_R^{0c}), and one very heavy neutrino, which is mainly the singlet (N_L^{0c}, N_R^0).

If one does allow $m_t \neq 0$ (but $\ll m_S$) then there are still two Majorana neutrinos with masses $|a - \frac{m_D^2}{m_S}|$ and m_S, respectively, while $\theta \sim m_D/m_S \ll 1$ still holds. (The minus sign in ν_R^{0c} is removed if $a - \frac{m_D^2}{m_S}$ is positive). In this case, however, one loses the natural explanation of why m_1 is so small, unless m_t is itself induced by the underlying physics and is of the same order as m_D^2/m_S.

The seesaw model is easily generalized to F families. One then has the general $2F \times 2F$ Majorana mass matrix in (15). Assuming that the eigenvalues of m_S are all much larger than any of the components of m_D or m_t (if it is non-zero) one can calculate the eigenvalues and mixing matrices to leading order in m_S^{-1}. One finds that there are F light Majorana neutrinos (consisting of the F doublets (ν_L^0, ν_R^{0c}), up to corrections of order $m_D m_S^{-1}$ and F heavy Majorana neutrinos (consisting of the singlets (N_L^{0c}, N_R^0), to $O(m_D m_S^{-1})$). That is, one can write

$$\begin{aligned} \begin{pmatrix} \nu_L^0 \\ N_L^{0c} \end{pmatrix} &= U_L \begin{pmatrix} n_{lL} \\ n_{hL} \end{pmatrix} \\ \begin{pmatrix} \nu_R^{0c} \\ N_R^0 \end{pmatrix} &= U_R \begin{pmatrix} n_{lR}^c \\ n_{hR}^c \end{pmatrix}, \end{aligned} \tag{30}$$

where n_{lL} and n_{hL} are F component vectors of light and heavy Majorana mass eigenstates, respectively, and similarly for n_{lR}^c, n_{hR}^c. As usual, U_L and U_R are $2F \times 2F$ unitary matrices which diagonalize M in (15), viz

$$U_L^\dagger \begin{pmatrix} m_t & m_D \\ m_D^T & m_S \end{pmatrix} U_R = m_d = \begin{pmatrix} m_l & 0 \\ 0 & m_h \end{pmatrix}, \tag{31}$$

where m_l and m_h are diagonal $F \times F$ matrices of the F light and F heavy eigenvalues, respectively. To leading order in m_S^{-1} one can write $U_L^\dagger$ and U_R in block diagonal form

$$U_L^\dagger = KU_R^T = \begin{pmatrix} K_1 & 0 \\ 0 & K_2 \end{pmatrix} \begin{pmatrix} A^T & -A^T m_D m_S^{-1} \\ D^T m_S^{-1\dagger} m_D^\dagger & D^T \end{pmatrix}, \tag{32}$$

where A^T and D^T are unitary (to leading order) $F \times F$ matrices defined by

$$\begin{aligned} m_l &= K_1 A^T (m_t - m_D m_S^{-1} m_D^T) A \\ m_h &= K_2 D^T m_S D \end{aligned} \tag{33}$$

i.e. the mass matrix for the light neutrinos is $m_t - m_D m_S^{-1} m_D^T$, which is diagonalized by A, while that for the heavy neutrinos is m_S, diagonalized by D. K_1 and K_2 are diagonal phase matrices which ensure that m_l and m_h are real and positive. We see from (30-33) that indeed there are F heavy states with masses of $O(m_S)$, and in the simplest case $m_t = 0$ there are F states which are naturally very light ($O(m_D^2 m_S^{-1})$). (For $m_t \neq 0$ one must separately assume m_t is small). Furthermore, the mixing between the light and heavy sectors is very small (of $O(m_D m_S^{-1})$), while the matrices A and D, which describe mixings within the two sectors, are in general arbitrary.

There are several classes of seesaw models [18], depending on the scale of m_S. In simple grand unified models one assumes that the scale is a typical GUT unification scale of around 10^{14} GeV. In many such models (e.g. SO_{10}) one has that the neutrino Dirac mass matrix m_D is the same as m_u/k where m_u is the u-quark mass matrix and $k \simeq 4.7$ represents the running of the Yukawa couplings between the GUT scale and low energies. If one makes the somewhat ad-hoc assumption that the matrix m_S is just $M_X I$, where $M_X \sim 10^{14}$ GeV is the unification scale and I is the identity matrix, one has (for $m_t = 0$) the light eigenvalues

$$m_{\nu_i} \sim \frac{m_{u_i}^2}{M_X k^2} \tag{34}$$

$\sim 10^{-11}$ eV, 10^{-6} eV, 10^{-3} eV, i.e. the neutrino masses are naturally expected to be extremely tiny, and to scale like the squares of the u, c, and t quark masses. (Equation (34) was computed for $m_{top} \sim 50$ GeV). Several caveats are in order: the assumption of $m_S \sim M_X I$ was quite arbitrary. One could easily imagine that the eigenvalues of m_S are smaller than M_X due to small Yukawa coupling couplings (increasing m_{ν_i}). Also, they need not be the same. For example, if the m_S eigenvalues followed the same family hierarchy as the ordinary fermions (i.e. $m_{S_i} \propto m_{u_i}$) then one would have m_{ν_i} scaling as m_{u_i} rather than $m_{u_i}^2$. (A similar linear hierarchy ensues in some variant GUTs in which m_S is zero at tree level but is generated at higher orders [31]. Of course, more complicated patterns for m_S and m_D (in (33)) are also possible. Furthermore, in many cases loop corrections to the (GUT) Higgs potential may induce [29] VEV's for Higgs representations that can yield a non-zero triplet terms m_t in (31). These are most likely to affect the smallest masses (e.g. m_{ν_e}). Equation (34) should therefore be regarded only as a typical order of magnitude.

If one does assume that $m_S = M_X I$, however, then m_u^2/M_X is diagonalized by the same transformations that diagonalize m_u. Since one also has equal electron and d-quark mass matrices (i.e. $m_e = m_d/k$) in most simple GUTs the final result is that flavor mixing in the lepton sector (analogous to (7)) is described by the same mixing matrix as the CKM quark mixing matrix. This result continues to hold [19] approximately for a far wider class of m_S than does the simple mass prediction in (34).

Lower mass scales for m_S imply larger values for the light neutrino masses (and generally less predictive power for m_D). Several authors [20] have suggested that the heavy Majorana scale could be the intermediate range $10^8 - 10^{12}$ GeV associated with invisible axions. For $m_{D_i} \sim m_{e_i}$ and $m_S \sim 10^9$ $GeV I$, for example, one obtains the values $\sim 10^{-7}$ eV, 10^{-2} eV, 10 eV for m_{ν_e}, m_{ν_μ}, m_{ν_τ}, respectively.

If m_S is in the several TeV range (as expected in some left-right symmetric [32] $SU_{2L} \times SU_{2R} \times U_1$, models [21], for example) one typically expects (for $m_{D_i} \sim m_{e_i}, m_S \propto I$) $m_{\nu_e}, m_{\nu_\mu}, m_{\nu_\tau}$ to have relatively large values 10^{-1} eV, 10 KeV, and 1 MeV, respectively. As we will see, such models run into severe cosmological difficulties unless the mass hierarchy is somehow modified or a fast decay mechanism is found for the ν_μ and ν_τ. Of course, one could also have m_S much smaller than the $SU_{2L} \times SU_{2R} \times U_1$ scale (e.g. in the 10 GeV – 100 GeV range), with corresponding larger masses for the light neutrinos. Similar statements apply to models with extra Z bosons in the 100 GeV – 10 TeV range, which usually also have heavy Majorana neutrinos.

Finally, one can consider light seesaw models, in which typically $m_S \ll 1\ GeV$. Such models are very artificial and abandon the principal advantages of the seesaw, because both m_D and m_S must be taken unnaturally small to obtain an acceptable ν_e mass. Their only virtue is that they yield strongly suppressed neutrinoless double beta decay rates, even though the neutrinos are Majorana.

Seesaw models were first introduced in GUT type models in which lepton number is explicitly violated by the gauge interactions. One can also consider non-gauge seesaw models [33] in which lepton number is spontaneously broken by the VEV of the Higgs field which generates m_S. Such models imply the existence of a massless Goldstone boson, the singlet-Majoron. [34] Unlike the triplet-Majoron in the Gelmini-Roncadelli model,[8] which can couple strongly to the ordinary neutrinos (coupling $\sim h_t$), the singlet-Majoron effectively decouples from ordinary particles. That is, it couples strongly to the heavy neutrino, with a coupling of order $m_D m_S^{-1}$ to off-diagonal $n_l n_h$ vertices, and with strength $(m_d m_S^{-1})^2$ to light neutrinos.

It is difficult to implement the seesaw model in most superstring inspired models, because there is no Higgs field available to generate a large m_S. It has been suggested [35] that m_S could be generated by a higher order effective operator, but such model may run into serious cosmological problems [36].

There have also been variant seesaw models constructed [37] in which the light neutrinos occur in degenerate pairs which can be combined from Dirac neutrinos with a conserved L.

Finally, I mention the charged Higgs models [23], in which small Majorana masses are generated by loop diagrams involving new charged Higgs bosons with explicit L-violating couplings (Fig. 1). Viable versions often lead to pseudo-Dirac neutrinos. The approximately conserved lepton number is typically $L_e - L_\mu + L_\tau$, for example, rather than L. The actual mass scale depends on unknown Yukawa couplings and masses.

4 Experimental Constraints

There are a number of excellent reviews [1] of the experimental status of neutrino mass. My major purpose in this section is to comment on the implications of the various theoretical models for the different types of experiments.

4.1 Kinematic Tests

Direct kinematic limits on the masses of the ν_e, ν_μ, and ν_τ are given in Table 2. The ITEP group [38] has long claimed evidence for a non-zero ν_e mass in the 20 eV range from tritium β decay, but this has not been confirmed by other groups, and in fact the Zurich-SIN measurement is on the verge of conflicting with the ITEP result. In addition the neutrinos from supernova 1987A observed by the Kamiokande [45] and IMB [46] experiments place upper limits in the 20 eV range on the ν_e mass (otherwise the arrival times of the detected neutrinos would be spread out more than is observed), but it is hard to make this limit precise because it depends on the details of the neutrino emission [42].

A 20 eV neutrino mass is just in the range that would be most interesting cosmologically, so clearly it is essentially to resolve the situation. Hopefully, the current and next generation of tritium β decay experiments will be sensitive down to a few eV, but it is doubtful whether experiments of this type will ever be able to probe to much lower scales. As can be seen in Table 1, none of the models really predict m_{ν_e} in the 20 eV range (the $SU_{2L} \times SU_{2R} \times U_1$ models come closest), but most can accomodate masses in this range by fine-tuning parameters.

Table 2: Kinematic limits/values on neutrino masses.

$17\ eV < m_{\nu_e} < 40\ eV$	ITEP [38]
$m_{\nu_e} < 18\ eV$	Zurich [39]
$m_{\nu_e} < 27\ eV$	LANL [40]
$m_{\nu_e} < 32\ eV$	INS-Tokyo [41]
$m_{\nu_e} < O(20\ eV)$	SN1987A [42]
$m_{\nu_\mu} < 0.25\ MeV$	SIN [43]
$m_{\nu_\tau} < 50\ MeV$	ARGUS [44]

As can be seen in Table 2, the direct kinematic limits on m_{ν_μ} (from $\pi_{\mu 2}$ decay) and on m_{ν_τ} (from $\tau \to \nu_\tau + 5\pi$) are relatively weak. The experiments are extremely difficult (the mass scales being probed are very much smaller than the energies released in the decays), so it is unlikely that these measurements will improve by much more than a factor of two.

4.2 Heavy Neutrinos

There are many limits [1,47] on possible small admixtures of heavy neutrino states in the ν_e or ν_μ, including universality tests in nuclear β decay, searches for secondary peaks or distortions of the lepton spectra in β, π, and K decay, searches for the decay products of heavy neutrinos (e.g. $\nu_h \to \nu_e e^+ e^-$) produced in beam dumps, e^+e^- annihilation, or neutrino scattering. The limits on the mixing angle U_{ai}, $a = e$ or μ, where $\nu_a^0 = \sum_i U_{ai}\nu_i$ are quite impressive, especially for m_i in the range $10\ MeV - 10\ GeV$, where they are comparable to the expectations in (29) of a seesaw model with $m_1 \sim 10\ eV$ and $m_2 = m_i$. The lower part of this range corresponds to the masses expected in the "light-seesaw" model (Table 1), while the $1\ GeV - 1\ TeV$ range is consistent with $SU_{2L} \times SU_{2R} \times U_1$ models. [21]

As has already been mentioned, heavy neutrinos in the $GeV - TeV$ range are likely to give too large m_{ν_μ} and m_{ν_τ} unless the typical seesaw hierarchy $m_{\nu_i} \propto m_{l_i}^n$ or $m_{u_i}^n$, $n = 1$ or 2, for the light neutrinos is avoided or new physics is invoked to ensure fast decays or annihilations for the ν_μ and ν_τ. On the other hand, if such new physics is present some of the limits (those based on decays) may no longer be valid, because in many cases the heavy neutrinos will decay rapidly into unobservable channels (e.g. $\nu_h \to \nu_l +$ Majoron) before reaching the detector.

4.3 Neutrino Oscillations

Neutrino oscillations are a beautiful example of a common quantum phenomenon: viz that if one starts at time $t = 0$ in a state that is not an energy eigenstate [48] then at later times it can oscillate into another (orthogonal) state. For example, suppose that the ν_e^0 and a second neutrino ν_a^0 (e.g. $\nu_a^0 = \nu_\mu^0$ or ν_τ^0) are mixtures of two mass eigenstates ν_1 and ν_2 with mixing angle θ. If at time $t = 0$ the weak eigenstate $\nu_e^0 = \cos\theta\ \nu_1 + \sin\theta\ \nu_2$ is produced (e.g. in the process $\pi^+ \to \pi^0 e^+ \nu_e^0$) then at time t it will have evolved into the state

$$\begin{aligned} \nu_e^0(t) &= \cos\theta\ \nu_1 e^{-iE_1 t} + \sin\theta\ \nu_2 e^{-iE_2 t} \\ &\simeq \cos\theta\ \nu_1 e^{\frac{-i m_1^2 t}{2p}} + \sin\theta\ \nu_2 e^{\frac{-i m_2^2 t}{2p}}. \end{aligned} \tag{35}$$

In the second form I have assumed relativistic neutrinos $E_i = \sqrt{p^2 + m_i^2} \sim p + m_i^2/2p$ with definite momentum [49] $p >> m_i$, and have neglected an irrelevant overall phase $\exp(-ipt)$. The state $\nu_e^0(t)$ has a non-trivial overlap with ν_a^0. After traveling a distance $L \sim t$, there will be a probability

$$\begin{aligned} P(\nu_e \to \nu_a) &= |\langle \nu_a^0 \mid \nu_e^0(t) \rangle|^2 \\ &= \sin^2 2\theta \sin^2\left(\frac{\Delta m^2 L}{4p}\right) \end{aligned}$$

$$= \sin^2 2\theta \sin^2 \left(\frac{1.27 \Delta m^2 (eV^2) L(m)}{p(MeV)} \right) \tag{36}$$

that the state will have evolved into ν_a^0 (as can be observed in the process $\nu_a N \rightarrow e_a N'$, for example), and a probability $P(\nu_e \rightarrow \nu_e) = 1 - P(\nu_e \rightarrow \nu_a)$ that the state will remain a ν_e^0. In (36), $\Delta m^2 = m_1^2 - m_2^2$, and the last form is valid for Δm^2 in eV^2, L in m, and p in MeV. It is seen that the $\nu_e \rightarrow \nu_a$ probability depends on both the mixing angle θ and on $\Delta m^2 L/p$. For moderate values of the latter quantity the probability oscillates as a function of L and p, while for very large values the oscillations are averaged by a finite-sized detector or non-monochromatic source, (the second factor in (36) averages to 1/2). It is easy to generalize [1] (36) to the case that the initial neutrino is a mixture of more than two mass eigenstates, $\nu_a^0 = \sum_i U_{ai} \nu_i$. One obtains

$$P(\nu_e \rightarrow \nu_a) = \sum_i |U_{ei} U_{ai}^*|^2 + Re \sum_{i \neq j} U_{ei} U_{ai}^* U_{ej}^* U_{aj} e^{\frac{-i(m_i^2 - m_j^2)L}{2p}} \tag{37}$$

Neutrino oscillations can be searched for in (a) appearance experiments, in which one looks for the interactions of ν_a in a detector, and (b) disappearance experiments, in which one looks for a reduced ν_e flux. In both cases one can compare the observed counting rate with the expectation from known backgrounds (appearance) or from the expected flux (disappearance) as determined, for example, by measuring the electron spectrum from $n \rightarrow pe^- \bar{\nu}_e$ in reactor $\bar{\nu}_e$ oscillation experiments. A much cleaner technique is to search for actual oscillations in the appearance or disappearance probabilities as a function of L or p, such as by using two detectors at different distances form the source.

There are many limits on neutrino oscillations from accelerator experiments [1] (e.g. counter and emulsion experiments and beam dumps, searching for $\nu_\mu \rightarrow \nu_e$, $\nu_\mu \rightarrow \nu_\tau$ and $\nu_e \rightarrow \nu_\tau$, as well as ν_μ disappearance), and reactors [1] ($\bar{\nu}_e$ disappearance), as well as on the oscillations of ν_μ produced in cosmic ray interactions in the atmosphere [50]. (Implications for the Solar neutrino problem are discussed below). The results of these searches [51] are summarized in Fig. 2. The

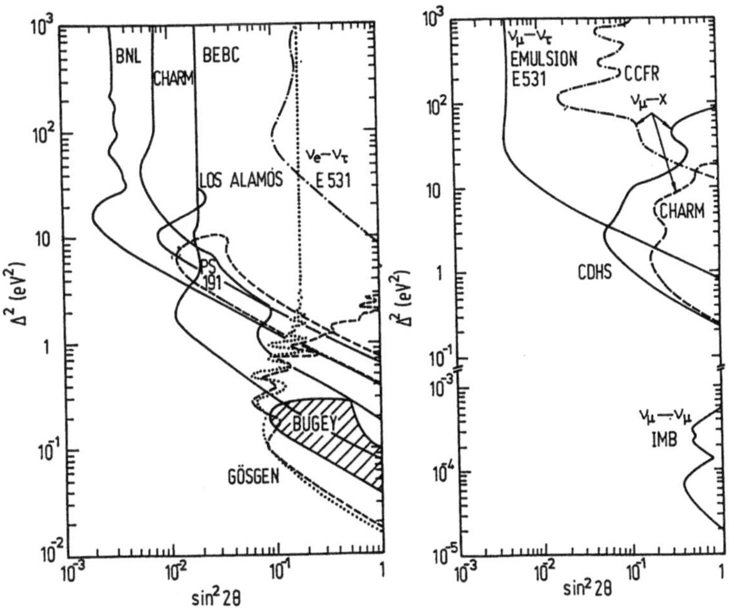

Figure 2: 90% $c.l.$ limits on neutrino oscillations, from [55]. (a) $\nu_\mu \rightarrow \nu_e$ (BNL, CHARM, BEBC, Los Alamos, PS-191), $\nu_e \rightarrow \nu_\tau$ (E531), and $\bar{\nu}_e \rightarrow \nu_X$ (Bugey, Gösgen). (b) $\nu_\mu \rightarrow \nu_\tau$, ν_X, ν_μ. The Bugey [52] and PS-191 [54] regions are allowed by positive results. The other contours are exclusion plots (the regions to the right are excluded).

Bugey reactor experiment [52] reports a positive signal for $\bar{\nu}_e$ disappearance, but their results are contradicted by the Gösgen experiment [53]. Similarly, the CERN PS-191 counter experiment [54] reports an excess of ν_e events in a ν_μ beam, but their signal is in conflict with several other $\nu_\mu \to \nu_e$ experiments [55]. Clearly, a clarification of the situation is essential .

From Fig. 2 it is clear that there are stringent limits on neutrino mixings for $|\Delta m^2|$ above $\simeq 1\ eV^2$. This should be contrasted with the suggested value $m_{\nu_e} \sim 17 - 40\ eV$ by the ITEP experiment [38]. If the ITEP result is correct then most likely the ν_e could not have any significant mixing with other neutrinos (the alternative possibility, that the ν_e is almost degenerate with another neutrino flavor so that $|\Delta m^2| << m^2_{\nu_e}$, seems rather contrived but cannot be excluded). A comparison of Fig. 2 with the expectations of various models (Table 1) suggests that $\nu_\mu \to \nu_\tau$ oscillations may be the most optimistic possibility for the future. Many of the seesaw-type models predict that the lepton mixing angles are roughly correlated with the corresponding quark mixing angles. This would suggest $\sin^2 2\theta \sim 10^{-4},\ 10^{-2},\ 10^{-1}$ for $\nu_e \leftrightarrow \nu_\tau$, $\nu_\mu \leftrightarrow \nu_\tau$, and $\nu_e \leftrightarrow \nu_\mu$, respectively.

Oscillations between ordinary SU_2 doublet neutrinos ($\nu_e^0, \nu_\mu^0, \nu_\tau^0$, and possible fourth family ν's), known as first class or flavor oscillations, occur for pure Dirac and pure Majorana neutrinos, as well as in the multi-family seesaw models. In models involving both Dirac and Majorana mass terms of comparable magnitude, however, there can be additional light neutrinos, and the mass eigenstates can have significant admixtures of both SU_2 doublets and singlets. In this case second class oscillations [16] can occur, in which the ordinary neutrinos oscillate into SU_2 singlets with negligible interactions. These "sterile" neutrinos are essentially undetectable, so second class oscillations can be observed [56] only in disappearance experiments. Of course, first and second class oscillations can occur simultaneously. For three families, for example, there could be oscillations between six Majorana neutrinos (3 doublets and 3 singlets).

Yet another possibility [57] are models in which the ordinary neutrinos have small mixings with heavy neutrinos. In that case the neutrinos actually produced in weak processes are the projections of the weak eigenstates onto the subspace of light or massless neutrinos. It can easily occur that the projections of the ν_e^0 and ν_μ^0, for example, are not orthogonal. The result is that a ν_μ^0 could produce an e^- in a subsequent reaction. Such a non-orthogonality would mimic the effects of oscillation appearance experiments, even if the masses of the light neutrinos are zero or negligible.

4.4 Cosmology

There are many limits on neutrino mass and decays from cosmology [58]. Ordinary light or massless neutrinos would have been produced by such weak processes as $e^+e^- \leftrightarrow \nu\nu^c$ in the early universe. As long as the weak reaction rate [59]

$$\Gamma_{weak} \sim \langle \sigma v \rangle n_T \sim G_F^2 T^5 \tag{38}$$

($\langle \sigma v \rangle \sim G_F^2 T^2$ is the thermally averaged cross section times relative velocity, and $n_T \sim T^3$ is the density of target particles, where T is the temperature) was large compared to the expansion rate $H \sim T^2/m_p$ (where $m_p = G_N^{-1/2} \sim 10^{19}\ GeV$ is the Planck scale) the number of neutrinos stayed in equilibrium. However, as soon as T dropped below the temperature $T_D \sim (G_F^2 m_P)^{-1/3} \simeq 3\ MeV$ for which $\Gamma_{weak} \sim H$, the weak rate became negligible and the neutrinos decoupled, i.e. effectively stopped interacting. According to most models these neutrinos should remain in the present universe, undisturbed from the first second of the big bang except for a redshifting of their momenta by the expansion of the universe. They are analogous to the $2.7°K$ microwave radiation (which decoupled later). If the neutrino masses are much less than 1 eV there should be $\simeq 50$ neutrinos/cm^3 of each type (ν_{eL}, ν_{eR}^c, etc) with momenta characterized by a thermal spectrum with temperature $\simeq 1.9°K$ ($10^{-4}\ eV$). Despite the large number of neutrinos ($\simeq 10^{10}$ per baryon) they are essentially impossible to detect [60] - [62] because their cross section $\sim G_F^2 E_\nu^2 \sim 10^{-62} cm^2$ is so low. [63]

The major cosmological bound is based on the energy density of the present universe. There are predicted to be so many relic neutrinos that even for a small mass in the 10 eV range they would be important. Limits on the energy density imply

$$\sum_i m_{\nu_i} < 40\ eV \tag{39}$$

where the sum extends over the light, stable (at least compared to the age of the universe) doublet neutrinos. Conversely, a neutrino with mass in this range would dominate the energy density and could account for the dark (missing) matter in galaxies and clusters [64]. In particular, for the ITEP value $m_{\nu_e} \sim (17-40)\ eV$, the ν_e would be an ideal candidate for the dark matter, but one would probably then have to find a mechanism to explain why the ν_e is the heaviest neutrino.

Similarly, the energy density associated with light or massless neutrinos for $T \sim T_D$ affects nucleosynthesis and leads to the limit $N_\nu \leq 4$ on the number of neutrino flavors with $m_\nu \leq 1\ MeV$. [65]

There are also a variety of constraints on unstable neutrinos. An ordinary doublet mass eigenstate neutrino ν_2 (with $m_{\nu_2} > m_{\nu_1}$) is expected to decay into

$$\begin{aligned} \nu_2 &\rightarrow \nu_1\gamma, \quad (m_{\nu_2} < 2m_e) \\ \nu_2 &\rightarrow \nu_1 e^+e^-, \quad (2m_e < m_{\nu_2} < m_\mu + m_e). \end{aligned} \tag{40}$$

The first decay occurs at one loop, while the second occurs at tree level. Both decays are very slow for small m_{ν_2} and the decay products are detectable. There are a large variety of cosmological and astrophysical constraints [66] on m_{ν_2} and τ_{ν_2} from the present energy density, the growth of galaxies, the distortion of the 2.7°K background radiation, the non-observation of the decay photons, supernovae, and nucleosynthesis and breakup. For reasonable mixing angles these limits exclude the range 40 eV – (20 – 40) MeV for ordinary neutrinos [67] decaying according to (40). Combined with laboratory limits this implies [66,68] that the ν_μ and ν_τ (i.e. their dominant mass eigenstate components) should be lighter than 40 eV. In particular, this poses serious problems for the TeV scale seesaw model.

Most of the cosmological limits can be evaded if new physics is invoked to allow fast and invisible (except for the relativistic energy of the decay products) decays or annihilation for the heavy neutrinos. One possibility is the decay $\nu_2 \rightarrow 3\nu_1$. However, the rate for this mode from off-diagonal Z couplings [69] is too slow, while models in which the couplings of a Higgs triplet [70] (present in $SU_{2L} \times SU_{2R} \times U_1$) are arranged to allow a fast decay generally run into problems [71] with $\mu \rightarrow 3e$.

More promising are models in which $\nu_2 \rightarrow \nu_1 G$, where G is a Goldstone boson [72]-[74] associated with a spontaneously broken global symmetry. Likely examples are the case that G is a familon [72] (a Goldstone boson associated with a broken family symmetry) or a triplet-Majoron [73]. In fact, for triplet-Majorons one expects the annihilation process $\nu\nu \rightarrow MM$ (which begins when T drops below v_t) to have removed any relic neutrinos from the present universe [8]. In familon models some care must be taken to avoid unacceptably large flavor changing neutral current effects. The decay $\nu_2 \rightarrow \nu_1 M$ is too slow in the simpler versions of the singlet-Majoron model [74] to avoid cosmological problems.

The role of spontaneous L violation in Majoron models in reducing possible initial large lepton asymmetries to cosmologically interesting values at the time of nucleosynthesis is discussed in [75].

4.5 Double Beta Decay

Another important source of information on the ν_e mass (if it is Majorana) is neutrinoless double beta decay ($\beta\beta_{0\nu}$).

First consider the lepton-number conserving two-neutrino ($\beta\beta_{2\nu}$) process $(Z,N) \rightarrow (Z+2, N-2)e^-e^-\nu_e^c\nu_e^c$, which can be thought of as two ordinary beta decays occurring in the same nucleus

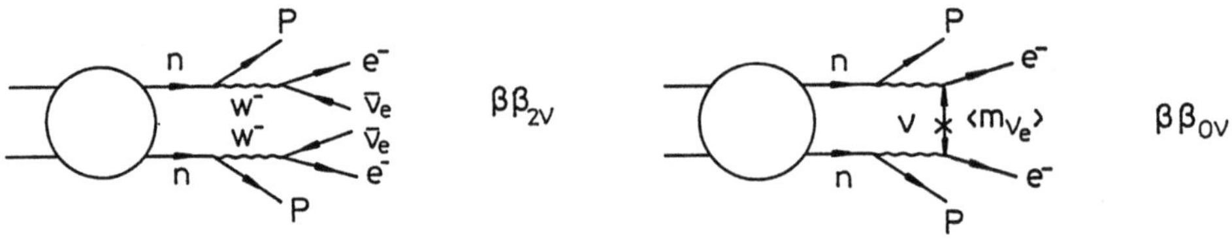

Figure 3: Diagrams for two neutrino ($\beta\beta_{2\nu}$) and neutrinoless ($\beta\beta_{0\nu}$) double beta decay.

(Fig. 3). In the context of neutrino mass this process is mainly of interest as a calibration of the calculated nuclear matrix elements that are needed for the neutrinoless case. There has long been a two order of magnitude discrepancy between the predicted rates [76], e.g. for $^{130}Te \to {}^{130}Xe$, and indirect measurements by geochemical techniques [77]. Within the last year, however, this discrepancy has gone away. The geochemical measurements were confirmed by the first laboratory observation of double beta decay (at Irvine [78].) In addition, several groups [76] have found that previously neglected ground state correlation effects could suppress the matrix element by the required order of magnitude. Furthermore, there is no analogous uncertainty in the $\beta\beta_{0\nu}$ case.

The neutrinoless double beta decay process $(Z,N) \to (Z+2,N-2)e^-e^-$, which violates lepton number by two units, can proceed through the second diagram [79] in Fig. 3. In the absence of mixing the quantity $\langle m_{\nu_e}\rangle$, the effective Majorana neutrino mass, is

$$\langle m_{\nu_e}\rangle = \begin{cases} 0, & \text{Dirac neutrino} \\ m_{\nu_e}, & \text{unmixed Majorana neutrino} \end{cases} \tag{41}$$

Although the matrix element is proportional to $\langle m_{\nu_e}\rangle$, which is necessarily very small, $\beta\beta_{0\nu}$ has an enormous advantage in phase space over $\beta\beta_{2\nu}$ and could be observable for $\langle m_{\nu_e}\rangle$ in the eV range. Of course, the sum of the electron energies should be a sharp peak in $\beta\beta_{0\nu}$ (and a continuum for $\beta\beta_{2\nu}$), so the principal difficulty is controlling the background. [80,1] Currently, the most sensitive experiments are for $^{76}Ge \to {}^{76}Se\ e^-e^-$. No evidence for $\beta\beta_{0\nu}$ has been observed, [81] and the lower limit on the lifetime is [55] $\tau_{1/2} > 9 \times 10^{23}\ yr$ (68% $c.l.$). According to several calculations of the nuclear matrix elements [82] this implies $\langle m_{\nu_e}\rangle \leq 1\ eV$. However, a recent estimate by Engel et al. [83] yielded a much weaker limit $\langle m_{\nu_e}\rangle \leq 11\ eV$, so caution is advisable.

Even the largest value $\langle m_{\nu_e}\rangle \leq 11\ eV$ is smaller than the range $m_{\nu_e} \sim (17-40)\ eV$ suggested by the ITEP experiment. If the latter is correct the simplest possibility is that the ν_e is Dirac. Another possibility [84] is that the ν_e^0 is a mixture of Majorana mass eigenstate neutrinos. Then, $\langle m_{\nu_e}\rangle$ becomes

$$\langle m_{\nu_e}\rangle = \sum_i m_i U_{Lei}^2 \xi_i F(m_i, A), \tag{42}$$

where $m_i \geq 0$ is the physical mass of the ith mass eigenstate, U_{Lei} is the mixing matrix element ($\nu_{eL}^0 = \sum U_{Lei}\nu_{iL}$) and $\xi_i = \pm 1$ is the CP parity of ν_{iL}. ξ_i is just K_{ii} in (13), and a negative value $\xi_i = -1$ means simply that the eigenvalue of M in (10) was negative before choosing K to redefine ν_R^c. In (42), $F(m_i, A)$ is a nucleus dependent propagator correction, [85] defined by

$$F(m_i, A) = \frac{\langle e^{-m_i r}/r\rangle}{\langle 1/r\rangle}. \tag{43}$$

It is ~ 1 for $m_i \ll 10\ MeV$. For $m_i \gg 10\ MeV$, $F(m_i, A) \ll 1$ (it falls as m_i^{-2}) and allows the possibility [86] of A dependence of $\langle m_{\nu_e}\rangle$.

Because of the possibility of negative contributions to $\langle m_{\nu_e}\rangle$ it is conceivable that there are cancellations so that $\langle m_{\nu_e}\rangle$ is much smaller than the mass of the dominant Majorana component of ν_e (e.g. $m_1 \sim (17-40)\ eV$). Such a cancellation is actually not so contrived as it might first appear. If all of the m_i are small enough that $F(m_i, A) = 1$ then from (12) $\langle m_{\nu_e}\rangle$ is just the M_{ee} component of the original Majorana mass matrix in (10). As we have seen, M_{ee} must be generated by a Higgs triplet and vanishes in many models. In fact, the light seesaw model of Table 1 automatically leads to $\langle m_{\nu_e}\rangle = 0$ for sufficiently small m_i. For two neutrinos, for example, $\langle m_{\nu_e}\rangle = m_1 \cos^2\theta - m_2 \sin^2\theta$, which vanishes by (28) and (29). However, the light seesaw model was devised just in order to give $\langle m_{\nu_e}\rangle = 0$. For seesaw models with more natural scales $m_2 \gg 10\ MeV$ one has that $F(m_i, A) \ll 1$ and $U_{e1} \sim 1$, so that $\langle m_{\nu_e}\rangle \sim m_{\nu_e}$. In most Majorana models, therefore, one expects $\langle m_{\nu_e}\rangle \sim m_{\nu_e}$ unless fine-tuned deviations from the seesaw formula are invoked.

Whether or not the cancellation of the terms in (42) is natural, one can consider whether it is phenomenologically viable. For two neutrinos, for example, the conditions $m_1 \sim 20\ eV$, and $\langle m_{\nu_e}\rangle \ll m_1$ imply

$$\tan^2\theta = \frac{m_1}{m_2 F(m_2, A)}, \tag{44}$$

where $m_1 \leq m_2$, $\tan^2\theta \leq 1$ since the ITEP experiment presumably measures the dominant component of ν_e. However, the reactor oscillation limits in Fig. 2 allow only two possibilities. One is that $m_1 \simeq m_2$, $0 \sim 45^0$. In that case ν_1 and ν_2 can be combined to form a Dirac neutrino (or pseudo-Dirac if the degeneracy is not exact), possibly with a non-canonical lepton number (such as $L_e - L_\mu + L_\tau$) conserved. Alternatively, one can have $m_2 \geq 450\ eV$. However, the various laboratory and cosmological limits exclude [22] almost all values of m_2 except for small windows around 40 MeV and 2 GeV. Hence, if the ITEP results turn out to be correct they would almost certainly imply either (a) the ν_e is Dirac, or (b) there is new physics (such as a Majoron) that evades the cosmological bounds.

There are additional contributions to neutrinoless double beta decay in $SU_{2L} \times SU_{2R} \times U_1$ models [87]. Typically, such models contain additional charged $W_R^\pm$ bosons which couple to right-handed currents $\bar{e}_R \gamma^\mu N_R$, where N_R is a heavy Majorana neutrino. The exchange of a N_R (rather than a ν_L in Fig. 3) yields a new contribution $M_N F(M_N, A)(M_{W_L}/M_{W_R})^4$ to $\langle m_{\nu_e}\rangle$, which sets non-trivial constraints [21] on M_N and M_{W_R}. Furthermore, mixed contributions involving one ordinary left-handed current $\bar{e}_L \gamma^\mu \nu_L$ and one right-handed current $\bar{e}_R \gamma^\mu N_R$ can yield contributions to to $0^+ \to 2^+$ decay amplitudes that are not directly proportional to a neutrino mass [88] However, the relevant amplitudes are of order [1,87]

$$\left(\frac{M_{W_L}}{M_{W_R}}\right)^2 \theta, \quad \zeta\theta, \tag{45}$$

where θ is a light-heavy neutrino mixing angle and ζ is the $W_L - W_R$ mixing angle. One typically expects $(M_{W_L}/M_{W_R})^2$ and ζ to be less than 10^{-3}. Since we expect $\theta \sim m_D/M_{W_R} \leq 10^{-4} - 10^{-5}$ in a typical TeV-seesaw, the expected values for the quantities in (45) are smaller than the experimental limits (of $\sim 10^{-6}$).

One typically has $\langle m_{\nu_e}\rangle \ll m_{\nu_e}$ for the charged Higgs models [23] because the antisymmetry of the relevant Yukawa coupling forces M_{ee} to vanish.

4.6 The Solar Neutrino Problem

For some years the event rate in the $^{37}Cl \to {}^{37}Ar$ Solar neutrino experiment [89] ($2.0 \pm 0.3\ SNU$ [90]) has been considerably below the prediction [91] $5.8 \pm 2.2\ SNU$ of the standard Solar model. The discrepancy has recently been confirmed by the Kamiokande group which reports [92] an upper limit on the ν_e flux (from $\nu_e e$ elastic scattering) that is less than half the expected event rate. One explanation for the discrepancy is the existence of vacuum oscillations of the ν_e into other neutrinos. These could be important for neutrino mass-squared differences [93] $\Delta m^2 \equiv m_1^2 - m_2^2$ as small as $\Delta m^2 \sim (10^{-11} - 10^{-10})\ eV^2$, but only if the mixing angles are large.

Another possibility [94] is that the ν_e is a Dirac particle with a magnetic moment in the range $\mu_{\nu_e} \sim (0.6 - 10) \times 10^{-10} \mu_B$. The ν_e spin could then process in the Solar magnetic field into a sterile right-handed ν_e, thus reducing the observed flux by a factor $\simeq 2$. The necessary value of μ_{ν_e} is barely consistent with laboratory limits [95] but is probably excluded by astrophysical constraints

Table 3: Limits on the neutrino magnetic moments. A value $\mu_{\nu_e} \sim (0.6 - 10) \times 10^{-10}\mu_B$ would be needed to resolve the Solar ν problem.

laboratory [95]	$\mu_{\bar{\nu}_e} < 1.5\times10^{-10}\mu_B$
	$\mu_{\nu_\mu} < 9.5\times10^{-10}\mu_B$
Stellar cooling [96] ($\gamma\rightarrow\nu\bar{\nu}$)	$\mu_\nu < 0.8\times10^{-11}\mu_B$
Nucleosynthesis [96] (ν_R produced by spin precession)	$\mu_\nu < 0.5\times10^{-10}\mu_B$
Standard model [97] (Dirac mass)	$\mu_\nu \sim 3\times10^{-19}\left(\frac{m_\nu}{1\ eV}\right)\mu_B$

from nucleosynthesis and stellar cooling [96] (Table 3). The worst objection, however, is that the necessary μ_{ν_e} is unnaturally high. In the standard model with a Dirac mass one expects [97]

$$\mu_\nu = \frac{3G_F m_\nu m_e}{4\pi^2\sqrt{2}}\mu_B \sim 3\times10^{-19}\left(\frac{m_\nu}{1eV}\right)\mu_B \tag{46}$$

which is many orders of magnitude too small. Non-standard models [98] can yield larger μ_ν, but to obtain a sufficiently large value appears highly contrived.

Other canonical explanations involve non-standard Solar models. The existing experiments are mainly sensitive to the relatively high energy (from 0.81 MeV up to 14 MeV) neutrinos from ^8B decay. The flux of these ^8B neutrinos depends very sensitively on the temperature of the Solar core and could be changed significantly by modifications of the standard Solar model. Recently, there has been much attention to the possibility that weakly interacting massive particles (WIMPs), which could form the dark matter, could carry energy out of the Solar core and lower the central temperature slightly. [58] Less exotic modifications of the standard model are also possible.

A $^{71}Ga \rightarrow {}^{71}Ge$ experiment could distinguish the nonstandard Solar model from the first two possibilities. Most of the expected ^{71}Ga event rate is from the low energy pp neutrinos, the flux of which can be inferred from the over-all Solar luminosity and is relatively insensitive to the temperature of the Solar core. The predicted ^{71}Ga event rate of $\simeq$ 107 SNU can be reduced at most to around 78 SNU in most non-standard Solar models [91,99]. The traditional view has been that a flux lower than this would imply large vacuum oscillations, which would reduce the ^{71}Ga rate by a factor comparable to the ^{37}Cl event rate reduction for most oscillation parameters (e.g. to around 40 SNU).

Yet another possibility, i.e. that neutrinos decay between the Sun and the Earth, is all but excluded by the survival of neutrinos from supernova 1987A, except in some two-component models with large mixing angles. [100]

Recently, Mikheyev and Smirnov [101] have proposed an elegant new solution to the Solar neutrino problem, in which even tiny vacuum mixing angles can be amplified by the coherent interactions of ν_e with matter.

Considering $\nu_e \leftrightarrow \nu_\mu$ oscillations for definiteness, the vacuum oscillation equation in (35) can be described in terms of the weak basis states $|\nu_e\rangle$ and $|\nu_\mu\rangle$ by

$$|\nu(t)\rangle = \nu_e(t)|\nu_e\rangle + \nu_\mu(t)|\nu_\mu\rangle, \tag{47}$$

where the coefficients satisfy the Schrödinger-like equation

$$i\frac{d}{dt}\begin{pmatrix}\nu_e(t)\\ \nu_\mu(t)\end{pmatrix} = M_0\begin{pmatrix}\nu_e(t)\\ \nu_\mu(t)\end{pmatrix}, \tag{48}$$

with

$$M_0 = \begin{pmatrix} \frac{\Delta m^2}{2p}\cos 2\theta & \frac{-\Delta m^2}{4p}\sin 2\theta \\ \frac{-\Delta m^2}{4p}\sin 2\theta & 0 \end{pmatrix}, \tag{49}$$

where an irrelevant term proportional to the identity (which only affects the overall phase) has been dropped. Wolfenstein pointed out [102] that in the presence of matter, M_0 is replaced by the M', where

$$M' = M_0 + \begin{pmatrix} \sqrt{2}G_F n_e & 0 \\ 0 & 0 \end{pmatrix} \tag{50}$$

and n_e is the density of electrons. The new term [102] - [104] is the effect of the coherent forward scattering amplitude for $\nu_e e^- \rightarrow \nu_e e^-$ via the charged current. The effects of neutral current scattering from e^-, p, and n have been neglected because they are the same for ν_e and ν_μ and only contribute to the overall phase. For $\Delta m^2 < 0$ (i.e. $m_{\nu_e} < m_{\nu_\mu}$ [104]) there is a critical density [105] $n_e^{crit} = -\Delta m^2 \cos 2\theta/(2\sqrt{2}G_F p)$ for which the diagonal elements of M' are equal (i.e. zero). At that density a resonance occurs, i.e. even a tiny off diagonal mixing term leads to large mixing effects.

In particular, if n_e in the Sun varies slowly an adiabatic approximation applies [101,106]. ν_e's produced in the core of the Sun (where $n_e > n_e^{crit}$) correspond to the larger mass eigenstate ν_2 of M' (Fig. 4). Outside the Sun, on the other hand, the higher energy state ν_2 corresponds to ν_μ for $\Delta m^2 < 0$. Hence, if the variation of n_e with the distance from the center of the Sun is sufficiently slow, the initial ν_e will be adiabatically converted to ν_μ as they pass through the resonance.

A number of authors [99],[101],[106] - [108] have analyzed the implications of the Mikheyev-Smirnov-Wolfenstein (MSW) effect for the Solar neutrinos quantitatively. It is found that there are three classes of parameters which can explain the reduction of ^8B neutrinos observed in the ^{37}Cl experiment. These roughly form the sides of a triangle, as is illustrated schematically in Fig. 5. For solution (a) corresponding to $|\Delta m^2| \sim 5 \times 10^{-5}\ eV^2$, $\sin^2 2\theta \geq 4 \times 10^{-4}$, the adiabatic hypothesis is valid and $\simeq 100\%$ conversion occurs. However, only the high energy ^8B neutrinos actually encounter a resonance layer (the central density is too low for the low energy neutrinos) and are converted. For this parameter range one expects little reduction in the counting rate for the gallium experiment (i.e. the effect is similar to non-standard Solar models in that respect).

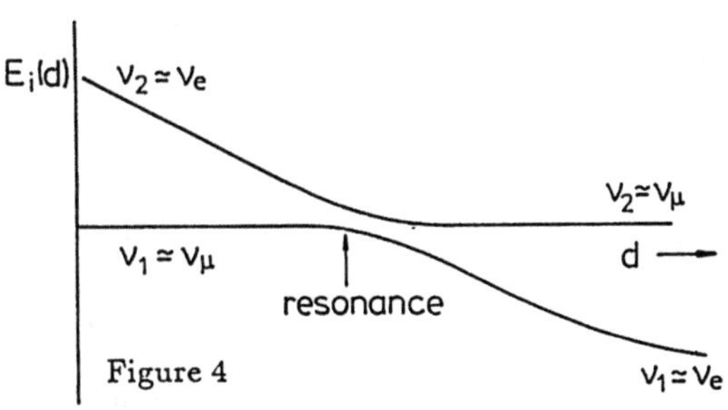

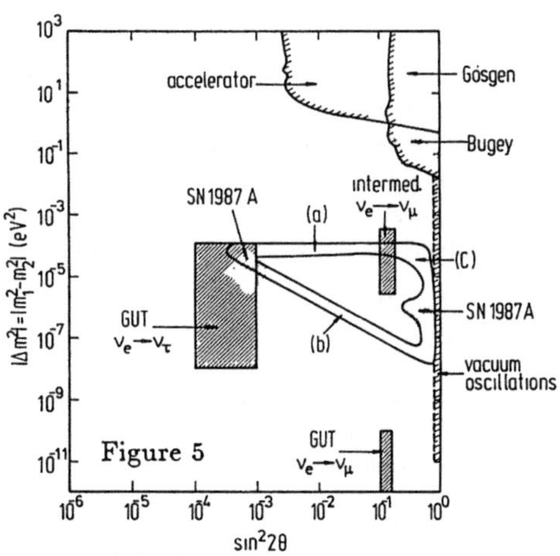

Figure 4: The energy eigenvalues of M' as a function of d, the distance from the center of the Sun.

Figure 5: A schematic view of the regions in the Δm^2 - $\sin^2 2\theta$ plane which can explain the Solar neutrino problem via the Mikheyev-Smirnov-Wolfenstein (MSW) effect.

For solution (b), extending down to $|\Delta m^2| \sim 10^{-8}\ eV^2$ one has [107] $|\Delta m^2| \sin^2 2\theta \sim 10^{-7.5}\ eV^2$. Here the adiabatic approximation starts to break down. All neutrino energies are affected, but the conversion probability is less than unity. For these solutions one expects a significant reduction in the gallium counting rate, similar to large vacuum oscillations or magnetic moments. Solution (c), corresponding to large vacuum mixing angles, is an extension of the vacuum oscillation solution. In the middle of region (c) one expects a large day-night asymmetry in the ν_e counting rate due to MSW regeneration in the Earth at night [99].

The MSW solution is very elegant, but it severely complicates the task of sorting out which if any of the proposed solutions to the Solar ν problem is correct. It will take an ambitious program of experiments [99,109] to clarify the matter.

The first two events from SN1987A observed by the Kamiokande experiment [45] point back towards the supernova. They are consistent with ν_e from the initial neutronization burst, scattering via $\nu_e e^- \rightarrow \nu_e e^-$. However, they could also be $\bar{\nu}_e$ from the subsequent thermal burst, scattering via $\bar{\nu}_e p \rightarrow e^+ n$, which has a much larger cross section (and which produces an isotropic distribution of positrons). If they are indeed ν_e they are problematic for the MSW mechanism because one expects $\nu_e \rightarrow \nu_\mu$ conversions on the way out of the supernova. However, there are two parameter regions (shown in Fig. 5), which would still be consistent [110], corresponding respectively to incomplete conversion in the supernova and reconversion in the Earth. Unfortunately there is no way to determine whether the two events are really ν_e's.

The MSW mechanism is consistent with the expectation of GUT [19] and intermediate scale [20] seesaws. As is illustrated in Fig. 5 the predictions of the GUT seesaw are consistent with $\nu_e \leftrightarrow \nu_\tau$ conversions in the Sun. In this case, the mass ranges are too small to ever see any direct laboratory effects of neutrino mass. The intermediate scale seesaw could account for the Solar ν problem via $\nu_e \leftrightarrow \nu_\mu$ conversions. In that case, $\nu_\mu \leftrightarrow \nu_\tau$ oscillations could well be observable in the laboratory.

It is also possible that the Solar ν problem could be explained by small neutrino masses, but that neutrino appearance experiments might nevertheless yield positive signals due to non-orthogonal neutrino states (induced by mixings between very light and very heavy neutrinos. [57])

5 Summary

- The question of whether the neutrinos have mass is vital for both particle physics and cosmology. However, there is at present no compelling evidence for neutrino mass. The ITEP result $m_{\nu_e} \sim (17-40)\ eV$ has not been confirmed by other experiments and is on the verge of being excluded. Although there are two positive indications of neutrino oscillations (with different parameters), these are contradicted by other experiments. The negative results suggest $m_{\nu_i} \simeq O(1\ eV)$ unless there are very small mixings or degeneracies. There are also stringent limits on incoherent mixing with heavy neutrinos.

- The MSW solution to the Solar neutrino problem would work for ν_μ or μ_τ in the $10^{-2}\ eV$ range, consistent with intermediate mass or GUT seesaws.

- The nonobservation of neutrinoless double beta decay implies $\langle m_{\nu_e} \rangle < 1 - 11\ eV$. If the ITEP result is correct this would most likely imply a Dirac neutrino or new physics to evade cosmological bounds.

- A ν mass in the 5 - 40 eV range would dominate the energy density of the universe and would be an excellent candidate for the dark matter, though other mechanisms would have to be invoked to explain the initial formation of galaxies. Conversely, the light stable neutrinos must be lighter than $\simeq 40\ eV$. A variety of astrophysical, laboratory, and cosmological bounds exclude unstable neutrinos up to $\sim 40\ MeV$ (unless new physics is invoked for fast, invisible decays or annihilations), implying that m_{ν_μ}, $m_{\nu_\tau} < 40\ eV$.

References

[1] There are several recent reviews of neutrino mass. See, for example, F. Boehm and P. Vogel, *Physics of Massive Neutrinos*, (Cambridge Univ. Press, Cambridge, 1987); J.D. Vergados, Phys. Rev. 133, 1 (1986); S.M. Bilenky and S.T. Petcov, Rev. Mod. Phys. 59, 671 (1987); R. Eichler, *1987 Int. Symp. on Lepton and Photon Interactions at High Energies.*

[2] Under a CP transformation $\psi_L(\vec{x},t) \to \gamma^0\psi_R^c(-\vec{x},t)$, so ψ_L and ψ_R^c are essentially CP conjugates.

[3] L. Wolfenstein, Nucl. Phys. B186, 147 (1981).

[4] E.S. Konopinski and M. Mahmoud, Phys. Rev. 92, 1045 (1953).

[5] Fermion mass matrices need not be Hermitian. One could generalize still further and introduce $K(\neq F)$ N_R^0 fields, in which case m_D would be $F \times K$ dimensional.

[6] We have chosen a basis for the charged leptons so that the analogue of V_L for the electrons is the identity matrix.

[7] $V_L^\dagger$ contains F^2 real parameters, of which $F(F-1)/2$ are angles, analogous to the Cabibbo angle, which describe transitions between families and lead to such effects as neutrino oscillations. $2F-1$ parameters are unobservable phases, which can be removed by appropriate choices of phases for the n_{iL} and e_{iL}^- fields (corresponding phases must be chosen for the n_{iR} and e_{iR} to ensure real masses). The remaining $(F-1)(F-2)/2$ parameters are phases which could lead to CP violation in the leptonic sector.

[8] G.B. Gelmini and M. Roncadelli, Phys. Lett. 99B, 411 (1981); H. Georgi et al., Nucl. Phys. B193, 297 (1983).

[9] More generally, K is block diagonal, the blocks being unitary symmetric matrices in the subspaces of degenerate eigenvalues.

[10] One has $n_{iR}^c = K_{ji}C\bar{n}_{jL}^T$, so that n_i is self-conjugate under the generalized C transformation $n_i \to K_{ji}C\bar{n}_j^T$.

[11] The counting of angles in $U_L^\dagger$ is the same as for the Dirac case. However, there are an additional $F-1$ CP violating phases associated with the fact that U_L and U_R are related by (13). For example, if one chooses K_L so that $K = I$, then the phases of the n_{iL} are fixed and cannot be redefined to remove phases from U_L. See S.M. Bilenky, J. Hosek, and S.T. Petcov, Phys. Lett. 94B, 495 (1980); M. Doi et al., Phys. lett. 102B, 323 (1981).

[12] W.Y. Keung and G. Senjanovic, Phys. Rev. Lett. 50, 1427 (1983); J.F. Gunion and B. Kayser, *Proc. 1984 Summer Study on the SSC*, eds R. Donaldson and J.G. Morfin (APS, 1984) p. 153.

[13] H. Goldberg, Phys. Rev. Lett. 50, 1419 (1983); L. M. Krauss, Phys. Lett. 128B, 37 (1983), Nucl. Phys. B227, 556 (1983).

[14] For a review, see B. Kayser, Comm. Nucl. Part. Phys. 14, 69 (1985).

[15] New interactions could in principle distinguish the two cases even for $m_\nu = 0$, depending on whether they conserved lepton number.

[16] S.M. Bilenky and B. Pontecorvo, Lett. Nuovo Cimento 17, 569 (1976). The implications of the existence of Dirac and Majorana neutrino mass terms were also studied in: V. Barger et al., Phys. Rev. Lett. 45, 692 (1976); S.M. Bilenky, J. Hosek and S.T. Petcov, Phys. Lett.

94B, 495 (1980); J. Schechter and J.W. F. Valle, Phys. Rev. D22, 2227 (1980); T.P. Cheng and L.F. Li, Phys. Rev. D22, 2860 (1980); T. Yanagida and M. Yoshimura, Progr. Th. Phys. 64, 1870 (1980).

[17] The relative sign of n_{1R} and n_{2R} will reemerge, however, in neutrinoless double beta decay, where it will be seen to be observable.

[18] M. Gell-Mann, P. Ramond, and R. Slansky, in *Supergravity*, eds. F. van Nieuwenhuizen and D. Freedman, (North Holland, Amsterdam, 1979) p. 315; T. Yanagida, Prog. Th. Physics B135, 66 (1978).

[19] P. Langacker, S. T. Petcov, G. Steigman, and S. Toshev, Nucl. Phys. B282, 589 (1987).

[20] R.N. Mohapatra and G. Senjanovic, Z. Phys. C17, 53 (1983); Q. Shafi and F.W. Stecker, Phys. Rev. Lett. 53, 1292 (1984); K. Kang and M. Shin, Nucl. Phys. B287, 687 (1987); K. Kang and A. Pantziris, Phys. Lett. 193B, 467 (1987); P. Langacker, R.D. Peccei, and T. Yanagida, Mod. Phys. Lett. A1, 541 (1986).

[21] R.N. Mohapatra and G. Senjanovic, Phys. Rev. D23, 165 (1981). For constraints on the heavy neutrino mass, see R.N. Mohapatra; Phys. Rev. D34, 909 (1986).

[22] P. Langacker, B. Sathiapalan, and G. Steigman, Nucl.Phys. B266, 669 (1986).

[23] A. Zee, Phys. Lett. 93B, 389 (1980); 161B, 141 (1985); Nucl. Phys. B264, 99 (1986); K. Tamvakis and J. Vergados, Phys. Lett. 155B, 373 (1985); G.K. Leontaris and K. Tamvakis, Phys. Lett. 191B, 421 (1987); M. Fukugita and T. Yanagida, Phys. Rev. Lett. 58, 1807 (1987).

[24] A. Masiero, D.V. Nanopoulos, and A.Sanda, Phys. Rev. Lett. 57, 663 (1986).

[25] U. Amaldi et al., Phys. Rev. D36, 1385 (1987).

[26] Unlike similar hierarchy problems in comparing the weak and unification scale, we know that there is no supersymmetry to control the radiative corrections.

[27] D.S.P. Dearborn, D.N. Schramm, and G. Steigman, Phys. Rev. Lett. 56, 26 (1986); H.Y. Cheng, Phys. Rev. D36, 1649 (1987); and references therein.

[28] R. Barbieri, J. Ellis, and M.K. Gaillard, Phys. Lett. 90B, 249 (1980).

[29] In many models which implement the seesaw mechanism Higgs triplet fields are also present. Even if these have zero VEV at tree level they may obtain VEV's due to higher order corrections to the Higgs potential. See M. Magg and C. Wetterich, Phys. Lett. 94B, 61 (1980); D. Chang and R.N. Mohapatra, Phys. Rev. D32, 1248 (1985); J. Bijnens and C. Wetterich, DESY 87-042 (1987).

[30] A simple diagonalization yields eigenvalues $-m_D^2/m_S$ and m_S. The sign of the smaller eigenvalue is reversed by choosing $K = \mathrm{diag}(-1\ 1)$, which accounts for the minus sign in ν_R^{0c}.

[31] E. Witten, Phys. Lett. 91B, 81 (1980).

[32] Some versions of these models [29] are expected to have induced values for m_t in (31).

[33] Y. Chikashige, R.N. Mohapatra, and R.D. Peccei, Phys. Lett. 98B, 265 (1981).

[34] In many of the intermediate mass seesaw models [20] the singlet-Majoron and the invisible axion are the same particle.

[35] S. Nandi and U. Sarkar, Phys. Rev. Lett. 56, 564 (1986); J. Derendinger, L. Ibanez, and H.P. Nilles, Nucl. Phys. B267, 365 (1986).

[36] J. Ellis et al., Phys. Lett. 188B, 415 (1987).

[37] M. Roncadelli and D. Wyler, Phys. Lett. 133B, 325 (1983); A.S. Joshipura et al., Phys. Lett. 156B, 353 (1985); P. Roy et al, Phys. Rev. D30, 2385 (1984); O. Shanker, Nucl. Phys. B250, 351 (1985); J. Oliensis and C.H. Albright, Phys. Lett. 160B, 121 (1985).

[38] S. Boris et al., Phys. Rev. Lett. 58, 2019 (1987).

[39] M. Fritschi et al., Phys. Lett. 173B, 485 (1986).

[40] J.F. Wilkerson et al, Phys. Rev. Lett. 58, 2023 (1987).

[41] H. Kawakami et al., *'86 Massive Neutrinos in Physics and Astrophysics*, ed. O. Fackler and J. Tran Than Van.

[42] See, for example, D. Schramm, *1987 Int. Symp. on Lepton and Photon Interactions at High Energies*; H. Meyer, this volume; E.W. Kolb et al., Phys. Rev. D35, 3598 (1987).

[43] R. Abela et al., Phys. Lett. B146, 431 (1984).

[44] ARGUS collaboration, quoted by R. Eichler, [1].

[45] K. Hirakata et al., Phys. Rev. Lett. 58, 1490 (1987).

[46] R.M. Bionta et al., Phys. Rev. Lett. 58, 1494 (1987).

[47] For reviews and a general discussion, see M. Gronau, C.N. Leung, and J.L. Rosner, Phys. Rev. D29, 2539 (1984); F.J. Gilman and S.H. Rhie, Phys. Rev. D32, 324 (1985). F.J. Gilman, Comm. Nucl. Part. Phys. 16, 231 (1986); R.E. Shrock, Phys. Rev. D24, 1232, 1275 (1981).

[48] A similar phenomenon occurs for a system of coupled classical oscillators.

[49] For a discussion of quantum mechanical subtleties, such as the uncertainty in p, see F. Boehm and P. Vogel, [1].

[50] IMB, J.M. Lo Secco et al., Phys. Lett. 184B, 305 (1987).

[51] The limits in Fig. 2 are obtained by analyzing the data in terms of the formula in (36) for oscillations between two types of neutrino only. Slightly weaker limits are obtained if one allows mixings between 3 neutrino flavors (ν_e, ν_μ, ν_τ). See K. Kleinknecht, Comm. Nucl. Part. Phys. 16, 267 (1986).

[52] J.F. Cavaignac et al., Phys. Lett. 148B, 387 (1984).

[53] G. Zacek et al., Phys. Rev. D34, 2621 (1986).

[54] G. Bernardi et al., Phys. Lett. 181B, 173 (1986).

[55] R. Eichler, [1].

[56] Another manifestation of the underlying physics would be off-diagonal neutral currents, so that a heavy neutrino could decay into three light neutrinos.

[57] D. Wyler and L. Wolfenstein, Nucl. Phys. B218, 205 (1983); B.W. Lee and R.E. Shrock, Phys. Rev. D16, 1444 (1977); P. Langacker and D. London, to be published.

[58] For a review, see G. Gelmini, this volume.

[59] Both Γ_{weak} and H have additional factors involving the number of types of relativistic particles at temperature T.

[60] P. Langacker, J.P. Leveille, and J. Sheiman, Phys. Rev. D27, 1228 (1983).

[61] N. Cabibbo and L. Maiani, Phys. lett. 114B, 115 (1982).

[62] For a recent review, see B. Müller, this volume.

[63] For masses in the 10 eV range the $\nu's$ would be clustered in the galaxy and their cross sections would be increased by 10^8, but direct detection still appears virtually impossible.

[64] Massive neutrinos could easily account for the scales of galactic clusters. However, it is hard to see how such structures could have fragmented into galaxies in the time available, unless another ingredient, such as cosmic strings, provided the seed.

[65] G. Steigman et al., Phys. Lett.176B, 35 (1986).

[66] For a recent review, see M. Roos, this volume; M. Turner, AIP Conf. Proc. 72, 266 (1980).

[67] Most of these limits apply to light SU_2-singlet neutrinos as well, because in many cases their small mixing with ordinary doublet neutrinos is sufficient to produce them prolifically in the early universe. See [22].

[68] S. Sarker and A.M. Cooper, Phys. Lett. B148, 347 (1984); L. Krauss, Phys. Rev. Lett. 53, 1976 (1984); R. Cowsik, Phys. Lett. 151B, 52 (1985); H. Harari and Y. Nir, Phys. Lett. 188B, 163 (1987), Nucl. Phys. B292, 251 (1987).

[69] Y. Hosotani, Nucl. Phys. B191, 411 (1981); M. Davis et al., AP.J. 250, 423 (1981); P. Hut and S. White, Nature 310, 637 (1984); P. Binetruy et al., Phys. Lett. 134B, 147 (1984); A.A. Natale, Phys. Lett. 141B, 323 (1984).

[70] M. Roncadelli and G. Senjanovic, Phys. Lett. 107B, 59 (1983).

[71] P.B. Pal, Nucl. Phys. B227, 237 (1983); H. Harari and Y. Nir, [68].

[72] D.B. Reiss, Phys. Lett. 115B, 217 (1982); F. Wilczek, Phys. Rev. Lett. 49, 1549 (1982); G. Gelmini, S. Nussinov, and T. Yanagida, Nucl. Phys. B219, 31 (1983).

[73] See [8].

[74] [33] and J. Schechter and J.W.F. Valle, Phys. Rev. D25, 774 (1982); G.B. Gelmini and J.W.F. Valle, Phys. Lett. 142B, 181 (1984); G.B. Gelmini et al., Phys. Lett. 146B, 311 (1984); A. Kumar and R. N. Mohapatra, Phys. Lett. 150B, 191 (1985).

[75] P. Langacker, G. Segre, and S. Soni, Phys. Rev. D26, 3425 (1982).

[76] For a recent review, see A. Faessler, this volume.

[77] T. Kirsten et al., Phys. Rev. Lett. 50, 474 (1983), Z. Phys. C16, 189 (1983).

[78] M.K. Moe, this volume.

[79] Neutrinoless double beta decay could conceivably occur through other mechanisms, such as Higgs exchange, or could be accompanied by the emission of a triplet-Majoron. [81] See G.K. Leontaris and J.D. Vergados, CERN-TH.4835/1987. J.D. Vergados, [1].

[80] F.T. Avignone, this volume.

[81] The PNL/USC group [80] has reported evidence for $^{76}Ge \rightarrow\ ^{76}Se\ e^- e^- +$ Majoron, but their results are in contradiction with other experiments.

[82] W.C. Haxton and G.J. Stephenson, Prog. Part. Nucl. Phys. 12, 409 (1984); T. Tomoda et al., Nucl. Phys. A452, 591 (1986); K. Grotz and H.V. Klapdor, Nucl. Phys. A460, 395 (1986); T. Tomoda and A. Faessler, to be published.

[83] J. Engel, P. Vogel, and M.R. Zirnbauer, to be published.

[84] L. Wolfenstein, Phys. Lett. 107B, 77 (1981); M. Doi et al., Phys. Lett. 102B, 323 (1981).

[85] See W.C. Haxton and G.J. Stephenson, [82], J.D. Vergados, [1].

[86] A. Halprin, S.T. Petcov, and S.P. Rosen, Phys. Lett. 125B, 335 (1983); C.N. Leung and S.T. Petcov, Phys. Lett. 145B, 416 (1984).

[87] M. Doi, T. Kotani, and E. Takasugi, Prog. Th. Phys. (Supp) 83, 1 (1985).

[88] However, in a gauge theory some neutrino must be massive. Otherwise the diagrams cancel. See B. Kayser, *7th Moriond Workshop on Searches for New and Exotic Phenomena*, Les Arcs, Jan. 1987.

[89] J.K. Rowley, B.T. Cleveland and R. Davis, AIP Conference Proc. 126, 1 (1985).

[90] A Solar neutrino unit (SNU) is defined as 10^{-36} interactions per target nucleus per second.

[91] J.N. Bahcall et al., Rev. Mod. Phys. 54, 767 (1982); J.N. Bahcall, AIP Conference Proc. 126, 60 (1985).

[92] H. Hirata et al, to be published.

[93] For recent discussions, see V. Barger et al., Phys. Rev. D24, 538, (1981); S.L. Glashow and L.M. Krauss, Phys. Lett. 190B, 199 (1987); L. Krauss and F. Wilczek, Phys.Rev. Lett. 55, 122 (1985).

[94] M.B. Voloshin and M.I Vysotsky, ITEP 86-1(1986); L.B. Okun et al., Zh. Eksp. Teor. Fiz. 91, 754 (1986) (JETP 64, 446 (1986)). For earlier references, see [19].

[95] A.V. Kyuldjiev, Nucl. Phys. B243, 387 (1987); K. Abe et al., Phys. Rev. Lett. 58, 636 (1987).

[96] For a recent discussion, see M. Fukujita and S. Yazaki, Kyoto preprint. RIFP-709.

[97] B.W. Lee and R. Shrock, Phys. Rev. D16, 1444 (1977); K. Fujikawa and R. Shrock, Phys. Rev. Lett. 45, 963 (1980).

[98] M.J. Duncan et al., Phys. Rev. Lett. 191B, 304 (1987); J. Liu, Phys. Rev. D35, 3447 (1987); M. Fukugita and T. Yanagida, [23].

[99] For a review, see W. Hampel, this volume.

[100] F.A. Frieman, H.E. Haber, and K. Freese, Phys. Lett. 200B, 115 (1988).

[101] S.P. Mikheyev and A. Yu Smirnov, Yad Fiz. 42, 1441 (1985) (Sov. Jour. Nucl. Phys. 42, 1441 (1985)); Nuo. Cim. 9C, 17 (1986).

[102] L. Wolfenstein, Phys. Rev. D17, 2369 (1968); D20, 2634 (1979).

[103] V. Barger et al., Phys. Rev. D22, 2718 (1980).

[104] A crucial sign error was corrected in [60].

[105] The scattering amplitude in M' changes sign for antineutrinos. Hence, resonance can occur for either ν_e or $\bar{\nu}_e$, depending on the sign of Δm^2.

[106] H.A. Bethe, Phys. Rev. Lett. 56, 1305 (1986).

[107] S.P. Rosen and S.M. Gelb, Phys. Rev. D34, 969 (1986).

[108] P. Langacker, S.T. Petcov, G. Steigman, and S. Toshev, Nucl. Phys. B282, 589 (1987).

[109] E.W. Beier et al., 1986 Snowmass Workshop, Pennsylvania UPR-0140E.

[110] J. Arafune et al., Phys. Rev. Lett. 59, 1864 (1987); P.O. Lagage et al., Phys. Lett. 193B, 127 (1987); L. Wolfenstein, Phys. Lett. 194B, 197 (1987); D. Nötzgold, Phys. Lett. 196B, 315 (1987); S. P. Rosen, Los Alamos LA-UR-87-1296.

Small Neutrino Masses in Gauge Theories*

R.N. Mohapatra

Department of Physics and Astronomy, University of Maryland,
College Park, MD 20742, USA

We discuss the various ways in which small neutrino masses can be understood in gauge theories such as left-right symmetric models, $SO(10)$ grandunified theories and superstring models.

1. Introduction

Since Pauli postulated the neutrino more than fifty years ago, it has played a key role in our understanding of the weak interactions. Today, as we we look beyond the standard model in search of clues to resolve the unsatisfactory aspects of the standard model such as the fermion masses, mixings and approximate symmetries of the model, the neutrino appears again to be holding the key. The neutrino mass is the focus of attention in this discussion. If the neutrino has a mass, there are two challenges to theorists: one is to understand why its mass is so tiny compared to the mass of the charged fermion of the corresponding generation and secondly, whether its mass is Dirac ($\Delta L = 0$) or Majorana ($\Delta L = 2$) type. In this article, I discuss various popular extensions[1,2,3)] of physics beyond the standard model in order to understand neutrino mass in those models. Before starting this discussion, a brief review is given of the observational constraints on the masses and decays of the neutrinos.

2. Observational Constraints on Massive Neutrinos:

There are two classes of observational constraints on massive neutrinos: one, implied by cosmological observations and second, implied by accelerator and reactor experiments attempting to detect neutrino stability. The cosmological constraints date back to the pioneering works of Cowsik and Mclelland[5] and Gershtein and Zeldovich[6] who pointed out that, there exists a neutrino sea all around us, with a temperature of $1.9^{o}\,K \simeq 1.6 \times 10^{-4}$ ev. Usual considerations of Fermi-Dirac statistics then leads to the conclusion that the present number density of each neutrino species is given by $n_\nu \simeq \frac{3}{11}\, n_\gamma \simeq 109/cm^3$. If neutrinos have mass m_{ν_i} and are stable, then their contribution to the mass density of the universe must be smaller than the critical mass density of the universe leading to an upper limit on the sum of the masses of the stable neutrinos:

$$\left(m_{\nu_e} + m_{\nu_\mu} + m_{\nu_\tau} + \cdots\right)\, n_\nu \leq 10^4\, h_o^2\, ev$$

where h_o denotes the uncertainty in the Hubble constant (i.e. $H_o^{-1} = 10^{10}/h_o$ years) (or the age of the universe) and we have $.4 \leq h_o \leq 1$, with smaller h_o corresponding to an older universe. For $h_o \approx .5$ to 1, we get,

$$\Sigma\, m_{\nu_i} \leq 25 \text{ to } 100\; ev\,. \tag{2.1}$$

*Work supported by a grant from the National Science Foundation.

Later on it was pointed out[7] that for stable heavier neutrinos, there is a lower bound of about 2 GeV. The situation is, however, very different for unstable neutrinos as was first pointed out by Dicus, Kolb and Teplitz.[8] The reason of course is that, if a neutrino species decays in the course of the evolution of the universe, it no longer contributes to mass density at present; the decay products, however, carry the information about the mass in their energy but energy, unlike the mass gets redshifted with the cosmological evolution and if the decay happens early enough, the energies could be sufficiently redshifted to be harmless. In this way, one gets a constrained joint bound on mass and lifetime as follows:

$$m_{\nu_H} \left(\frac{\tau_{\nu_H}}{\tau_U}\right)^{1/2} \leq 25\, ev \ \text{ to } \ 100\, ev\ . \tag{2.2}$$

There are three principal decay modes of heavier neutrinos that one may consider to avoid the bound in eqn. (2.1):

a) $\nu_H \to \nu_e \gamma$: Cosmology implies the following bound on this decay mode. Unless the half-life is less than[8a] 10^4 sec., the decay photons will disintegrate the deuterium and affect the isotropy of the observed black body radiation. In gauge models, these decay times are too long to be useful in avoiding the cosmological bounds. (See later). Therefore, much attention has been focussed in the following two decay modes:

b) $\nu_H \to \nu_e + \chi$, where χ is the Majoron,[9] which represents the Goldstone boson corresponding to spontaneous breaking of global $B - L$ quantum number. These decay modes are known[10] to lead to life-times much shorter than the photonic decay mode.

c) $\nu_H \to 3\nu_e$: This mode, in certain models such as the left-right symmetric ones, is important for consistency of neutrino masses with cosmology but often requires fine tuning of parameters in gauge models[11] since by an $SU(2)_L$ rotation, $\nu_\mu \to 3\nu_e$ gets related to $\mu \to 3e$ decay, which is known to have a tiny upper bound on its branching ratio ($\lesssim 10^{-11}$). In some models with Majorana neutrinos $\nu_H \to 3\nu_e$ can be induced via flavor changing neutral current coupling Z , but these effects are small.

It is important to point out that arbitrary neutrino mass can be generated by including the right-handed neutrino in the standard model; but the decay modes in cases (b) and (c) involve new physics beyond this and are therefore extremely important. For instance, evidence for neutrino mass in the forbidden range (10^{-7} to 2 GeV) combined with cosmological constraints will imply the existence of the massless Majoron or an ultralight neutral Higgs boson (in the Gev range).[12]

Another kind of constraint on the neutrino mass spectrum has recently been in the news a great deal having to do with the so-called Mikheyev-Smirnov-Wolfenstein solution to the solar neutrino puzzle. According to this proposal, the neutrino mass matrix gets altered due to interaction with a dense medium thereby enabling the electron-neutrino emitted from the solar core to get oscillated away to ν_μ or ν_τ. This will explain the deficit of neutrinos observed by Davis. The resolution of the solar neutrino puzzle requires $10^{-7} \leq m^2_{\nu_i} - m^2_{\nu_e} \leq 10^{-4}$ ev^2 with the corresponding mixing angle $\theta_{ei} \geq 10^{-2}$. Barring any accidental degeneracy, this would lead one to conclude that $m_{\nu_e} << m_{\nu_\mu} << m_{\nu_\tau}$ with either ν_μ or ν_τ weighing $\simeq 10^{-2}$ eV. We will describe a unique $SO(10)$ model that leads to a neutrino mass of the above kind without any unnatural finetuning of parameters.

Laboratory experiments also have led to constraints on neutrino masses as follows:[13] $m_{\nu_e} \leq 18$ ev, $m_{\nu_\mu} \leq .25$ Mev and $m_{\nu_\tau} \leq 70$ Mev. There exist constraints on the mixing angles from various accelerator experiments, which we do not include in this article, since we will not be using it in the subsequent sections.

3. Majorana, Dirac or Pseudo-Dirac Neutrino Masses:

Unlike a boson or an electrically charged fermion, a neutral spin half object such as the neutrino can have several kinds of masses. To understand this, we consider 2-component Weyl spinor ν to denote the chiral spin 1/2 neutrino emitted in the process of β-decay. It transforms as a $\left(\frac{1}{2},0\right)$ representation of the Lorentz group and admits a Lorentz invariant mass term such as:

$$\mathcal{L}^{(1)}_{mass} = i\,m\,\nu^T\,\sigma_2\nu + h.c. \tag{3.1}$$

If we assume that the neutrino carries a global quantum number such as lepton number (under which $\nu \to e^{i\alpha}\,\nu$), then $\mathcal{L}^{(1)}_{mass}$ violates lepton number by two units. In this case, the neutrino is said to have a Majorana mass and is called a Majorana particle. If the only spin half neutral fermion (with properties similar to the neutrino) around is ν, then no other mass term is possible for the neutrino. So, the neutrino will either remain massless or become a Majorana particle. On the other hand, if there is another spin 1/2 neutral lepton N present, then, one can envisage a generalized form for the mass term:

$$\mathcal{L}^{(2)}_{mass} = i\left(\nu^T\,N^T\right)\begin{pmatrix} m_{LL} & m_{LR} \\ m_{LR} & m_{RR} \end{pmatrix}\sigma_2\begin{pmatrix} \nu \\ N \end{pmatrix} \tag{3.2}$$

(We will denote this 2×2 mass matrix in eqn. (3.2) by M). If all elements of M are not-vanishing, again no global symmetry will be respected by $\mathcal{L}^{(2)}_{mass}$; therefore, in this case, the neutrino will also be a Majorana particle and lepton number will be violated by two units. One can, however, have special situations:

a) $m_{LL} = m_{RR} = 0$: In this case, one can define the lepton number $U(1)$ symmetry as follows: $\nu \to e^{i\alpha}\,\nu$ and $N \to e^{-i\alpha}\,N$. $\mathcal{L}^{(2)}_{mass}$ will then respect lepton number. The ν and N combine to form a four-component Dirac neutrino. In the two component notation, one has the following eigenstates:

$$\nu_\pm = \frac{\nu \pm N}{\sqrt{2}} \tag{3.3}$$

with mass eigenvalues $\pm m_{LR}$; making the mass positive changes ν_- to $i\nu_-$. In the four component language, we write,

$$\psi = \begin{pmatrix} \nu \\ i\sigma_2\,N^* \end{pmatrix}. \tag{3.4}$$

Then,

$$\psi^c = C\bar{\psi}^T = -\gamma_2\psi^* = \begin{pmatrix} N \\ i\sigma_2\nu^* \end{pmatrix}. \tag{3.5}$$

We then define:

$$\psi_\pm = \frac{1}{\sqrt{2}}\left(\psi \pm \psi^c\right) = \begin{pmatrix} \nu_\pm \\ i\sigma_2\nu_\pm^* \end{pmatrix}. \tag{3.5}$$

From (3.5), it is obvious that, ψ_+ and ψ_- are even and odd under charge conjugation respectively.

b) $m_{LR} >> m_{LL} = m_{RR}$: In this case, the neutrino is called a pseudo-Dirac particle.[14] In this case there is a slight mixing between the CP-odd and CP-even eigenstates $\psi_\pm$; although the neutrino is a predominantly Dirac particle, it has a very slight admixture of Majorana character. This situation is of particular interest in the study of evolution of a chiral left-handed neutrino state, such as that emitted in beta decay. When m_{LL} or m_{RR} are nonzero (and small, m_{LL} , $m_{RR} << m_{LR}$), the eigenstates of M are still $\nu_\pm$; but their masses are split by an amount δm , i.e.

$$m_\pm = \mp m_{LR} + m_{LL} \,. \tag{3.7}$$

The left-handed neutrino oscillates into the extra sterile lepton N , with a probability given by

$$P_{\nu N}(t) \simeq sin^2 \frac{1}{2} \left(\frac{m_{LL}}{R}\right) \left(\frac{m_{LR}}{E}\right)^t \,. \tag{3.8}$$

If $m_{LL} \times m_{LR} \simeq 10^{-11}$ eV^2, this can provide an alternative solution to the solar neutrino problem.[15]

c) $m_{LL} = 0$; $m_{LR} << m_{RR}$: In this case, the neutrino is a Majorana particle. But unlike cases (a) and (b) where the two neutrino states are nearly degenerate, in this case, there are two neutrinos, one heavy ($m_H \simeq m_{RR}$) and one light ($m_L \simeq m_{LR}^2/m_{RR}$). This is well known as the "see-saw" mechanism and provides a "naturally" small value for the neutrino mass without any finetuning of parameters. The eigenstates are also nearly pure, i.e.

$$\nu_{Light} \simeq \nu + \left(\frac{m_{LR}}{m_{RR}}\right) \cdot N$$

$$\nu_{Heavy} \simeq N - \left(\frac{m_{LR}}{m_{RR}}\right) \nu \,. \tag{3.9}$$

Before closing this section, we wish to reemphasize the point that Majorana mass terms require the breaking of Lepton number, L. Since in general gauge theories the only gauge-anomaly free combination of quantum numbers is $B-L$, we will talk about breaking of $B-L$ symmetry. In the context of gauge theories, there are three possible ways to achieve this breaking:

(i) Explicit breaking of $B-L$ symmetry.
(ii) Spontaneous breaking of local $B-L$ symmetry[16]
(iii) Spontaneous breaking of global $B-L$ symmetry.[9]

In the rest of the article, we will focus on the local $B-L$ symmetry models and its connection to neutrino mass. In case (iii), a Goldstone boson (the Majoron[9]) appears in the theory coupled to the neutrino. This particle leads to many interesting phenomenological and cosmological implications for neutrino decay and neutrino scattering. For instance, heavier neutrinos can now decay via the emission of Majoron such as $\nu_H \to \nu_e + \chi$; there can also be neutrino annihilation such as $\nu\nu \to \chi\chi$. In the laboratories, one can have new modes for neutrinoless double β-decay such as $(N,Z) \to (N-2, Z+2) + 2e^- + \chi$. These modes are under search at various laboratories around the world.[17]

4. The Left-Right Symmetric Models:

In the standard model of Glashow, Weinberg and Salam, only one helicity component is introduced into the theory to fit observations in muon and beta decay. Since the model has exact $B-L$ symmetry, it follows from the discussions of sec. 3 that neutrinos must be massless. In this model, the left-handed nature of weak interactions is also built in by hand, because of which no fundamental understanding of the origin of parity violation emerges. In order to provide an understanding of parity violation, the left-right symmetric theories of weak interactions were developed in a series of papers[1] in 1974-75. In these models, the basic interactions are invariant under parity transformations: the observed near maximal parity violation at low energies is then understood to be a consequence of spontaneous symmetry breaking. Before proceeding to the discussion of neutrino masses in this model, we first present the basic ideas of the model.

Motivation and description of the Model

There is a rather compelling reason why left-right symmetry ought to be a fundamental and natural symmetry of the quark-lepton world. It is well-known that weak interactions are symmetric between quarks and leptons. We can use this symmetry to write a unified formula for electric charge in terms of weak isospin and $B-L$ quantum number, in a manner analogous to the Gell-Mann-Nishijima formula for the hadronic world, i.e.

$$Q = I_{3W} + \frac{B-L}{2} \,. \tag{4.1}$$

Since at low energies electroweak processes involve only the left-handed weak isospin, we must rewrite the eqn. (4.1) as

$$Q = I_{3L} + I_{3R} + \frac{B-L}{2} \,. \tag{4.2}$$

It is then natural to assume that Nature realizes the full right-handed weak isospin in the same way as it realizes the left-handed one. This then leads to the left-right symmetric model of weak interactions, based on the gauge group $SU(2)_L \times SU(2)_R \times U(1)_{B-L}$.[1] The quarks and leptons are assigned to the gauge group as follows: (denote $Q \equiv (u,d)$ and $\psi \equiv (\nu, e^-)$)

$$Q_L : (2,1,1/3)\,; \quad Q_R : (1,2,1/3)\,; \quad \psi_L : (2,1,-1)\,; \quad \psi_R : (1,2,-1)\,. \tag{4.3}$$

It is clear that gauge interactions are symmetric between left and right-handed fermions; therefore, prior to spontaneous breaking of gauge symmetry, weak interactions like strong, electromagnetic and gravitational interactions conserve parity. This can be seen explicitly, by writing the gauge interactions

$$\mathcal{L}_{WK} = \frac{g}{2}\left[\vec{J}_{\mu,L}\cdot\vec{W}^{\mu}_{L} + \vec{J}_{\mu,R}\cdot\vec{W}^{\mu}_{R}\right] + g' V^{B-L}_{\mu}\cdot B^{\mu} \,. \tag{4.4}$$

The $\vec{W}_L$, $\vec{W}_R$ and B are the gauge bosons corresponding to $SU(2)_L$, $SU(2)_L$ and $U(1)_{B-L}$ gauge symmetries. As long as the W_L and W_R have the same mass, the weak interactions conserve parity. The observed parity violation at low energies is then attributed to vacuum being parity asymmetric, which manifests itself in $m_{W_R} >> m_{W_L}$. The symmetry breaking pattern responsible for the parity violation observed at low energies can be obtained using the following generic pattern for the Higgs multiplets: (we drop the reference to color gauge group from now on) ϕ: $(2,2,0)$, $H_L\,(a,1,b) + H_R\,(1,a,b)$ where a and b will be fixed later. The first stage of the symmetry breaking is implemented by the neutral component of the right-handed multiplet H_R acquiring a v.e.v. i.e., $\langle H^{\circ}_{R}\rangle = v_R$. This reduces the electroweak symmetry to $SU(2)_L \times U(1)_Y$, which is subsequently broken by ϕ acquiring a v.e.v.

$$\langle\phi\rangle = \begin{pmatrix} \kappa & 0 \\ 0 & \kappa' e^{i\alpha} \end{pmatrix} \,. \tag{4.5}$$

The second stage of the symmetry breaking can induce[18] a v.e.v. for $\langle H^{\circ}_{L}\rangle = v_L \simeq \gamma\,\kappa^2/v_R$ assuming $\kappa' >> \kappa$; the quarks and leptons acquire a mass at the second stage, as do the W_L and Z-bosons. The form of the low energy charged current interaction in this model is as follows:

$$H^{c.c}_{WK} = \frac{G_F}{\sqrt{2}}\left[\left(\cos^2\zeta + \eta\sin^2\zeta\right) J^{+}_{\mu_L} J^{\mu^-}_{L} + \left(\eta\cos^2\zeta + \sin^2\zeta\right) J^{+}_{\mu_R} J^{\mu^-}_{R}\right.$$

$$\left. + e^{i\alpha}\cos\zeta\,\sin\zeta\,(1-\eta)\, J^{+}_{\mu_L} J^{\mu^-}_{R} + \text{h.c.}\right] \tag{4.6}$$

where $\eta = \left(\frac{m_{W_L}}{m_{W_R}}\right)^2$ and ζ is the mixing angle between the charged W-bosons and is given in terms of symmetry breaing parameters as $\zeta \simeq \left(\kappa\kappa'/m_{W_R}^2\right)$. The currents $J_{\mu L,R}$ are given as follows:

$$J_{\mu L}^{+} = \overline{P}\,\gamma_\mu\,(1+\gamma_5)\,U_L N + \overline{E}^o\,\gamma_\mu\,(1+\gamma_5)\,V_L E^- \tag{4.7}$$

where P, N, E^o and E^- denote the up quark, down quark, neutrino and electron column vector involving all generations and U_L and V_L denote the charged current mixing matrices for the quark and lepton sector. The corresponding right-handed currents are obtained by replacing L by R and γ_5 by $-\gamma_5$.

The form of the neutral current interaction of Z_1 and Z_2 can be written in these models in the form[18,19]

$$\mathcal{L}_{WK}^{N.C.} = \frac{g}{\cos\theta_W}\left[Z_{1\mu}\left\{J_{ZL}^{\mu} - \eta_Z J_\mu\right\} + \frac{1}{(\cos 2\theta_W)^{1/2}}\,Z_{2\mu}J_\mu\right] \tag{4.8}$$

where $\eta_Z = (M_{Z_1}/M_{Z_2})^2$ and $J = \sin^2\theta_W\,J_{ZL} + \cos^2\theta_W\,J_{ZR}$; $J_{ZL,R} = \left(I_{3L,R} - Q\sin^2\theta_W\right)$. We, then, see that the effects of right-handed bosons vanish as $M_{Z_2} \to \infty$ and we obtain the neutral current interaction of the standard model. Typical precision of neutral current experiments is about 10%, which allows one to have a rather light Z_2; a recent analysis[19] yields $M_{Z_2} \geq 275$ Gev for these models.

As far as the charged W_R-boson is concerned, we consider both the light Dirac and Majorana cases. And, we will further assume that the left and right handed mixing matrices are equal. This is the case of manifest left-right symmetry[20] which emerges for the simplest choice of Higgs multiplets ϕ. In this case, if the neutrino is a Dirac particle with ν_L and ν_R forming this particle or is a Majorana particle such that $m_{\nu_R} \leq 10$ Mev, the most model independent bounds on m_{W_R} and ζ come from the analysis of a recent TRIUMF muon decay experiment.[21] In this experiment the endpoint spectrum of the positrons moving along a direction opposite to the spin of a stopped 100% polarized μ^+ is measured. In pure $V-A$ theory, this spectrum vanishes at the end point, thus any non-vanishing would be an indication of the presence of $V+A$ currents. The quantity measured this way, to be called R is given in terms of the muon decay parameters δ, ξ and ρ as

$$R = 1 - \frac{\delta\xi}{\rho}\,P_\mu \tag{4.9}$$

where in terms of the η and ζ defined after eqn. (4.6) we get

$$R = 4\eta^2 + 2\zeta^2 + 4\eta\zeta\,. \tag{4.10}$$

The experimental result for $1-R = .99863 \pm .00046$ (stat) $\pm .00075$ (syst.). This implies that $m_{W_R} \geq 432$ Gev for arbitrary ζ and $m_{W_R} \geq 514$ Gev for $\zeta = 0$; for arbitrary m_{W_R}, $-0.050 \leq \zeta \leq .035$ and $|\,\zeta\,| \leq .035$ for $m_{W_R} \to \infty$. These analyses have recently been extended[22] to the case of non-manifest left-right symmetry,[23] where bounds on m_{W_R} become weaker.

Turning now to the case of the heavy right-handed neutrino (i.e. m_{ν_R} in the Gev range), the only bound comes from non-leptonic decays such as K-decays[24] and $K_L - K_S$ mass difference[25] (ΔM_K), the latter case leading to the most stringent bound. The reason for this bound is that calculation of the one loop $\Delta S = 2$ effective Hamiltonian arising from the $W_L - W_R$ exchange leads to an enhanced contribution to ΔM_K as follows

$$\Delta m_K = \Delta m_{K_{LL}}\,[1 - 430\,\eta]\,. \tag{4.11}$$

Thus we see that unless $\eta \leq 2.3 \times 10^{-3}$ or $m_{W_R} \geq 1.6$ Tev, eqn. (4.11) will give the wrong sign and contradict observations. This is the most stringent bound on m_{W_R}.

Coming now to the bound on the mixing parameter ζ, we have already seen that for Dirac neutrinos muon decay experiments provide a bound $\zeta < 3.5\%$. For heavy right-handed neutrinos, the most model independent bound comes from the study[26] of y-distribution in deep inelastic scattering of antineutrinos off nuclei. In the absence of $V + A$ current, the valence quarks lead to $(1-y)^2$ type distribution, whereas the left-right mixing leads to flat y-distribution proportional to ζ. Analysis of existing experiments[27] lead to $\zeta < .1$. Other more model dependent bounds can be obtained by considering deviations from current algebra relations in non-leptonic decays,[26] ($\zeta \leq 4 \times 10^{-3}$) as well as from beta decays[28] of nucleus combined with unitarity of the KM-matrix ($\zeta < .005$). Finally, we note that with the simplest Higgs structure (one ϕ and $\Delta_{L,R}$), one has the following theoretical upperbound[29] on the mixing parameter $\zeta \leq \left(\frac{m_{W_L}}{m_{W_R}}\right)^2$. If we use the bound from the $K_L - K_S$ mass difference, it implies $\zeta \leq 2 \times 10^{-3}$, which is the most stringent, though model dependent bound on this parameter.

Bound on the mass of the Majorana Neutrino ν_R:

The see-saw model for neutrino masses implies that the m_{ν_R} is in the tens of Gev range. An interesting limit that correlates $(m_{\nu_R})_{min}$ with m_{W_R} arises from present experimental limits[30] on neutrinoless double β-decay.[31] The point is that a heavy right-handed neutrino contributes to $(\beta\beta)_{ov}$ decay via the exchange of W_R boson with an amplitude proportional to $G_F^2 \eta^2 m_{\nu_R}$ $\langle e^{-m_{\nu_R} r}/r^o \rangle_{nuc.}$ and this contribution adds incoherently with the usual light neutrino contribution. Using the available nuclear matrix element calculations,[32] we obtain the bound shown in fig. 1. An upper limit on $m_{\nu_R} \leq m_{W_R}$ arises from considerations of vacuum stability, which is the right-hand line in fig. 1.

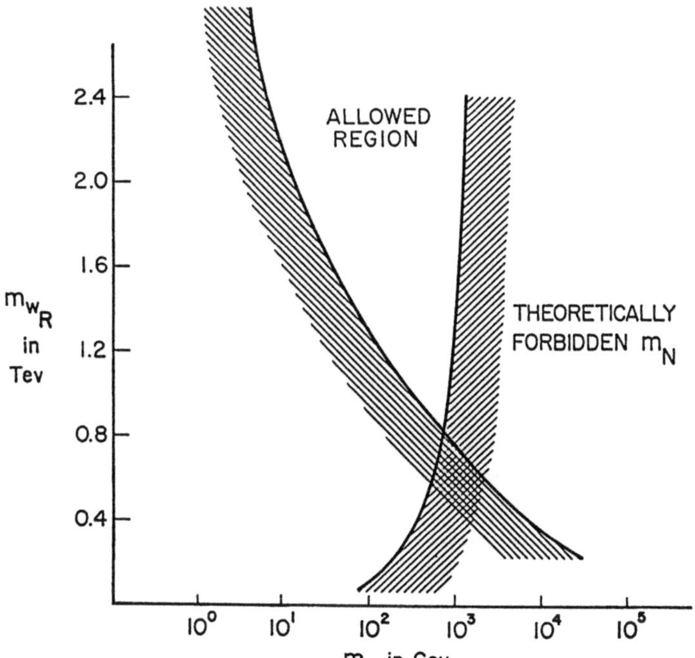

Fig. 1.
Correlated bounds on m_{N_R} and m_{W_R} from experiments on neutrinoless double beta decay (left arm) and considerations of vacuum stability. The shaded region is forbidden.

5. Dirac vs. Majorana Neutrino in Left-Right Models:

Due to left-right symmetry, ν_L is accompanied by ν_R leading to a rich set of possibilities for neutrino masses. The detailed nature, however, depends on the pattern of symmetry breaking which must lead to the standard model at low energies. We see that a crucial role is played by the Higgs boson that is responsible for breaking the $SU(2)_R$ symmetry, i.e. the nature of the Higgs multiplets H_L and H_R. Two possible choices of H_L and H_R can be considered: the first one which we will call the canonical choice was introduced in ref. 18 (case i) and it leads to Majorana neutrinos and provides an understanding of the smallness of the neutrino mass being related to the suppression of $V+A$ currents. The second one we mention here leads to light Dirac neutrinos (case ii) at the price of expanding the model to include singlet leptons.[33]
Case (i): Here, H are chosen to be triplets under the weak gauge group and are denoted by $\vec{\Delta}_L(3,1,2) + \vec{\Delta}_R(1,3,2)$. They couple to leptons only as follows:

$$\mathcal{L}'_Y = h\left\{\psi_L^T c^{-1}\tau_2\vec{\tau}\cdot\vec{\Delta}_L\psi_L + \psi_R^T C^{-1}\tau_2\vec{\tau}\cdot\vec{\Delta}_R\psi_R\right\} + \text{h.c.} \tag{5.1}$$

Choosing $\langle\Delta_R^\circ\rangle = v_R$ and assuming that the neutrino Dirac mass coming from the $SU(2)_L$ breaking is m_e, we obtain the following 2×2 mass matrix for $(\nu, N)'$

$$\begin{array}{cc} & \begin{array}{cc}\nu & \quad N\end{array} \\ \begin{array}{c}\nu \\ N\end{array} & \begin{pmatrix} h\gamma\,\kappa^2/v_R & m_e \\ m_e & h\,V_R \end{pmatrix} \end{array} . \tag{5.2}$$

This mass matrix provides a qualitative explanation for the small neutrino mass, which comes out to be $m_\nu \simeq (h\gamma\kappa^2 - m_e^2)/v_R$; according to this formula, $m_\nu \to 0$ as $v_R \to \infty$ in the absence of right-handed weak interactions; it is, therefore, aesthetically pleasing that if neutrino has a small mass, it is not just a freak accident of nature but is related to a new intrinsic property of weak interactions being ultimately parity conserving. Turning this question around, we could say that once mass of the neutrino is known, that would provide a rough idea about the mass scale for right-handed interactions. There is, however, one technical difficulty with eqn. (5.2); to get acceptable values for neutrino mass (i.e. ev or less), we will have to tune the parameter $\gamma \simeq 10^{-6}$ or so. It turns out that such a small value arises automatically,[34] if parity symmetry is broken at a scale M_P higher than the breaking scale of $SU(2)_R$ gauge symmetry.[35] In fact, in this case, we find, $v_L \simeq \frac{\kappa^2 v_R}{M_P^2}$ which is of order 10^{-10} or so for v_R in the Tev range and $M_P \simeq 10^9$ Gev. This has the impact that at $\mu \simeq m_{W_R}$, $g_L \neq g_R$ and the multiplet Δ_L decouples from low energies. Alternatively, it is worth pointing out if γ is set to zero at the tree level, then it receives contributions at the one loop level only from Yukawa couplings, which are much smaller than one ($\sim 10^{-2}$ or so). So if we assume that such one loop graphs are cut off in momentum at $\mu \simeq M_{Planck}$, we would expect $\gamma \simeq \frac{h^4}{16\pi^2}\,\ell n\,(M_P/m_W) \lesssim 10^{-8}$ or so, which may justify the finetuning adopted earlier.

A further point worth discussing in connection with the mass matrix in eqn. (5.2) is that if the scale v_R is high, then for arbitrary γ, we get $m_{\nu_i} \simeq h\gamma\cdot\kappa^2/v_R$. We find $m_{\nu_i} \approx 10$ eV requires for $h \simeq 10^{-3} - 10^{-4}$, $\gamma \simeq 10^{-1}$, $v_R \simeq 10^8 - 10^9$ Gev. An important characteristic of this point of view is that all light neutrinos can be nearly degenerate in mass, which could manifest themselves in substantial neutrino oscillations.

The mass eigenstates for neutrinos in this case (ignoring mixings between generations) is given by (setting $\gamma \simeq 0$)

$$\begin{aligned} \nu_e &\simeq \nu + \xi N \\ N_e &\simeq N - \xi\nu \end{aligned} \tag{5.3}$$

where

$$\xi \simeq (m_{\nu_e}/m_{N_e})^{1/2} .$$

Due to this admixture of the right-handed neutrinos, the GIM cancellation in this sector does not occur; as a result, one obtains flavor changing neutral current decays of Z such as $Z \to \nu_e \bar{\nu}_\mu$. This, then, opens up a channel for heavier neutrinos to decay via Z-exchange, e.g. $\nu_\mu \to \nu_e \bar{\nu}_e \nu_e$. This decay amplitude is, however, small in practice (see below).

Turning to the neutrino masses, we find (in the simplest approximation), the following mass formula for the neutrinos, i.e.

$$m_{\nu_i} \simeq m_{e_i}^2 / m_{N_i} \tag{5.4}$$

where i denotes the generation and ℓ_i denotes the charged lepton. Assuming m_{N_i} to be generation independent, we find that, $m_{\nu_e} : m_{\nu_\mu} : m_{\nu_\tau} = m_e^2 : m_\mu^2 : m_\tau^2$. This relation was first proposed in ref. 18. If we assume, $m_{\nu_e} \simeq 1$ to 5 ev, we obtain from the above scaling law, $m_{\nu_\mu} \simeq 40$ - 200 kev and $m_{\nu_\tau} \simeq 10$ - 50 mev. Now, recalling our discussion in sec. 2, we see that, for stable neutrinos, these values are in the forbidden range. Therefore, for the model to be acceptable, ν_μ and ν_e must decay with typical life times of about $\tau_{\nu_\mu} \simeq 10^{12}$ sec. or less. The photonic decay mode, which occurs via a one loop graph[37] is bigger than 10^{16} sec. for $m_{\nu_\mu} \simeq 200$ kev. This is clearly much too long for our purpose. Let us, therefore, turn to the $.3\nu_e$ decay mode of ν_μ. This decay mode can arise in two ways in left-right symmetric models: one, via the flavor changing Z-couplings to neutrinos and two, via the exchange of neutral Higgs boson Δ_L^o. The first mechanism leads to an effective $\nu_\mu \to 3\nu_e$ interaction with effective couplings strength $\sim G_F \left(m_{\nu_\mu}/m_{N_e}\right)^{1/2}$; since $G_F \left(m_{\nu_\mu}/m_\mu\right)^{1/2} \simeq 10^{-8}\ \mathrm{Gev}^{-2}$. This leads to a lifetime of order

$$\tau_{\nu_\mu} \geq 10^{17}\ sec.. \tag{5.5}$$

In the second mechanism, suggested in ref. 11 the strength of the four neutrino interaction is of order $\sim \left(h_{e\mu} h_{ee}/m_{\Delta_L^o}^2\right)$, where $m_{\Delta_L^o}$ is the mass of the neutral Higgs boson, $h_{\mu e}$ and h_{ee} are the Yukawa couplings of Δ_L with leptons. Since $m_{\Delta_L^o}$ is unknown, it can be arbitrary. To see how big it is, we note that, to satisfy eqn. (2.2), we find the following constraint:[12]

$$h_{e\mu} h_{ee}/M_{\Delta_L^o}^2 \geq 6 \times 10^{-6}\ Gev^{-2} \left(\frac{m_{\nu_\mu}}{40\ kev}\right)^{-3/2}. \tag{5.6}$$

By an $SU(2)_L$-rotation, the Δ_L^{++} gives rise to the highly suppressed decay mode $\mu \to 3e$, leading to the constraint that,

$$h_{\mu e} h_{ee}/m_{\Delta_L^{++}}^2 \leq 10^{-10}\ Gev^{-2}. \tag{5.7}$$

Since Δ_L^{++} and Δ_L^o are members of the same $SU(2)_L$ multiplet, their mass difference can at most be 250 GeV. This, together with eqn. (5.6) and (5.7) implies that,[12] $m_{\Delta_L^o} \simeq$ few Gev. This can therefore be tested once the Z-width is measured since $B\left(Z \to \Delta_L^o \Delta_L^{o*}\right)/B\left(Z \to \nu\bar{\nu}\right) \simeq 2$.

Finally, we turn to the Majoron decay mode of ν_μ and ν_τ in left-right symmetric models.[38] If the left-right symmetric model is extended by the inclusion of the left-right symmetric doublets $\chi_L\left(\frac{1}{2},0,1\right)$ and $\chi_R\left(0,\frac{1}{2},1\right)$, then, the model can admit an extra global symmetry, (similar to the $B-L$ quantum number). Spontaneous breaking of this symmetry leads to the existence of a mass less Goldstone boson χ. The Goldstone boson couples to neutrinos and allows for the decay mode $\nu_\mu \to \nu_e \chi$. The typical strength of this decay mode can be big enough[37] to avoid the constraints from cosmological mass density.

Case (ii) Now, we turn to the Dirac neutrinos in left-right symmetric models. As mentioned earlier, this requires the introduction of new singlet fermions θ_1 and θ_2 and new set of Higgs bosons $\chi_L\left(\frac{1}{2},0,1\right) \oplus \chi_R\left(0,\frac{1}{2},1\right)$. The existence of these multiplets enables one to write the following Yukawa coupling:

$$\mathcal{L}_Y = f\left(\bar{\psi}_L\,\chi_L\,\theta_1 + \bar{\psi}_R\,\chi_R\,\theta_2\right) + \mu\bar{\theta}_1\theta_2 + h.c.\,. \tag{5.8}$$

The first point we wish to note about eqn. (5.8) is that the parameter μ does not receive infinite contributions from higher loops and is, therefore radiatively stable. The significance of this observation is that if we fix μ to be $\simeq 1 - 10$ Mev it will not change in higher orders. Minimization of the potential leads to $\langle\chi_R\rangle \simeq v_R\,,\ \langle\chi_L\rangle \simeq \frac{\mu' \kappa}{V_R} \approx 10^{-2}\,\mu'$ where we choose $\mu' \approx \mu$. (This is because as $\mu \to 0\,,\ \mu' \to 0$). Then, $\langle\chi_L\rangle \simeq 10^{-4} - 10^{-5}$ Gev. In the subsequent discussion, we will ignore this. We then obtain the following 4×4 mass matrix:

$$\begin{array}{c} \\ \nu \\ \theta_1 \\ N \\ \theta_2 \end{array}
\begin{array}{c} \begin{array}{cccc} \nu & \theta_1 & N & \theta_2 \end{array} \\ \begin{pmatrix} 0 & 0 & m_D & 0 \\ 0 & 0 & 0 & \mu \\ m_D & 0 & 0 & f\,V_R \\ 0 & \mu & f\,V_R & 0 \end{pmatrix} \end{array}\,. \tag{5.9}$$

This matrix leads to two Dirac neutrinos, one heavy with mass $\sim f\,v_R$ and another light, with mass, $m_\nu \sim m_D\mu/f\,v_R$. This light four component spinor has the correct weak interaction properties to be identified as the neutrino.

Coming to the mass spectrum, we find that for $f \simeq 1$ and $v_R \sim$ Tev, $m_\nu \simeq 1$ eV. To extend this model to include th higher generations, we include 3 pairs of $(\theta_1\,,\,\theta_2)$. If we assume f and μ to be generation independent, then neutrino masses scale linearly with charged lepton masses. Thus, if $m_{\nu_e} \sim 10$ eV, we would have $m_{\nu_\mu} \sim 2$ kev and $m_{\nu_\tau} \sim 40$ kev. We again see the need for ν_μ and ν_τ to decay to avoid cosmological mass constraints.

Let us discuss the magnetic moment of neutrino. It is well-known that a Dirac neutrino can have a magnetic moment, μ_ν. In fact in the case conventional left-right models with Dirac neutrinos, μ_ν has been calculated.[38] The interest in this parameter is due to recent observations by Okun, Voloshin and Vysotski[39] that the observed deficit in the number of solar neutrinos reaching the earth can be explained if $\mu_\nu \simeq 10^{-11}\,\mu_e$. In conventional left-right[1] symmetric models with no extra singlet neutrinos, the principal diagram contributing to μ_ν is shown in fig. 2 and has the magnitude

$$\mu_\nu \simeq \frac{\alpha\, m_e}{8\pi\, sin^2\theta_W\, m_{W_L}^2}\,\xi \simeq 8\times 10^{-11}\,\xi\cdot\mu_e\,. \tag{5.10}$$

If we take the phenomenological upperbounds of 10^{-2} on ξ, $\mu_\nu \simeq 10^{-12}\,\mu_e$, which is smaller than that required for solar neutrino puzzle. Note that in a model like this, the smallness of the neutrino mass is put in by hand. If, however, we consider a model like the one proposed in this section, (where small m_ν arises without finetuning of parameters), we estimate

$$\mu_\nu \simeq \frac{\alpha}{8\pi\, sin^2\theta_W}\cdot\left(\frac{m_e}{m_W^2}\right)\left(\frac{m_D}{m_{W_R}}\right)\cdot\xi\,. \tag{5.11}$$

Due to the additional suppression factor (m_D/m_{W_R}), which is of order 10^{-5}, implying that, $\mu_\nu \simeq 10^{-17}\,\mu_e$, making it completely irrelevant for any discussion of solar neutrino puzzle.

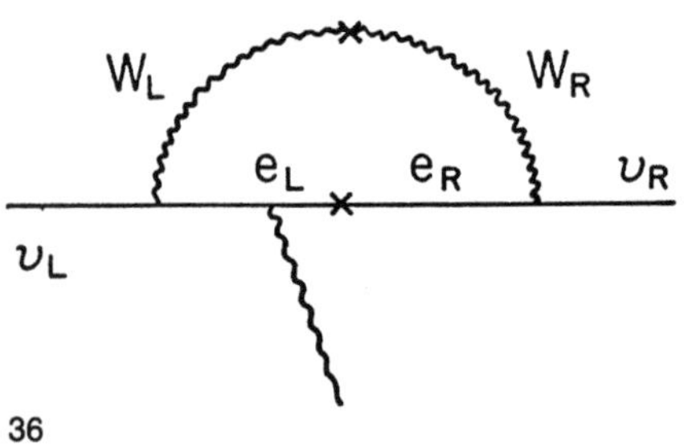

Fig. 2.
The diagram contributing to $\nu_H \to \nu_e\,\gamma$ decay as well as the magnetic moment of the neutrino in left-right symmetric models.

Finally, we wish to note that one can construct a left-right symmetric model by addition of weak iso-singlet quarks and leptons such that the Dirac mass of the neutrino vanishes at the tree level and arises at the one-loop level via $W_L - W_R$ mixing.[40] In this case, the Dirac neutrino mass is given by:

$$m_{\nu_i} = \left(\frac{\alpha}{4\pi \, sin^2 \theta_W}\right)^2 \left(\frac{m_b m_t}{m_{W_R}^2}\right) \cdot m_{\ell_i} \, . \tag{5.12}$$

For $m_{W_R} \simeq 10$ Tev (say), this will lead to $m_{\nu_\mu} \simeq 5 \times 10^{-4}$ ev, (and $m_{\nu_e} \simeq 5 \times 10^{-6}$ eV) which is in the range of interest for the MSW solution to the solar neutrino puzzle.

6. $SO(10)$ Embedding:

In this section, we discuss the $SO(10)$ embedding of the left-right symmetric models discussed in the previous section and any new possibilities for neutrino masses, that may arise in these models. It is of course well known that, the simplest grandunification model that leads to left-right symmetric models at low energies is the $SO(10)$ model. The fermions (both quarks and leptons) belong to the 16-dimensional spinor representation of $SO(10)$. The symmetry breaking pattern of interest to us is the following one:

$$\begin{array}{c} SO(10) \longrightarrow G \longrightarrow SU(3)_C \times SU(2)_L \times SU(2)_R \times U(1)_{B-L} \\ \downarrow \\ G_{std.} \end{array}$$

We will first consider the scenario that leads to Majorana neutrinos: this is achieved by including the Higgs multiplets, which transform as $\{54\}$ (S), $\{45\}$ (A), $\{126\}$ (Σ), and $\{10\}$ (H) dimensional representations. (The symbols within the parenthesis denote those representations in what follows). The $\{54\}$-dimensional representation breaks $SO(10)$ down to $SU(4)_C \times SU(2)_L \times SU(2)_R$, whereas $\{45\}$ breaks this group further down to $SU(3)_C \times SU(2)_L \times SU(2)_R \times U(1)_{B-L}$ without the discrete D-parity symmetry that relates the left and right-handed gauge couplings. The $\{126\}$-dimensional representation contains the Δ_L and Δ_R defined earlier and therefore serves to break not only $SU(2)_R$ symmetry but also gives a heavy mass to the right-handed neutrinos. Due to the difference between the scales M_P of D-parity breaking and M_R of $SU(2)_R$-breaking, the Δ_L^o v.e.v. is suppressed by an additional factor of $(v_R/M_P)^2$ with respect to κ^2/v_R. This gives the real see-saw form to the neutrino mass matrix.[34]

Having outlined the overall scenario, we mention two interesting specific cases that emerge in this model: in one case, the D-parity breaking scenario enables the $sin^2 \theta_W$ constraints to be satisfied even with a low M_{W_R} (in the Tev range) leading neutrino mass spectrum of ev-kev-Mev type as discussed in sec. 5; in the second case, if the $B-L$ breaking scale (which coincides with $SU(2)_R$ scale in the cases we consider) is identified with the Peccei-Quinn scale,[41] then a unique $SO(10)$ model emerges, that simultaneously provides a solution to solar neutrino problem via the MSW mechanism.

In order to comment on these two possibilities, we first explain the meaning of D-parity. When $SO(10)$ breaks down to one of its maximal subgroups, $SO(4) \times SO(6)$, there is an additional discrete symmetry which can also remain unbroken. We call this discrete symmetry as D-parity. Under D-parity, $q \to q^C$ and $\vec{W}_L \to \vec{W}_R$; so as far as the quark sector is concerned it acts as charge conjugation or parity; but its effect on Higgs bosons is more complicated; therefore we prefer to call it D-parity rather than C or P. Its practical impact is to maintain the equality of gauge couplings constant of the $SU(2)_L$ and $SU(2)_R$ groups. Until the work of ref. 35, it was not realized that $SU(2)_R$ and D-parity could be broken separately. If $SU(2)_R$ and D-parity were broken together, the observed value of $sin^2 \theta_W$ could only be consistent with $SO(10)$ model if the value of m_{W_R} is rather high ($\sim 10^{10} - 10^{12}$ Gev). On the other hand, if D-parity was broken "strongly" (i.e. if g_L/g_R differed very much from 1 at the scale m_{W_R}), one could have $m_{W_R} \simeq$ from TeV's without spoiling the agreement with $sin^2\,\theta_W$. Since in the neutrino mass matrix, m_{W_R} determines the mass of the right-handed neutrino, m_{RR}, we would get an ev-Kev-Mev type neutrino spectrum. The question, then, is: do we have a built-in mechanism to achieve sufficiently fast neutrino decay? This is now more subtle due to the existence of D-parity breaking; since the mass of the entire $\vec{\Delta}_L$ triplet is now lifted to the D-parity breaking scale M_P. Thus, $\nu_\mu \to 3\nu_e$ decay modes are suppressed. On the other hand, one can have the Majoron decay modes present by including Higgs multiplets $\{16\}$, if we forbid the $\{16\} \cdot \{16\} \cdot \{126\}$ Higgs coupling a là ref 37. In such a case, a consistent $SO(10)$ model with m_{ν_e} in the ev range could be constructed.

Let us now turn to the second scenario for the neutrino mass suggested in ref. 41. The theoretical motivation for the model is the observation that while the vectorial $B-L$ symmetry is anomaly-free and is, in fact, a part of $SO(10)$ symmetry, the axial $B-L$ is plagued with anomalies arising from the QCD gluon contributions and is the source of the well-known strong CP-problem. The suggestion of Peccei and Quinn was of course to make axial $B-L$ into a continuous global symmetry of the gauge model (operating both on quarks and Higgs bosons) and use it to convert θ into symmetry parameter, in which case $\theta = 0$ is as physical as $\theta \neq 0$. The breaking of this symmetry leads to axion and the presently consistent axion models[42] seem to require that 10^9 Gev $\leq M_{PQ} \leq 10^{12}$ GeV. Since PQ symmetry and vectorial gauge $B-L$ symmetry as so closely related, it was suggested in ref. 41 that perhaps they may be broken at the same scale. This has the following profound implications: that $SO(10)$ must break via a two-step process in either of the following ways:

$$SO(10) \xrightarrow{M_U} \begin{cases} SU(2)_L \times SU(2)_R \times SU(4)_C \\ SU(2)_L \times SU(2)_R \times SU(3)_C \times U(1)_{B-L} \end{cases} \xrightarrow{M_{B-L}} G_{Std.}$$

with $M_{B-L} \simeq 10^{12}$ Gev and $M_U \simeq 10^{15}$ Gev[43] after including the two-loop effects.

An important consequence of this is that

$$m_{\nu_i} \simeq m_{u_i}^2/9\,M_{BL} \qquad (6.1)$$

where m_{u_i} is the up-quark mass of the corresponding generation and with $M_{BL} \simeq 10^{13}$ Gev, we get, $m_{\nu_e} \simeq 10^{-8}$ ev; $m_{\nu_\mu} \simeq 10^{-3}$ ev, $m_{\nu_\tau} \simeq 4$ ev. We then see that m_{ν_μ} has mass in the right range to provide a solution to the solar neutrino problem a là MSW. Secondly, it leads to a proton lifetime of around 10^{35} years.[41,43] This could therefore be tested in the next generation of proton decay experiments.

7. Neutrinos in Superstring Inspired Models:

In this section, we consider the properties of neutrinos in the superstring model. To make this discussion useful, we first give a miniature review of the gross features of superstring inspired models. These models imply a supersymmetric E_6-grandunified model with matter fields belonging to $\{27\}$-dimensional representations of the E_6-group and with Higgs fields also transforming as $\{27\} + \{\overline{27}\}$ fields under E_6. There are additional E_6-singlet fields. In particular note the absence of any E_6 representation such as $\{351\}$, which contains the $\{126\}$-

dimensional representation of $SO(10)$ that was essential to the see-saw mechanism for small neutrino masses. Therefore, the nature of neutrino masses are likely to be more complicated in superstring models. To facilitate this discussion, we first write down the decomposition of the {27}-dim. representation of E_6 under $SO(10)$ as well as $SU(3)_c \times SU(3)_L \times SU(3)_R$.

$$E_6 \longrightarrow SO(10) \times U(1)_X \ :$$

$$\{27\} = \{16\}_1 \oplus \{10\}_{-2} \oplus \{1\}_4$$

$$\begin{aligned} \text{Particle Content}: \quad \{16\}\,; \quad & \begin{pmatrix} u_1 & u_2 & u_3 & \nu \\ d_1 & d_2 & d_3 & e^- \end{pmatrix}_{L+R} \\ \{10\} \ : \quad & g\,,\, g^c \ \text{singlet quark} \\ & \begin{pmatrix} H_u^o & H_u^+ \\ H_d^- & H_d^o \end{pmatrix} \\ \{1\} \quad & n_o \end{aligned} \tag{7.1}$$

$$E_6 \supset SU(3)_c \times SU(3)_L \times SU(3)_R \ :$$

$$\{27\} \supset (3,3,1) \oplus (\bar{3},1,\bar{3}) \oplus (1,\bar{3},3)$$

$$(3,3,1) \ : \ \begin{pmatrix} u \\ d \\ g \end{pmatrix} \ ; \ (\bar{3},1,\bar{3}) \ : \ \begin{pmatrix} u^c \\ d^c \\ g^c \end{pmatrix}$$

$$(1,\bar{3},3) \ : \ \begin{pmatrix} H_u^o & H_u^+ & e^+ \\ H_d^- & H_d^o & \nu^c \\ e^- & \nu & n_o \end{pmatrix} \equiv L \tag{7.2}$$

We see from eqn. (7.1) and (7.2) that aside from the known fields, there are eleven extra fermions out of which three are neutral. In total, each generation have 5 neutral leptons. Add to this the neutral gauginos (4 colorless ones) and many Higgsinos plus E_6 singlet fermions, which makes a total of more than 20 neutral fermions. All these in general can mix making any understanding of the nature of the neutrino next to impossible in general. We will therefore simplify the picture by first ignoring generation mixing and secondly by assuming that there exist intermediate mass scales, which decouple the neutral leptons $H_u^o - H_d^o$. We also further assume that, in these models, $B-L$ symmetry scale is either an intermediate scale such as $10^{11} - 10^{12}$ Gev or a Tev. In this simplified approximation, four mechanisms have been proposed to understand neutrino masses, which we briefly summarize below.

(i) *R*-parity violation Mechanism:[44]

In this mechanism, it is assumed that the low energy gauge group is given by $SU(3)_c \times SU(2)_L \times U(1)_{I_{3R}} \times U(1)_{B-L}$ and that the superpartner of ν^c (the anti right-handed neutrino) acquires a v.e.v. of order V_R to break I_{3R} and $B-L$. Because of this, the gaugino corresponding to the broken generator (denoted by $\lambda_{Y'}$) couples to the ν^c and leads to the following mass matrix gaugino-neutrino mass matrix:

$$\begin{array}{c} \\ \nu \\ \nu^c \\ \lambda_{Y'} \end{array} \begin{array}{c} \begin{array}{ccc} \nu & \nu^c & \lambda_{Y'} \end{array} \\ \begin{pmatrix} 0 & m_D & 0 \\ m_D & 0 & g\,v_R \\ 0 & g\,v_R & \mu \end{pmatrix} \end{array} . \tag{7.3}$$

Here μ is the Majorana mass for λ_Y-gaugino that can arise in the two-loop order. This matrix can be diagonalized leading to a pseudo-Dirac 4-component heavy lepton ($m \simeq g\,v_R$) and a light 2-component Majorana neutrino

$$\nu_e \simeq \nu + \frac{m_D}{g\,v_R}\,\nu^c \tag{7.4}$$

with mass

$$m_{\nu_e} \simeq \frac{m_D^2 \cdot \mu}{(g\,v_R)^2} . \tag{7.5}$$

Usually, in supergravity models, $\mu \simeq 100$ Gev or so; assuming $(g\,v_R) \simeq 1$ Tev, we get $m_{\nu_e} \simeq 10^{-1}$ ev. If, on the other hand, we choose $v_R \simeq 10^{12}$ Gev, $m_{\nu_e} \simeq 10^{-9}$ ev. Thus, smallness of neutrino mass is understood without any unnatural fine tuning of parameters. It has been pointed out[45] that, in these models, the H_u^o-Higgsino mixes with the ν with mixing proportional to $h_\nu^D\,v_R$; this modified eqn. (7.5) and m_ν has a component which grows with v_R. Thus, if we want to understand the small m_ν, we must have an upperbound on v_R of a few Tev. A limitation of this method so that it works only for one generation.

E_6-Singlet Mechanism:

Use of E_6-singlets (S) for studying the neutrino mass problem was hinted at by Witten[46] and was analyzed in detail in ref. 47. here, one has to rely on possible symmetries that may arise in superstring models on specific Calabi-Yau spaces to restrict the Yukawa couplings in such a way that we have couplings of the following type in the superpotential: using the notation of (7.2),

$$W = W_o + \lambda_1\,LLL_H + \lambda_2\,L\bar{L}_H\,S \tag{7.6}$$

where subscript H stands for the Higgs field. If we then give v.e.v. to $\tilde{\nu}_H^c$, and $(H_u^o)_{Higgs}$, then, for each generation we get the following type of 3×3 mass matrix:

$$\begin{array}{c} \\ \nu \\ \nu^c \\ S \end{array} \begin{array}{c} \begin{array}{ccc} \nu & \nu^c & S \end{array} \\ \begin{pmatrix} 0 & m_D & 0 \\ m_D & 0 & \lambda_2\,v_R \\ 0 & \lambda_2\,v_R & 0 \end{pmatrix} \end{array} . \tag{7.7}$$

This, as discussed in sec. 6, leads to a massless neutrino and a massive Dirac neutrino. If we include possible a Majorana mass term for the S-fermion of order of 100 Gev, the mass matrix looks like that in eqn. (7.3) with similar values for m_ν. Note that these values fall outside the range needed to solve the solar neutrino puzzle for $v_R \simeq 1$ Tev or 10^{11} Gev.

(iii) Higher Dimensional Operators:

It was proposed in ref. 48 that one may solve the neutrino mass problem in superstring models with two intermediate mass scales corresponding to $\langle\tilde{\nu}_H^c\rangle \approx \langle\tilde{n}_{oH}\rangle \approx 10^{11} - 10^{12}$ Gev. In this case, a higher dimension term in the superpotential of the form $L\bar{L}_H \cdot \bar{L}_H/M_{Pl}$ leads to an effective see-saw-like mechanism with right-handed neutrino mass $m_{N_R} \simeq \langle\tilde{\nu}^c\rangle^2/M_{Pl} \simeq 10^4 - 10^6$ Gev. Using the standard see-saw formula, we get $m_{\nu_e} \simeq m_e^2/m_{N_R} \simeq 10^{-2}$ ev $- 10^{-4}$ ev $m_{\nu_\mu} \simeq 400$ ev $-$ 4 ev and $m_{\nu_\tau} \simeq 160$ Kev $-$ 1.6 Kev. Since in the Superstring model, there is no known way for ν_μ and ν_τ to decay, this scenario will be ruled out by cosmology.

However, there exist arguments from proton decay that $\langle \nu_H^c \rangle \simeq \langle n_{oH} \rangle$ may be of order 10^{14} Gev, leading to $m_{N_R} \simeq 10^{10}$ Gev in which case, $m_{\nu_e} \simeq 10^{-8}$ ev, $m_{\nu_\mu} \simeq 4 \times 10^{-4}$ ev and $m_{\nu_\tau} \simeq .16$ ev. This kind of a spectrum is cosmologically acceptable and with slight variation of coupling parameters could be useful for the solution of solar neutrino puzzle via the MSW mechanism.

(iv) One Loop Induced Neutrino Mass:[49]

It has also been proposed that existence of some accidental symmetries may force the tree level Dirac mass of the neutrino to vanish. As a result, the dominant contribution to neutrino masses can arise at the one loop level from diagrams of the type shown in fig. 3.

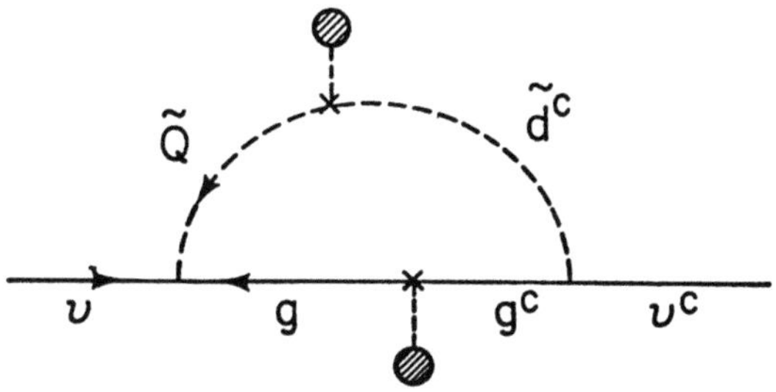

Fig. 3.
One loop graph for neutrino Dirac mass in some versions of superstring inspired models.

8. Conclusion:

To summarize, we have given an extensive review of the basic ideas that go into understanding the smallness of neutrino masses in general gauge and superstring inspired models. We see that various kinds of mass spectra can emerge in different models. Testing between these various possibilities may be possible if the MSW solution to the solar neutrino puzzle is backed up by future experiments and other considerations. For instance, this would rule out a low mass m_{W_R} in an $SO(10)$ type model as well as the mechanisms (i) and (ii) in superstring inspired models.

References

[1] J.C. Pati and A. Salam, Phys. Rev. D10, 275 (1974); R.N. Mohapatra and J.C. Pati, Phys. Rev. D11, 566 (1975); G. Senjanović and R.N. Mohapatra, Phys. Rev. D12, 1502 (1975). For a review, see R.N. Mohapatra in "Quarks, Leptons and Beyond," ed. by H. Fritzsch et al, Plenum (N.Y.), p. 219.

[2] R.N. Mohapatra and G. Senjanović, Phys. Rev. Lett. 44, 912 (1980); Phys. Rev. D23, 165 (1981).

[3] For a review, see "Superstring Theory" Vols. I and II by M. Green, J. Schwarz and E. Witten, Cambridge University Press (1986).

[4] For a review of standard model and grandunification see "Unification and Supersymmetry: The Frontiers of Quark-Lepton Physics," by R.N. Mohapatra, Springer-Verlag (N.Y.) (1986).

[5] R. Cowsik and J. Mclelland, Phys. Rev. Lett. 29, 699 (1972).

[6] S.S. Gershtein and Ya B. Zeldovich, JETP Lett. 4, 120 (1966).

[7] B.W. Lee and S. Weinberg, Phys. Rev. Lett. 39, 165 (1977); P. Hut, Phys. Lett. 69B, 85 (1977); K. Sato and M. Kobayashi, Prog. Theor. Phys. 58, 1775 (1977); M.I. Vysotskii, A.D. Dolgov and Ya. B. Zeldovich, JETP Lett. 26, 188 (1977).

[8] D. Dicus, E. Kolb and V. Teplitz, Phys. Rev. Lett. 39, 169 (1977).

[8a] S. Sarkar and A.M. Cooper, Phys. Lett. B148, 347 (1984).

[9] Y. Chikashige, R.N. Mohapatra and R.D. Peccei, Phys. Lett. 98 B, 265 (1981).

[10] Y. Chikashige, R.N. Mohapatra and R.D. Peccei, Phys. Rev. Lett. 45, 1926 (1980); G. Gelmini and J.W.F. Valle, 142B, 181 (1984); A. Kumar and R.N. Mohapatra, Phys. Lett. B150, 191 (1985).

[11] M. Roncadelli and G. Senjanović, Phys. Lett. B107, 59 (1983); P. Pal, Nucl. Phys. B227, 237 (1983).

[12] R.N. Mohapatra and Palash Pal, Phys. Lett. 179B, 105 (1986).

[13] Particle Data Table, Phys. Lett. B , April (1987).

[14] S. Petcov, Phys. Lett. B1101, 245 (1982); J.W.F. Valle, Phys. Rev. D27, 1672 (1983); M. Doi, T. Kotani, M. Kenmoku, and E. Takasugi, Phys. Rev. D30, 626 (1984).

[15] R.N. Mohapatra and J.W.F. Valle, Phys. Lett. B177, 47 (1986).

[16] R.N. Mohapatra and G. Senjanović, Phys. Rev. Lett. 44, 912 (1980); R.N. Mohapatra and R.E. Marshak, Phys. Rev. Lett. 44, 1316 (1980).

[17] For a recent review, see D. Caldwell, Proceedings of "XXIIIrd International Conference on High Energy Physics," ed. S. Loken, World Scientific Publishing Co. (1986), p. 951.

[18] R.N. Mohapatra and G. Senjanović, Phys. Rev. D23, 165 (1981).

[19] L. Durkin and P. Langacker, Phys. Lett. B166, 436 (1986); R. Robinett and J. Rosner, Phys. Rev. D25, 3036 (1982); V. Barger, E. Ma and K. Whisnant, Phys. Rev. D26, 2378 (1982).

[20] M.A.B. Bég, R.V. Budny, R.N. Mohapatra and A. Sirlin, Phys. Rev. Lett. 38, 1252 (1977). For further discussion, see J. Maalampi, K. Mursula and M. Roos, Nuc. Phys. B207, 233 (1982).

[21] J. Carr et al, Phys. Rev. Lett. 51, 627 (1983); D.P. Stoker, et al., Phys. Rev. Lett. 54, 1887 (1985).

[22] P. Herczeg, Los Alamos preprint LA-UR-85-2761 (1985).

[23] R.N. Mohapatra, New Frontiers in High Energy Physics, ed. A. Perlmutter and L. Scott (Plenum, New York, 1978), p. 337.

[24] J.F. Donoghue and B. Holstein, Phys. Lett. B113, 382 (1982).

[25] G. Beall, M. Bander and A. Soni, Phys. Rev. Lett. 48, 848 (1982). Inclusion of QCD corrections improve this bound; see G. Ecker and W. Grimus, Z. Phys. C30, 293 (1986).

[26] I.I. Bigi and J.M. Frere, Phys. Lett. B110, 255 (1982).

[27] H. Abramowicz, et al., Z. Phys. C12, 225 (1982).

[28] L. Wolfenstein, Phys. Rev. D29, 2130 (1984).

[29] E. Masso, Phys. Rev. Lett. 52, 1956 (1984).

[30] D.O. Caldwell, et al., Phys. Rev. D33, 2737 (1986); E. Ejiri, et al., Nucl. Phys. A448, 27 (1986); E. Fiorini, et al., Phys. Lett. B146, 450 (1984); F.T. Avignone III, et al., Phys. Rev. Lett. 54, 2309 (1985).

[31] R.N. Mohapatra, Phys. Rev. D34, 909 (1986). For limits on m_{ν_R} from other considerations, see M. Gronau, C. Leung and J. Rosner, Phys. Rev. D29, 2539 (1984).

[32] For reviews, see, W. Haxton and G. Stephenson, Jr., Progress in Nuclear and Particle Physics, 12, 409 (1984); J. Vergados, Phys. Rep. 133, 1 (1986); M. Doi, T. Kotani and E. Takasugi, Prog. Theor. Phys. Supp. 83, 1 (1985).

[33] R.N. Mohapatra, Proceedings of "The International Symposium on Weak and Electromagnetic Interactions in Nuclei," ed. H.V. Klapdor, Springer-Verlag (1986); S. Bertolini and J. Liu, CMU-HEP 87-09 (1987).

[34] D. Chang and R.N. Mohapatra, Phys. Rev. D32, 1248 (1985).

[35] D. Chang, R.N. Mohapatra and M.K. Parida, Phys. Rev. Lett. 52, 1072 (1984).

[36] U. Chattopadhyay and P.B. Pal, Phys. Rev. D34, 3444 (1986).

[37] A. Kumar and R.N. Mohapatra, Phys. Lett. B150, 191 (1985); R.N. Mohapatra and P. Pal, University of Massachusetts Preprint (1987).

[38] J.E. Kim, Phys. Rev. D14, 3000 (1976); M.A.B. Bég, W. Marciano and M. Ruderman, Phys. Rev. D17, 1395 (1978).

[39] L.B. Okun, M.B. Voloshin and M. Vysotski, ITEP Preprint 86-20 (1986).

[40] D. Chang and R.N. Mohapatra, Phys. Rev. Lett. 58, 1600 (1987).

[41] R.N. Mohapatra and G. Senjanović, Zeit. für Phys. C17, 53 (1983); F. Stecker and Q. Shafi, Phys. Rev. Lett. 50, 928 (1983); P. Langacker, R.D. Peccei and T. Yanagida, DESY Preprint 86-141 (1986).

[42] For reviews of strong CP-problem, see J.E. Kim, Phys. Rep. 150, 1 (1987); H.Y. Cheng, Indiana Preprint IUHET-125 (1986).

[43] D. Chang, R.N. Mohapatra, J. Gipson, R.E. Marshak and M.K. Parida, Phys. Rev. D31, 1718 (1985); R.N. Mohapatra, Proceedings of "Eighth Workshop on Grandunification," ed. by K.C. Wali, (World Scientific Publishing, Singapore, 1987).

[44] R.N. Mohapatra, Phys. Rev. Lett. 56, 561 (1986).

[45] J.W.F. Valle, private communication (1987).

[46] E. Witten, Nucl. Phys. B268, 79 (1986).

[47] R.N. Mohapatra and J.W.F. Valle, Phys. Rev. D34, 1642 (1986).

[48] S. Nandi and U. Sarkar, Phys. Rev. Lett. 56, 564 (1986).

[49] A. Masiero, D.V. Nanopoulos and A.I. Sanda, Phys. Rev. Lett. 57, 663 (1986); G.C. Branco and C.Q Geng, Phys. Rev. Lett. 58, 969 (1987); E. Ma, Phys. Rev. Lett. (to appear) (1987); J. Liu, Carnegie-Mellon Preprint (1987).

[50] M. Doi, T. Kotani, T. Kurimoto, H. Nishiura and E. Takasugi, Osaka preprint OS-GE-87-05 (1987).

Neutrinos in Cosmology

G. Gelmini

SISSA/ISAS and ICTP, P.O.Box 586, I-34100 Trieste, Italy

1. Introduction

We only know about the left handed neutrinos $\nu_{e_L}, \nu_{\mu_L}, \nu_{\tau_L}$ that they exist, they have weak interactions and they are much lighter than their corresponding charged leptons. They may be massive or massless, stable or not. If they are massive, they may be the left handed part of Majorana or Dirac fermions. In the first case, the right handed neutrinos may or may not exist and may be heavy or light, in the second case they must exist. Neutrinos may have new interactions, other than weak, which may allow them to annihilate or decay faster. Not only laboratory experiments can limit these possibilities. Because neutrinos were present in large numbers in the early universe, their properties affect the evolution and the present structure of the universe. Our knowledge of the history of the universe is thus used as a laboratory for placing limits on the mass, lifetime and number of flavours of neutrinos. Also astrophysics leads to limits on neutrinos, being the recent observations of neutrinos from the supernova SN 1987A an outstanding example. On the other hand, the knowledge of some properties of neutrinos may be crucial for cosmology and astrophysics, For example, the solution to a major cosmological problem would be given by a neutrino mass of around 30eV. this would mean that the "dark matter" consists of light neutrinos.

The standard model of cosmology, the hot big bang, provides a reliable description of the evolution of the universe from the epoch of primordial nucleosynthesis (at a time $t \sim 10^{-2}$ sec after the bang and temperature $T \sim$ 10MeV) until the present ($t_0 \sim 1.5 \times 10^{10}$ y, $T_0 = 2.7^o$K). It rests on three major empirical pieces of evidence [1]. One of them is the Hubble expansion of galaxies, i.e. the recession of galaxies with velocities proportional to their distances. The proportionality is given by the Hubble constant H (constant in space but not in time). The value of H is now $H = h \times 100$ Km/(Mpc.sec), with h a number between 0.4 and 1 [2]. Another piece of evidence is the existence of the 2.7^oK blackbody background radiation [3]. It consists of photons that interacted for the last time when atoms became stable, at $t \sim 5 \times 10^5$ y, $T \sim 0.3$ eV. The blackbody character and the high degree of isotropy [4] of this radiation give bounds on radiative neutrino decay modes. The third impressive piece of evidence, as already mentioned, is the nucleosynthesis of light elements. The cosmic abundance of ^{4}He and of the trace elements D, ^{3}He and ^{7}Li are well accounted for in terms of nuclear reactions that must have occurred at $t \sim 10^{-2} - 10^2$ sec, $T \sim$ 10-0.1 MeV [5,6].Nucleosynthesis gives bounds on any new particle that makes a non negligible contribution to the energy density of the universe during the synthesis. It provides also bounds on radiative decays of neutrinos after the synthesis.

There is a most useful experimental bound on the total energy density of the universe now (for a universe matter or radiation dominated) [7,8]

$$(\Omega h^2)_{now} < 1 \; . \qquad (1)$$

Here $\Omega = \rho/\rho_c$ is the energy density in units of $\rho_c = 10.5\ h^2\ KeV/cm^3$, the critical density necessary to have a flat universe ($\rho > \rho_c$ for a closed universe, $\rho < \rho_c$ for an open universe). The limit (1) can be obtained just from the empirical bound on the lifetime of the universe, $t_o > 1 \times 10^{10}$ y [9] in a matter or radiation dominated universe. H, t_o and ρ are not independent. The content of the universe determines the relation between them. In a universe matter dominated by particles of mass m, the density is $\rho_{matt.} \sim m/R^3 \sim mT^3$. R is the scale factor of the universe, $H = (dR/dt)/R$, and for most of the history of the universe, $R \sim T^{-1}$. The distance d between two far galaxies can be written as $d = R(t) d_c$ being d_c a constant. d_c is the distance measured in comoving coordinates, the coordinates that follow the expansion of the universe (as a grid painted on an expanding baloon). Eintein's equations and $\rho_{matt.}$ imply that $R \sim t^{2/3}$ on a matter dominated universe. For a radiation dominated universe, the energy of each particle is proportional to the temperature T thus $\rho_{Rad.} \sim T/R^3 \sim T^4$ and $R \sim t^{1/2}$. In this case the bound on Ωh^2 is 0.3, slightly more restrictive than eq.(1). However if the cosmological constant, i.e. the vacuum energy, is non zero, this bound is degraded. Direct observations indicate that Ω is of order one, what combined with the value of h tell us that Ωh^2 is not much larger than one in any case. We will consider a zero cosmological constant in the following. Ω of order one is a very surprising result, unless a dynamical reason exists for this value, such as a period of inflation [10] in the early universe. In this case Ω should be exactly 1.

Another very surprising fact is that most of the mass of the universe does not consists of stars, gas or dust. The main mass component of the universe is "dark", it does not emit or absorb light [11]. A fascinating possibility is that this dark matter, D.M., may be neutral elementary particles. From all the many D.M. candidates proposed in recent years [12], neutrinos are the only that exist for sure, and even so, we do not know if they are massive, as they should.

In the following we start by reviewing the implications of the condition (1) on the energy density in neutrinos, $\Omega_\nu \leq \Omega$. It provides an upper bound, $m_\nu < 100$ eV, on the mass of stable light ($m_\nu < 1$ MeV) neutrinos, a lower bound on the mass of heavy stable neutrinos ($m_\nu >$ few GeV) and upper bounds on lifetimes τ_ν. (By stable we mean $\tau > t_o$). We call neutrinos not only the three known left handed $\nu_{e_L}, \nu_{\mu_L}, \nu_{\tau_L}$ but also the neutrinos of possible further generations, 4th.., light or heavy, and the possible, light or heavy, right handed neutrinos. We will, then, briefly mention bounds derived from nucleosynthesis and bounds on radiative and invisible decay modes and annihilations. The last sections are devoted to neutrinos as the D.M.. Light or heavy neutrinos are good candidates for hot or cold D.M. respectively. The possibility of detecting cold D.M. from the halo of our galaxy and the attempts under way will be our last topic.

Besides a general review of the mentioned topics, there are some news that the reader will find in the following. With the experimental bounds known until recently, the neutrinos of the standard model (but massive) had the possibility of being unstable. A tau neutrino of mass around $m_\nu \sim 50$ MeV could decay into $e^+e^-\nu$ with a lifetime around $\tau_\nu \sim 10^2$ sec, within the standard electroweak model, respecting all known bounds. The observation that the neutrino burst from the supernova SN87A was not followed by photons, seems to have closed this window. This window has, in any case, been closed by the new bound on the ν_τ mass presented at this conference [13]. The exclusion of this window is very important because it means that either all three neutrinos ν_e, ν_μ, ν_τ are lighter than 100 eV or they have non standard interactions.

With respect to invisible neutrino decay modes, the observation of neutrinos from the supernova does not prelude (against a naive belief) the decay of solar neutrinos while arriving to the earth as a solution of the solar neutrino problem. Among the news we have that light neutrinos, for sometime out of favour as the D.M., have been resurrected as possible D.M. candidates, not alone but with some other mechanism to help forming galaxies, such as cosmic strings. Finally, interesting bounds on heavy neutrinos as components of the halo of our galaxy are being obtained, from both indirect and direct D.M. detection attempts.

2. Relic abundances

In the big bang model, at high T left handed neutrinos were in equilibrium with charged leptons and photons, due to their weak interactions. The number density of neutrinos n_ν in equilibrium is given by the Fermi-Dirac distribution, while the number density of photons is given by the Bose-Einstein distribution. Thus, in equilibrium the ratio n_ν / n_γ is calculable:

- for relativistic neutrinos, $m_\nu < T$, $\frac{n_\nu}{n_\gamma} = \frac{3}{4}\left(\frac{g_\nu}{2}\right)$ (2)

- for non-relativistic neutrinos, $m_\nu > T$, $\frac{n_\nu}{n_\gamma} = \left(\frac{\pi}{8}\right)^{1/2} \frac{g_\nu}{2\zeta(3)} \left(\frac{m_\nu}{T}\right)^{3/2} e^{-m_\nu/T}$ (3)

Here g_ν is the multiplicity of spin states, $g_\nu = 2$ for Weyl or Majorana neutrinos. Particles are in equilibrium when their interaction rate Γ is larger than the expansion rate of the universe, which is the Hubble constant H or, equivalently, when their interaction time Γ^{-1} is smaller than the age of the universe, H^{-1}. When $\Gamma^{-1} < H^{-1}$ at high T and there is a temperature such that $\Gamma^{-1} > H^{-1}$ for any smaller T, it is said that at that temperature $T_{f.o.}$ the particles "freeze-out" [14]. For any $T < T_{f.o.}$ the particles are out of equilibrium, they do not interact, they are neither created nor destroyed. The existence of equilibrium and then freeze out, depends on the different behaviour of Γ and H with T. For relativistic neutrinos (with full weak interactions, i.e. ν_L) $\Gamma \sim T^5$ while $H \sim T^2$, and $T_{f.o} \sim 1$ MeV. The number per comoving volume of these light, $m_\nu < 1$ MeV, neutrinos is constant after the freeze-out. The number of photons, however, is not. When T drops below the mass threshold for the production of e^+e^-, these annihilate, increasing the number of photons but not that of non interacting particles, such as neutrinos. Thus the number of neutrinos (plus antineutrinos), of every light species with respect to photons is now smaller than (2):

$$n_{\nu_i}/n_\gamma = (3/11)(g_{\nu_i}/2) = (108/\mathrm{cm}^3)\,(g_{\nu_i}/2) \tag{4}$$

The total contribution of the three light neutrinos to the density of the universe now is $\rho_\nu = \sum m_{\nu_i} n_{\nu_i} = \Omega_\nu \rho_c$. Thus the sum of the masses is [15]

$$\sum_i \left[\frac{g_{\nu_i}}{2}\right] m_{\nu_i} = 97 \text{ eV } (\Omega_\nu h^2) \tag{5}$$

The sum runs actually over all neutrino species lighter than 1 MeV. This bound applies to the left handed neutrinos, ν_L, those with full weak interactions, i.e. $g_\nu = 2$. Even in the case of Dirac neutrinos, in the standard electroweak model the right handed components ν_R are never in equilibrium for $m_\nu < 1$ MeV [16]. Since the ν_R do not have gauge interactions, they are produced through Dirac mass insertions that flip ν_L into ν_R, thus the production rate is suppressed at least by $(m_\nu/T)^2$. In extensions of the standard model in which ν_R do have gauge interactions the freeze-out occurs at $(T_{f.o.})_R > M_Z$ (the Z^0 boson mass), thus $n_{\nu_R} < (n_{\nu_L}/27)$ [17]. The situation is different for heavy neutrinos ($m_\nu >>$ MeV), for which $g_\nu = 4$ if they are Dirac and $g_\nu = 2$ if they are Majorana.

For heavy neutrinos that become non-relativistic while annihilations are in equilibrium (i.e. $\Gamma_{annih.} > H$), the number of neutrinos per comoving volume decreases with the Boltzmann factor, $n_\nu R^3 \sim n_\nu/T^3 \sim e^{-m_\nu/T}$ (see eq. 3, where $n_\gamma \sim T^3$). After annihilations cease (because $\Gamma_{annih.} < H$), n_ν/T^3 departs for its exponential equilibrium function at $T_* = m_\nu/x_*$, and becomes constant soon after [14,18]. The final density ρ_ν is proportional to the Boltzmann factor at T_*. Thus, an upper bound on ρ_ν implies a lower bound on m_ν. For weakly interacting relics $x_* \sim 20$ and for $\Omega_\nu = 0.1$ to 1, m_ν results in the few GeV range [14,18,19]. The value of x_* is larger for larger annihilation cross sections, annihilations remain longer in equilibrium, and the final density is smaller for the same mass. This explains the difference between the relic abundances of heavy Dirac and Majorana neutrinos [20] (see fig. 1 re-drawn with data from [21]). The annihilation of non-relativistic Majorana neutrinos is suppressed due to the presence of two identical particles in the initial state,since particle and antiparticle coincide $\nu^c \equiv \nu$. Selection rules, then, imply that the annihilation occurs from an s-wave (with amplitude proportional to the mass of the annihilation products) or from a p-wave. Thus we have [20],

$$(\Omega_\nu h^2)^{\rm DIRAC}_{\rm MAJORANA} = \begin{Bmatrix} 5.4 \\ 18.0 \end{Bmatrix} \left(\frac{\rm GeV}{m_\nu}\right)^2 \left[1 + \frac{3}{25}\ \ell n \left(\frac{m_\nu}{\rm GeV}\right)\right] \qquad (6)$$

For $(\Omega_\nu h^2) < 0.25$ for example, $m_\nu^{\rm DIRAC} > 3.3$ GeV and $m_\nu^{\rm MAJORANA} > 6.8$ GeV [20]. The annihilation cross-section for heavy neutrinos has a resonance at $m_\nu = m_{Z^0}/2$. Correspondingly ρ_ν has a dip. For even larger values of m_ν the cross section decreases with m_ν, $\sigma_{annih} \sim m_\nu^{-2}$ (instead of $\sigma_{annih} \sim m_\nu^2/M_Z^4$ for $m_\nu < M_Z/2$). Accordingly, the relic density increases with m_ν and $m_\nu >$(few TeV) is again forbidden by the condition (1) (fig. 1). As a curiosity let us notice that the annihilation cross section is also large for special values of m_ν such that a vector resonance as ψ and ψ' is formed in the s-channel. Thus some special discrete values of m_ν are also allowed [22].

Here we are assuming stable neutrinos. However, neutrinos of mass of order GeV or larger are very unlikely to be stable, unless they carry an exactly conserved particle number.

Unstable neutrinos may have masses in the range forbidden for stable neutrinos, between 100 eV and a few GeV. The condition (1) has to be applied

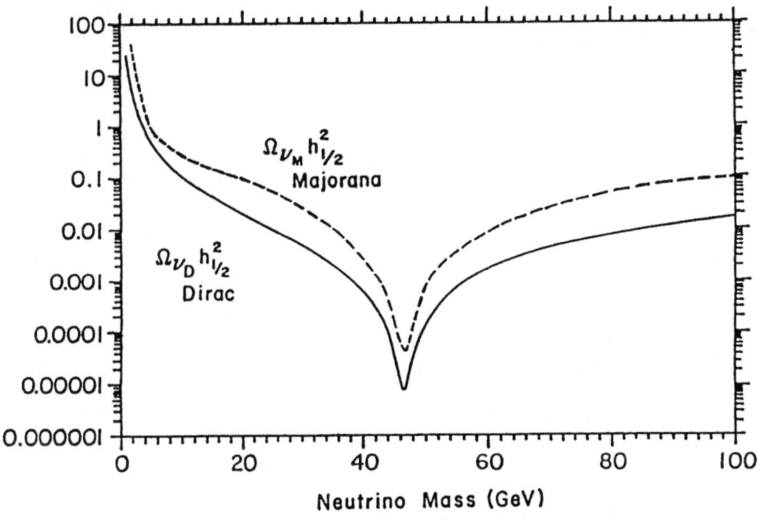

Fig. 1 - The vertical axis shows $\Omega_\nu h^2_{1/2}$, where $h_{1/2}$ is the Hubble constant in units of 50 km/Mpc sec and Ω_ν is the relic density for Dirac (solid curve) and Majorana (broken curve) neutrinos with $m_\nu > 1$ MeV

to the present density of the decay products. If a heavy neutrino, that matter dominates the universe, decays, the decay products, that are relativistic after the decay became dominant. Assuming that the universe is still now radiation dominated by the decay products, their density now is $\Omega_{D.P.} = \Omega_\nu \times (R_{at\ decay}/R_{now})$ being Ω_ν the density of the parent neutrino had it not decayed. The condition (1) and

$$(\Omega_{D.P.})_{now} \leq \Omega_{now} \qquad (7)$$

lead to an upper bound on the lifetime into any mode [8, 23]. For m_ν<1 MeV it is

$$\tau_\nu < 5 \times 10^{14} \text{ sec } (\text{KeV}/m_\nu)^2 (2/g_\nu)^2 (\Omega h^2/0.3)^2 \quad . \qquad (8)$$

This bound is shown for all masses in fig. 2 (large hatched up-left down-right area). The bound is more stringent if the decay products become non-relativistic and matter dominate the universe. Galaxy formation provides a bound more stringent than (8) by a factor 10^5 [24]. For m_ν < 1 MeV it is

$$\tau < 2 \times 10^9 \text{ sec } (\text{KeV}/m_\nu)^2 (2/g_\nu)^2 (\Omega h^2)^{3/2} \quad . \qquad (9)$$

In fact, in most models of galaxy formation structures grow, due to gravitation, from originally small density fluctuations. This growth cannot happen in a radiation dominated universe. Thus, the relativistic decay products should not be dominant when density perturbations in baryonic matter can start forming galaxies. This happens at recomination. That is $(\Omega_{D.P.}) < \Omega_B$ at recombination [24]. Some people does not consider this bound on the same footing of (7), since the problem of galaxy formation has not been solved anyway.

3. Primordial nucleosynthesis

The light nuclei, mainly ^{4}He are formed using all the neutrinos available when the universe cools to T~0.1 MeV. At that moment the number of photons per nucleon that have enough energy to dissociate the nuclei falls below unity. The number of neutrons per proton is just $(n/p) = e^{-(m_n - m_p)/T}$ while protons and neutrons are kept in equilibrium by weak interaction. The n/p ratio freezes out when the rate of weak interactions becomes equal and then smaller than H. H depends on the total energy of the universe at that moment, $H \sim \sqrt{\rho}$. If ρ is larger, the freeze out occurs at a higher temperature, the remaining n/p ratio is larger and so is the amount of ^{4}He synthetized. This amount increases with the number of baryons per proton, $\eta = n_B/n_\gamma$, besides ρ. Thus, in order to get bounds on any new contribution to ρ, such as new flavours of neutrinos, ρ_ν, the abundances of at least two elements are needed (since also η has to be fitted). YANG et al [6] concluded that one range of η fixes three very different abundances (^{4}He, (D+^{3}He), ^{7}Li), η= 4-7 $\times 10^{-10}$, what implies a baryon density

$$0.014 < \Omega_B h^2 < 0.035 \; . \qquad (10)$$

Recently, various alternative scenarios of nucleosynthesis have been proposed [25] in which the bounds on Ω_B are less stringent, but all of them have partial failures up to now. YANG et al. find that the number of flavours of 2-component relativistic neutrinos during the nucleosynthesis, i.e. m_ν < 0.1 MeV, cannot be larger than 4, N_ν <4. If during the nucleosynthesis there are three 2-component (g_ν=2) relativistic neutrino species and a 4th non relativistic one, the 4th neutrino should have a mass m_ν > 20 MeV [26].

Nucleosynthesis implies bounds on visible decays (which produce photons or charged particles) that occur after the synthesis. They alter the value of η (since new γ's are produced) and the decay products destroy the synthetized nuclei.

4. Bounds on lifetimes

Unstable neutrinos that decay into visible channels, i.e., into charged particles (for $m_\nu > 1$ MeV) or photons are severely constrained by various observations. Fig. 2 (partially re-drawn from E. Kolb ref. [27]) shows several of these bounds. The large hatched up-left to down-right region shows the lifetimes forbidden by eq. 7, for any decay mode (it assumes the universe is radiation dominated by the decay products). The wider hatched up-right to down-left region is forbidden for visible modes. The blackbody microwave background radiation, MBR, consists of photons that last interacted at the moment of electromagnetic decoupling or recombination ($t \sim 5 \times 10^5$ y, T~0.3 eV). At that moment ions and electrons, that formed a plasma in which photons had a short mean free path, combine into atoms. The universe becomes transparent to photons. The radiative decay of a non relativistic neutrino $m_\nu > T$, after the recombination (in the region A) would produce a visible energy density larger than the MBR, in excess of observed limits [28,29]. The blackbody spectrum of the MBR must have been established before recombination, in the thermalization era (that starts at $t \sim 10^5$ sec, T~5 KeV). Any extra photons produced in this era would result in distorsions of the MBR (area B of Fig. 2) [28] Photons produced before would have time to thermalize and become part of the blackbody spectrum. Areas C and D are excluded by nucleosynthesis: in the region C photons produced after the sysnthesis decrease η too much in the region D the contribution of the heavy neutrino to the density of the universe during nucleosynthesis is too big, leading to overproductions of ^{4}He [26]. The region E (of Fig. 2) is excluded by astrophysics: the neutrinos

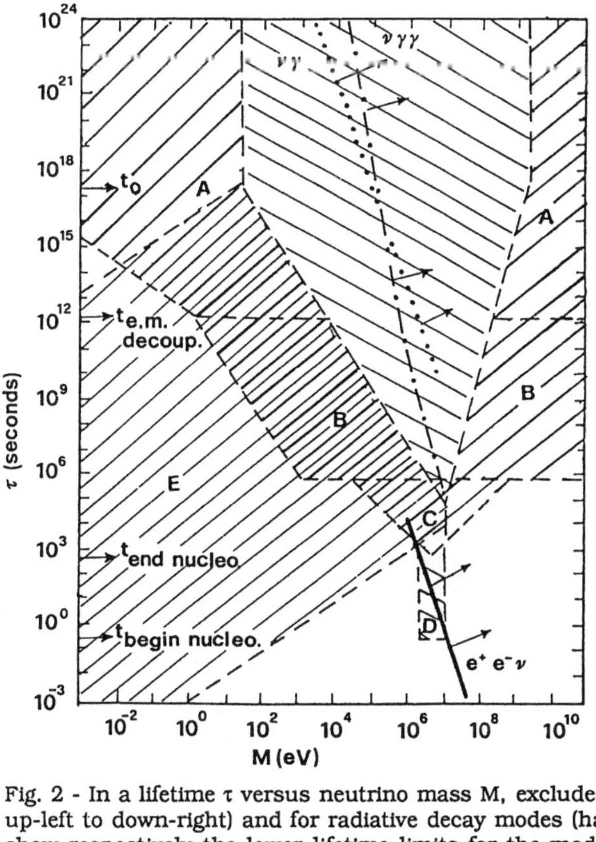

Fig. 2 - In a lifetime τ versus neutrino mass M, excluded regions for any decay mode (hatched up-left to down-right) and for radiative decay modes (hatched down-left to up-right). The lines show respectively the lower lifetime limits for the modes $\nu\gamma$, $\nu\gamma\gamma$ and νe^+e^- (M >1 MeV) in the standard electroweak model.

emitted by supernovas and white dwarfs decaying into visible modes would produce too large a background of high energy photons [29,31]. These bounds apply to the decays $\nu\to\nu'\gamma$, $\nu\to\nu'\gamma\gamma$ and $\nu\to e^+e^-\nu'$ (the e^+ and e^- end up producing energetic photons when annihilating or scattering with interstellar matter or radiation). The lower bounds of the respective lifetimes of these modes in the standard electroweak model are shown in Fig. 2 The mode $\nu'\gamma\gamma$ [32] can be faster than $\nu'\gamma$ [33], because the former is not GIM suppressed but both are far too slow. For $m_\nu > 1$ MeV, i.e. for the ν_τ, the dominant mode, $e^+e^-\nu$ [34], seemed to be able to respect all bounds, for certain values of m_ν and τ. There was a window [35] for tao neutrino masses between 40 and 70 MeV and lifetimes between 10^1 and 10^3 sec. The window was limited by astrophysical bounds (on the observation of γ-ray background radiation from supernovaes [36]) and laboratory bounds (on the mass [37] and on mixing angles [38]).

The non observation of photons following the burst of neutrinos from the SN87A seems to have closed this window [39,40] (TAKAHARA and SATO [39] arrive to this conclusions, provided the radius of the progenitor star was smaller than 10^{13} cm; now it is known to have been 3×10^{12}cm). This window has, in any case, been closed by the new upper bound of 35 MeV on the mass of the ν_τ presented in this conference by the ARGUS collaboration [13].

The decays of neutrinos into invisible particles, light neutrinos or goldstone bosons can be constrained only by dynamical arguments, such as condition (7) (or that, 10^5 times more restrictive in τ, coming from galaxy formation). The decay $\nu\to3\nu'$, through flavour changing neutral currents, F.C.N.C., or Higgs mediated in extensions of the standard model, cannot satisfy this bound without problems to avoid conflict with laboratory experiments and to suppress related radiative decays [41]. F.C.N.C. are possible in the standard model when Dirac and Majorana mass terms are both present. Harai and Nair [42] have recently confronted many neutrino models with the bound (7) and concluded that $\nu_\tau\to3\nu'$ was allowed, through F.C.N.C., for $m_\nu > 35$ MeV (a range of mass now excluded [13]).

The decays $\nu\to\nu'\phi$, with ϕ a Nambu-Goldstone boson, $m_\phi = 0$, may lead to very different lifetimes, according to the models. Lifetimes of cosmological scale are obtained with singlet Majorons [43] or familons [44]. A Majoron is a N.G. boson that appears together with Majorana neutrino masses (the reason for its name) when a global lepton number symmetry is spontaneously violated. There are two different models, the "singlet" [43] and the "triplet" Majoron [45,46]. Familons are N.G. bosons associated to the spontaneous breaking of an interfamiliar (or "horizontal") symmetry. The breaking associated with singlet Majorons and with familons occurs at a large scale V.

The neutrino lifetime is of order

$$\tau = 10^{15}\,\mathrm{sec}\left[\frac{V}{10^{10}\,\mathrm{GeV}}\right]^2\left(\frac{\mathrm{keV}}{m_\nu}\right)^3 . \tag{11}$$

Familons induce other non diagonal decays, $\mu\to e\phi$, $\kappa\to\pi\phi$, $\tau\to e\phi$ that have not been observed. The bounds $V>2\times10^9$ GeV, $V>2\times10^{10}$ GeV and $V>1\times10^7$GeV are respectively obtained [44]. That is, for familons the bound (7,8) can barely be respected, and the bound (9) could only be fulfilled in the odd case just the 3rd generation could decay into the 1st, the second generation being stable. The triplet Majoron [45,46] is the only model in which neutrinos have an interaction among themselves that is almost as strong as electromagnetism among charged particles. In the original model neutrinos could not decay (just annihilate). In a slightly modified version [47], neutrinos may decay very fast. For example, solar neutrinos may decay while arriving to the earth. This mechanism has been proposed as a solution to the solar neutrino problem, with $\tau_\nu\sim5\times10^{-5}(m_\nu\,/\mathrm{eV})$ sec [48]. The observation of electron neutrinos from the supernova means that $\tau_\nu\sim5\times10^{5}(m_\nu\,/\mathrm{eV})$ sec . Still, due to mixings between flavour and mass neutrino eigenstates both things are not incompatible, if the mixing is large [49]. Taking for simplicity just two neutrinos, if $\nu_e=\nu_1\cos\theta+\nu_2\sin\theta$, and $\nu_2\to\nu_1\phi$, the solar neutrino solution requires $45^\circ\le\theta\le90^\circ$ while the signal from the supernova should have been the same for $\theta\le60^\circ$.

Not only unstable neutrinos, but also fast annihilating neutrinos can evade the mass bounds on standard stable neutrinos. In the triplet Majoron model neutrinos annihilate completely into Majorons, leading to a neutrino free universe [46, 50]. A slower annihilation mediated by a light massive Higgs field [51] may lead to any value of Ω_ν with any value of m_ν [50].

5. Neutrinos as the Dark Matter

The "dark matter", the gravitationally dominant mass component of the universe is only detected through its gravitational effects [11]. One of the most striking examples is given by the orbital velocity v of stars and gas clouds orbiting spiral galaxies, as a function of the distance r of these objects to the centre. The function v(r) is called the "rotation curve". If M(r) is the mass contained within the radius r, the third Kepler's law says that $v(r)^2 = G\, M(r)/r$. The surprising fact is that v(r) is constant, thus M(r)~r, well outside the radius of the visible matter in the galaxy [52]. This indicates that most of the mass of spiral galaxies is in a dark approximately spherical halo, whose density falls off as r^{-2} and extends to a distance more that 10 times the radius of the disk. The D.M. is seen at all cosmological scales. The light emitting regions account for Ω ~0.01, the haloes of galaxies for Ω~ 0.10, at the largest scales (superclusters of galaxies) the detected D.M. density reaches $\Omega = 0.2 \pm 0.1$ (while inflation predicts that Ω = 1!) [53].

Could the D.M. be dark normal nucleonic matter? Big bang nucleosynthesis implies (eq. 10) that Ω_B may be as large as 0.22 (for h = 0.4) [6]. Thus primordial nucleosynthesis does not preclude the possibility that the D.M. are baryons. There are however arguments against a universe composed only of baryonic matter [54].Cosmologists are examining the exciting possibility that the D.M. consists of neutral elementary particles. In this case, the structures in the universe (galaxies, clusters) start their formation as lumps of D.M., into which baryonic matter (atoms, molecules) fall, after the electromagnetic decoupling. Density perturbations in baryonic matter cannot grow before the e.m. decoupling (due to their interactions with photons).

Cosmologists classify the D.M. into three types, according to which structures form first (galaxies or clusters of galaxies): hot, warm, cold [55]. With hot D.M. large clusters form first. With cold D.M. (and also warm, but barely) galaxies form first. We know that physical linear scales d, scale as $d(t) = R(t) d_c$ due to the Hubble expansion, where d_c is the linear scale measured in comoving coordinates. Taking the scale factor now $R(t_o) = 1$, the size d_c of a given structure is constantly equal to the size it has now. Since in a radiation or matter dominated universe (see the introduction above) $R(t) \sim t^n$ with n < 1, the horizon ct grows with time even in comoving coordinates, encompassing more material as time goes. Gravitation can act only within ct, obviously. Let us call t_{GAL} the moment a perturbation of the size of a typical galaxy (that contains 10^{11} M_{SUN}, 10^{12} M_{SUN} with the halo) is first encompassed by the horizon.. The temperature at that moment turns out to be T~ 1 KeV. Hot D.M. particles are moving with velocity ~ c at t_{GAL} and until much later. Galaxy size density fluctuations are erased, because every particle in the fluctuation moves from its original position to a distance much larger than the size of the fluctuation. The first fluctuation that survives has the size of the horizon when the hot D.M. particles become non relativistic. It is, thus, larger than a galaxy. Cold D.M. particles move from their original positions in the fluctuations a distance much smaller that ct_{GAL}.

Light neutrinos are hot D.M.. The structure that they form first has a mass

$$M = 3 \times 10^{15}\, M_{SUN} / \, (m_\nu / 30\ \mathrm{eV})^2. \qquad (12)$$

With m_ν ~ 30 eV superclusters form first and galaxies form later by fragmentation. Numerical simulations showed that galaxies form too late in this scenario and light neutrinos fell in disfavour as D.M. candidates. The situation has recently changed. Large structures, at the scale of superclusters have been observed, large voids, with galaxies in walls and filaments [56], large regions of the universe moving with large velocities [57]. Thus, neutrinos are reconsidered, because they naturally produce large structures. Some other mechanism, such as the presence of cosmic strings, is necessary to help forming galaxies faster. Loops of cosmic strings would act as seeds for galaxies [58]. Cosmic strings are linear defects that may remain from the Grand Unified phase transition [59].

The mass of 30 eV is preferred for a light neutrino due to a kinematic constraint by TREMAINE and GUNN [60]. They noticed that quantum mechanics implies that the number of D.M. particles which can fit within a gravitationally bound object is finite. Thus a minimum mass of the particles (corresponding to the maximal occupation number) is required to get a certain total mass. For rich clusters of galaxies this minimum mass is 5eV, for a large galaxy is 30 eV.

Particles of m~1 KeV are just becoming non relativistic at t_{GAL}. Galaxies form first but barely. They are warm D.M. The scenario of galaxy formation with unstable neutrinos, warm D.M. has been proposed [61] and analyzed at length. It does not work. Neutrinos are not good warm D.M. candidates [62].

Cold D.M. particles are non relativistic since much earlier than t_{GAL} galaxies form first and aggregate into clusters and superclusters. Heavy neutrinos of m_ν ~ few GeV have a relic abundance $\Omega \sim 0.1\text{-}1$, and are therefore good cold D.M. candidates. Cold D.M. was definitely the preferred type of D.M., until the recent observation of the large scale structures mentioned above. Several ideas are in consideration to allow cold D.M. to account for the observations.

Thus either light 30 eV stable nutrinos as hot D.M. or heavy, few GeV, Dirac or Majorana stable neutrinos as cold D.M. are both good D.M. candidates.

Light relativistic neutrinos with a very short mean free path may behave as cold D.M., since the total distance they move by diffusion is much shorter than ct at any given moment (i.e. they behave, in this respect, as non relativistic particles). A variation of the triplet Majoron was recently proposed to get the strong interaction needed, with a marginal success [63].

6. May we detect D.M. particles from the halo of our Galaxy?

We live in a spiral galaxy that, as all the others, has a dark halo, which contains most of its mass. The local density of our halo has been estimated in 0.4 GeV/cm^3 (with a factor of two of uncertainty) and the characteristic velocity of the halo particles in 10^{-3} c [64]. Thus, the flux on earth is large ~ 10^7/cm^2 sec (m/1 GeV) for particles of mass m. The possibility of detecting 30 eV neutrinos with this velocity is out of reach at present. Heavy D.M. particles, of a few GeV of mass may, instead, be detected either indirectly, through their annihilation products, or directly, through the energy they may deposit in collisions within detectors.

Annihilations may occur in the halo [65], if the D.M. particles can be trapped from the halo by the sun or the earth, annihilations would occur in the sun [66] or the earth [67]. The possibility of detecting the products of these annihilations has been extensively discussed recently. Heavy neutrinos annihilating in the halo may account for the entire spectrum of cosmic rays antiprotons, (and contribute to the electrons and photons) [68]. Heavy Dirac or Majorana neutrinos with masses larger than 4 GeV would be trapped by the sun and then annihilate there, producing an excess of neutrinos coming from the sun. This is an active subject of research right now, because the data from large proton decay experiments, IBM, Frejus, Kamiokande, are at the point of providing interesting bounds [66,69]. If Dirac neutrinos are heavier than 12 GeV they may be trapped by the earth and annihilating would produce too many neutrinos [67]. Particles lighter than those mentioned "evaporate": the rate at which they escape due to thermal velocities is larger than the capture rate. They never accumulate.

While the scattering cross sections of Dirac and Majorana neutrinos on protons are similar, those on heavier nuclei are very different. For a small momentum transfer, non relativistic Dirac neutrinos have coherent interactions with heavy nuclei, with a N^2 coherent factor (N=number of neutrons). Majorana neutrinos have spin dependent interactions: the cross section is proportional to J(J+1), where J is the spin of the nucleus. This is why Majorana neutrinos in the halo cannot be captured by the earth. The difference in scattering cross sections are very relevant for direct detection. Interesting bounds are obtained on heavy Dirac neutrinos from Ge spectrometers, but not on Majorana neutrinos.

The idea of direct detection is that an incident halo particle could collide with a nucleus within a detector and transfer an energy that could be measured [70]. A recoiling nucleus in a crystal gives part of its kinetic energy to the lattice to produce phonon waves which, after many scatterings within the crystal, thermalize originating an increase in temperature. The detection of these effects requires cryogenic techniques still under development [71]. Another effect of a recoiling nucleus in a crystal is the production of ionization, the excitation of electrons. This mechanism allows the use of existing ultralow background germanium detectors to search for galactic D.M.

A search for D.M. has been initiated with the Ge-detector of the PNL/USC collaboration [72] (see F.T. Avignone in these proceedings) (Similar results have been obtained by the UCSB/LBL collaboration with another Ge-detector [73]). The energy threshold of the detector was of 4 KeV of electronic energy, corresponding to an initial nuclear recoil energy of 15 KeV. At best this threshold could be reduced in the future by a factor of four. The signal obtained when some energy is deposited on an electron is larger than when the same energy is deposited on a nucleus, by a factor called R.E.F. in ref. [72], relative efficiency factor. The Ge-detectors were designed to detect the electrons emitted in double beta decay. The signal of a recoiling nuclus in these detectors had to be estimated. The R.E.F. is an energy dependent quantity which was determined, in ref. [72], to be ~ 3.75 in the 10-100 KeV region.

In order to obtain bounds the observed interactions rate R_o has to be compared with the predicted rate, R_p , if the D.M. candidates under study are in the galactic halo. The predicted rate depends on the local density, ρ_{LOCAL} of the D.M. particles, its mass m, its cross section σ on Ge nuclei and the velocity v, distribution in the halo, $R_p \sim (\rho_{LOCAL}) \sigma v/m$. In ref. 72, a Gaussian v-distribution was used with $v_{r.m.s.}$ = 250 Km/sec and v_{max} = 550 km/sec. These are very conservative,values. Also to be conservative, it was assumed that the halo slowly rotates with a local velocity of 70 Km/sec, in the direction of motion of the sun, like the galactic spheroid. The halo may not rotate at all, the rotation of the spheroid is the upper bound. A non zero rotation velocity of the halo, reduce the velocities of the D.M. particles with respect to the earth, thus R_p is reduced. After integrating over all velocities, an upper bound on $(\rho_{LOCAL}\sigma)$ is obtained by requiring $R_p < R_o$. To fix ideas the σ of standard heavy neutrinos was used in ref. 72. By fixing σ an area in (ρ_{LOCAL}, m) space is excluded. For Dirac neutrinos this area is shown in fig. 3.

Dirac neutrinos with mass between 20 GeV and 2 TeV have been excluded as main components of the halo. There are no bounds on Majorana neutrinos because σ is too small. These experimental bounds may reach at most to masses 8-10 GeV. New cryogenic techniques will be necessary to reach smaller masses and to test particles with spin independent interactions.

With a further assumption, bounds on the cosmological density of Dirac neutrinos may be obtained from the bounds on the local density [74]. With the plausible assumption that the ratio of total D.M. mass over total baryonic mass in our galaxy has the same value in the whole universe, and with improvements in the data expected soon, the very existence in the universe of heavy Dirac neutrinos may be disproved.

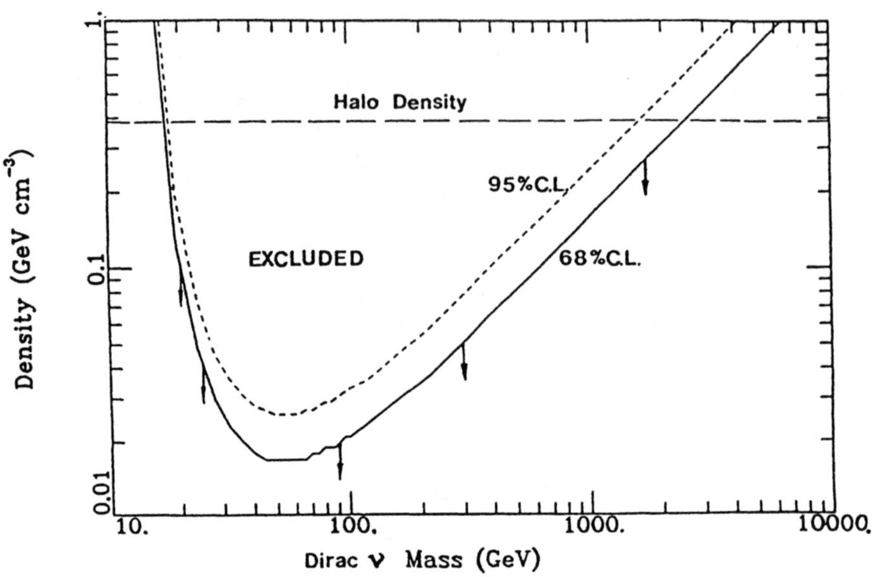

Fig. 3 - Bounds obtained with the PNL/USB Ge-detector (Ref. 72). Maximum halo density of standard heavy Dirac neutrinos consistent with the observed count rate, at the 68% and 95% confidence levels

References

1. For example S. Weinberg: Gravitation and Cosmology (J. Wiley, 1972).
2. A.R. Sandage and G. Tammann: Nature 307, 326 (1984); G. de Vaucouleurs and G. Bollinger, Ap. J. 233, 433 (1979).
3. P.L. Richards, in Inner Space/Outer Space, ed. by E. Kolb et al. (Univ, of Chicago Press, 1986).
4. D.T. Wilkinson, in Inner Space/Outer Space, ed. by E. Kolb et al. (Univ, of Chicago Press, 1986).
5. R.V. Wagoner, W.A. Fowler and F. Hoyle: Ap. J. 148, 3 (1967).
6. J. Yang, M.S. Turner, G. Steigman, D.N. Schramm and K. A. Olive: Ap. J. 281, 493 (1984).
7. J. Bernstein and G. Feinberg: Phys. Lett 101B, 39 (1981)
8. P.B. Pal: Nucl. Phys. B227, 237 (1983).
9. F.K. Thielemann, J. Metzinger and H.V. Klapdor: Astr. Astrophys. 123, 162 (1983); I. Iben and A. Renzini: Phys. Rep. 105C, 329 (1984).
10. A. Guth: Phys. Rev. D23 (1981); A. Linde: Phys. Lett. 108B, 389 (1982); A. Albrecht and P. Steinhardt, Phys. Rev. Lett 48, 1220 (1982).
11. Dark Matter in the Universe eds. J. Knapp and J. Kormendy (Reidel, 1986), Proceedings of the International Astronomical Union Symposium No. 117, Princeton, June 1985.
12. see for example M. Turner in ref. 11.
13. K. Schubert (ARGUS collaboration): these proceedings.
14. G. Steigman: Ann. Rev. Nucl. Part. Sci. 29, 313 (1979).
15. R. Cowsik and J.McClelland: Phys. Lett. 29, 669 (1972) (however they assumed g_ν=4).
16. F. Antonelli, R. Konoplich and D. Fargion, Lett. Nuovo Cim. 32, 289 (1981).
17. K. Olive and M. Turner: Phys. Rev. D25, 213 (1982).
18. S. Wolfram: Phys. Lett. 82B, 65 (1979).
19. B.W. Lee and S. Weinberg: Phys. Rev. Lett 39, 165 (1977). P. Hut: Phys. Lett. 69B, 85 (1977). K. Sato and M. Kobayashi: Prog. Theor. Phys. 58, 1775 (1977); M.I. Vysotskii, A.D. Dolgov and Ya. B Zel'dovich: JETP Lett. 26, 188 (1977).

20. E. Kolb and K. Olive: Phys. Rev. D33, 1202 (1986) and (erratum) Phys. Rev. D34, 2531 (1986).
21. T.K. Gaisser, G. Steigman and S. Tilav: Phys. Rev. D34, 2206 (1986).
22. G.L. Kane and I. Kani: Nucl. Phys. B277, 525 (1986).
23. D. Dicus, E. Kolb and V. Teplitz: Phys. Rev. Lett 39, 169 (1977).
24. G. Steigman and M.S. Turner: Nucl. Phys. B253, 375 (1985).
25. J.H. Applegate, et al.: Phys. Rev. D35, 115 (1987); C. Alcock et al: LLL preprint 1987; J. Audouze et al.: Ap.J. 293, L53 (1985); R.Dominguez-Terneiro; Ap.J. 313, 523 (1987); S. Dimopoulos et al.: SLAC preprint PUB-4356 (1987).
26. E. Kolb and R. Scherrer: Phys. Rev. D25, 1481 (1982).
27. E. Kolb: Proceedings of the 1986 Theoretical Advanced Studies Institute, Santa Cruz, California, July 1986, Fermilab preprint Conf. 86/146-A.
28. D. Dicus, E. Kolb and V. Teplitz; Ap. J. 221, 327 (1978).
29. R. Cowsik: Phys. Rev. Lett. 39, 784 (1977).
30. D. Dicus, E. Kolb, V. Teplitz and R. Wagoner: Phys. Rev. D17, 1529 (1978).
31. S.W. Falk and D.N. Schramm: Phys. Lett. 79B, 511 (1978).
32. J.F. Nieves: Phys. Rev. D28, 1664 (1983).
33. P.B. Pal and L. Wolfenstein, Phys. Rev. D25, 766 (1982).
34. K. Sato and M. Kobayashi, Prog. Theor. Phys. 58, 1775 (1977); E. Kolb and T. Goldman, Phys. Rev. Lett. 43, 897 (1979); R. Shrock, Phys. Rev. D24, 1232 (1981).
35. N. De Leener-Rosier, J. Deutsch and R. Prieels: Phys. Rev. Lett. 59, 1868 (1987).
36. M. Takahara and K. Sato: Phys. Lett. B174, 373 (1986).
37. H. Albrecht et al. (ARGUS collaboration): Phys. Lett 163B, 404 (1985).
38. N. De Leener-Rosier et al: Phys. Lett. B177, 228 (1986): G. Azuelos et al.:Phys. Rev. Lett. 56, 2241 (1986); F. Bergsma et al. (CHARM collaboration): Phys. Lett. 128B, 361 (1983).
39. M. Takahara and K. Sato: Modern Phys. Lett. A 2, 293 (1987).
40. A. Dar and S. Dado: preprint TECH-PH-40 (1987).
41. Y. Hosotani: Nucl. Phys. B191, 411(1981); M. Roncadelli and G. Senjanovich: Phys. Lett 107B, 59 (1981); A. Natale: Phys. Lett 141B, 323 (1984); P. Pal, Nucl. Phys. B227, 237 (1983); M. Gronau and R. Yahalom, Phys. Rev. D30, 2422 (1984).
42. H. Harari and Y. Nair: SLAC preprint PUB-4224 (1987).
43. Y. Chicashige, R.N. Mohapatra and R.D. Peccei: Phys. Lett. 98D, 265 (1981) and Phys. Rev. Lett. 45, 1926 (1980).
44. D.B. Reiss: Phys. Lett. 114B, 217 (1982); F. Wilczek; Phys. Rev. Lett. 49, 1549 (1982); G. Gelmini, S. Nussinov and T. Yanagida, Nucl. Phys. B219, 31 (1983); D. Dicus and V. Teplitz; Phys.Rev. D28, 1778 (1983); H.Y. Cheng; Indiana Univ. preprint IUHET-129 (1987).
45. G. Gelmini and M. Roncadelli: Phys. Lett. 99B, 411 (1981).
46. H. Georgi, S.L. Glashow and S. Nussinov: Nucl. Phys.B193, 297 (1981).
47. J.W.F. Valle: Phys. Lett. 131B, 87 (1983); G. Gelmini asnd J.W.F. Valle: Phys. Lett. 142B, 181 (1984).
48. J.N. Bahcall, S.T. Petcov, S. Toshev and J.W.F. Valle: Phys. Lett. 181B, 369 (1986); J.N. Bahcall, N. Cabibbo and A. Yahilk; Phys.Rev. Lett. 28, 316 (1972).
49. J. Frieman, H. Haber and K. Freese: SLAC preprint, PUB-4261 (1987).
50. E. Kolb and M. Turner: Phys. Lett. 159B, 102 (1985).
51. P.B. Pal, Phys. Rev. 30D, 2100 (1984).
52. S.M. Faber and J.S. Gallagher: Ann. Rev. Astron. and Astrophys., 17, 135 (1979); A. Bosma; Astron. J. 220, 1825 (1981); V.C. Rubin, W.K. Ford and N. Thonnard: Ap. J. 238, 471 (1980); D. Burstein and V.C. Rubin: Astroph. J. 297, 423 (1985).
53. P.J. Peebles, Ap. J. 284, 439 (1984).
54. Lectures of J. Silk at First ESO-CERN School on Particle Physics and Cosmology, Erice, Italy, January 1987.

55. S.J.Bond and A. Szalay: Ap. J. 274, 443 (1983): J.R. P[rimak and G.R. Blumenthal: Moriond conference (1983), Formation and Evolution of Galaxies and Large Structures in the universe, ed. by J. Audouze and J. Tran Thanh Van (Reidel, Dordrecht) 1983.
56. V. De Lapparent, M. Geller and J. Huchra: Ap. J. 302, L1.
57. A. Dressler et al.: Ap. J. 313, L37 (1987).
58. R. Brandenberger, N. Keiser and N, Turok, DAMTP preprint, June 1987, R.Brandenberger, N. Kaiser, D. Schramm and N. Turok, FERMILAB preprint, Pub. 87/126-A; E. Berschinger and P. Watts, MIT preprint 1987.
59. A. Vilenkin, Phys. Rep. 121, 263 (1985) and references therein.
60. S.Tremaine and J. Gunn: Phys. Rev. Lett. 42, 407 (1979).
61. M. Turner, G. Steigman and L. Krauss: Phys. Rev. Lett. 52, 2090 (1984); G. Gelmini, D. Schramm and J.W.F. Valle, Phys. Lett. 146B, 443 (1984).
62. See R. Flores, G. Blumenthal, A. Dekel and J. Primack: Santa Cruz preprint SCIPP 86/64, R. Flores: XXIII Int. Conf. on H.E.P., Berkeley, July 1986 and references therein.
63. G. Raffelt and J. Silk: U.C. Berkeley preprint 1987.
64. J. Caldwell and J. Ostriker: Ap. J. 251, 61 (1981); J. Bahcall and R. Soneira, Ap. J. Suppl. 55, 67 (1984) and Ap. J. Suppl. 44, 73 (1980); J. Bahcall, M. Schmidt and R. Soneira, Ap. J. 265, 730 (1983). See useful discussions in M. Turner: Phys. Rev. D33, 889 (1986) and A.K. Drukier, K. Freese and D. Spergel: Phys. Rev. D33, 34954 (1986).
65. J. Silk and M. Srednicki; Phys. Rev. Lett. 53, 624 (1984); J. Hagelin and G.L. Kane, Nucl. Phys. B263, 399 (1986); M. Srednicki, S. Theisen and J. Silk, Phys. Rev. D33, 2079 (1986); M. Turner, Phys. Rev. D34, 1921 (1986).
66. J. Silk, K. Olive and M. Srednicki, Phys. Rev. Klett. 55, 257 (1985); M. Srednicki, K. Olive and J. Silk, Nucl. Phys. B279, 804 (1987); J. hagelin, K.W. Ng and K. Olive, Phys. Lett. 180B, 375 (1986); K.W. Ng, K. Olive and M. Srednicki, Phys. Lett. 188B, 138 (1986).
67. K. Freese, Phys. Lett. B167, 295 (1986); L. Krauss, M. Srednicki and F. Wilczek, Phys. Rev. D33, 2079 (1986).
68. S. Rudaz and F. Stecker, Lab., for High Energy Astroph. NASA/Goddard Space Flight Center preprint 87-015 (1987).
69. K. Olive and M. Srednicki, in preparation (private communication).
70. M.W. Goodman and E. Witten: Phys. Rev. D31, 3059 (1985); I. Wasserman, Phys. Rev. D33, 2071 (1985); A.K. Drukier et al in ref. 64. See also A. Drukier and L. Stodolski: Phys. Rev. D30, 2295 (1984).
71. P.F. Smith, review at 2nd ESO/CERN Symposium on Cosmology, Astronomy and Fundamental Physics (ESO, Garching, March 1986), Rutherford Laboratory preprint RAL-886-029.
72. S.P. Ahlen, F.T. Avignone III, R.L. Brodzinski, A.K. Drukien, G. Gelmini and D.N. Spergel.
73. D.O. Caldwell, R.M. Eisberg, D.M. Grumm, M.S. Witherell, F.S. Goulding and A.R. Smith, Santa Barbara preprint, USCB-HEP-87-3.
74. G. Gelmini et al., Quarks and Galaxies workshop, L.B.L. Berkeley, California, 1986; Enrico Fermi Institute preprint EFI 86-77.

Radiative Massive Neutrino Decay

M. Roos

Department of High Energy Physics, University of Helsinki,
Siltavuorenpenger 20 C, SF-00170 Helsinki, Finland

1. Introduction

In the not too distant past, neutrinos were considered massless, and in consequence stable. Subsequently it was recognized that such an ideal situation would be a fortuitous accident, and that in general one could expect all neutrinos to be massive (although light) and mixing among themselves (although feebly). In consequence, all neutrinos save the lightest species would be unstable giving rise for instance to radiative decays

$$\nu_2 \to \nu_1 + \gamma \tag{1}$$

when $m_2 > m_1$.

To date, no such radiative decays have been observed. Searches for them in terrestrial particle physics data and in astrophysical processes permit to set limits to the mass and decay rate of massive neutrinos. Interesting limits can also be derived from the standard cosmological model. The subject of Sec.3 of this paper is to survey all such limits.

Before doing that, however, one should understand the properties of massive neutrinos expected in the standard model of electroweak interactions, as well as in some more general models. These include the standard model without GIM mechanism, some versions of left-right symmetric models, subconstituent models, and mirror fermion models. This is the subject of Sec.2.

Section 4 contains conclusions about possible massive neutrino ranges, possible theories, possible dark matter candidates, and possibilities for observations in the extragalactic photon flux.

2. Theories of Radiative Neutrino Decay

The effective matrix element for the decay (1) compatible with gauge invariance is of the general form [1]

$$M = \varepsilon^\mu q^\nu \bar{\nu}_1(p_1)\sigma_{\mu\nu}(a+b\gamma_5)\nu_2(p_2) , \tag{2}$$

where ε^μ and $q^\nu = (p_2-p_1)^\nu$ are the polarization vector and the momentum of the photon, respectively. From (2) one derives the width

$$\tau^{-1} = \Gamma(\nu_2 \to \nu_1\gamma) = \frac{1}{8\pi}\left(\frac{m_2^2-m_1^2}{m_2}\right)^3 (|a|^2+|b|^2) = \pi^{-1}E_0^3\ (|a|^2+|b|^2), \tag{3}$$

where E_0 is the energy of the photon.

The decay (1) in the standard model proceeds (in the unitary gauge) through the two one-loop diagrams shown in Fig.1. A full standard model calculation [2] in the case of ν_1 and ν_2 being Dirac particles gives (under the assumption that $m_i \ll m_\ell, M_W$)

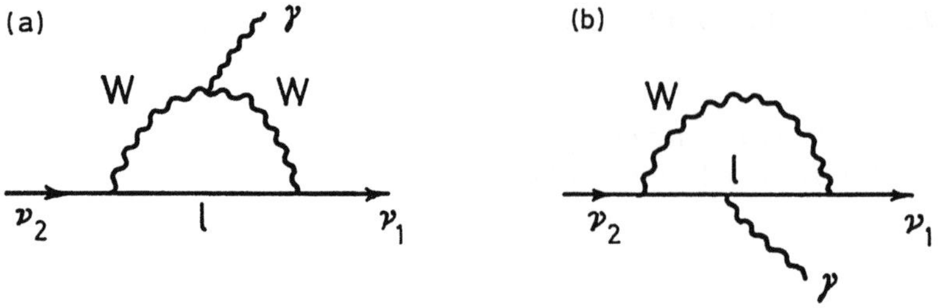

Fig.1: The two unitary gauge diagrams contributing to the radiative decay (1) in the standard model.

$$a_D = -\frac{eG_F}{8\sqrt{2}\pi^2}(m_1+m_2)\sum_\ell U_{1\ell}U^*_{2\ell}F(r_\ell),$$
$$b_D = -\frac{eG_F}{8\sqrt{2}\pi^2}(m_2-m_1)\sum_\ell U_{1\ell}U^*_{2\ell}F(r_\ell), \qquad (4)$$

where U is the leptonic mixing matrix and the sum runs over the charged leptons . The $F(r_\ell)$ is a smooth function of $r_\ell = (m_\ell/M_W)^2$ defined in Ref.[2]. For Majorana neutrinos one has $a_{Maj} = 0$, $b_{Maj} = 2b_D$, $b_{Maj} = 0$, depending whether the relative CP quantum number of ν_1 and ν_2 is +1 or -1. For $m_1 = 0$ one obtains from (3) and (4)

$$\tau^{-1}_{stand} = \Gamma(\nu_2 \to \nu_1\gamma) = \frac{K\alpha G_F^2}{128\pi^4} m_5^2 \left|\sum_\ell U_{1\ell}U^*_{2\ell}F(r_\ell)\right|^2 , \qquad (5)$$

where K = 1 (or 2) for Dirac (or Majorana) neutrinos. With the presently known three light charged leptons ($r_\ell \ll 1$), the leading term in the sum in cancelled because of the unitarity of the matrix U, and the rate is thus additionally suppressed by the factor r_ℓ^2.

The lower limit to τ_{stand} in the 3-family case with $m_1=0$, $K=1$, $\ell=\tau$, is then obtained by setting $U_{1\tau} \equiv \sin\Theta_{\tau 1} = 1$ in

$$\tau^{-1}_{S3} \equiv \tau^{-1}_{stand} = (1.2\cdot 10^{44}s)^{-1}\left(\frac{m_2}{1\ eV}\right)^5 \sin^2\Theta_{\tau 1} . \qquad (6)$$

This corresponds to the line S_3 in Fig.4.

If the Glashow-Iliopoulos-Maiani cancellation were avoided by the presence of additional heavy generations [2] or by making the unitarity sum in (5) incomplete (e.g., with an extra singlet ν), the lower bound for the radiative lifetime of the ν_2 is orders of magnitude less. To be specific, if the masses of the heavy lepton doublet (ℓ,ν_2) are bounded by [3]

$$m_\ell - m_2 < 310\ GeV ,$$

equation (6) is replaced by

$$\tau^{-1}_{S4} \approx (7.3\cdot 10^{37}s)^{-1}\left(\frac{m_2}{1\ eV}\right)^5 \sin^2\Theta_{21} . \qquad (7)$$

This corresponds to the limit S_4 in Fig.4. The sensitivity to m_ℓ is very small.

The essential feature of (5) is that the scale is (apart from G_F) set solely by the mass of the neutrino. This is a direct consequence of the pure left-handed structure of $W^\pm$-lepton interactions, a built-in property of the standard model. Indeed, if the two $W\ell\nu_i$ vertices in the decay amplitude (see Fig.1) are both of the same chirality, only the momentum part of the internal charged-lepton propagator $(p_\ell - m_\ell)^{-1}$ contributes to the matrix element. All terms of the gauge-invariant form (2) are then proportional to the neutrino mass, and thus yield the result $\Gamma \propto m_2^5$.

If, on the other hand, the $W\ell\nu_i$ vertex should contain a right-handed piece, also the mass part of the lepton propagator would contribute to the amplitude. This has a dramatic effect on the decay rate as has been pointed out before [4].

There are two types of electroweak models where right-handed interactions occur: left-right-symmetric models [5] and mirror models [6]. In both kinds of models the leptonic charged weak current is of the general form

$$L = W_\mu^\pm \frac{g}{2\sqrt{2}} \sum_{i=1,2} \sum_\ell \bar{\nu}_i \gamma^\mu (V_{\ell i} - A_{\ell i}\gamma_5)\ell \; . \tag{8}$$

The parameters a and b appearing in (2) are then in leading order given by [1,4,7]

$$\begin{aligned} a &= -\frac{eG_F}{8\sqrt{2}\pi^2} \sum_\ell m_\ell f(r_\ell)\,(V_{\ell 1}V^*_{\ell 2} - A_{\ell 1}A^*_{\ell 2}) \; , \\ b &= -\frac{eG_F}{8\sqrt{2}\pi^2} \sum_\ell m_\ell f(r_\ell)(V_{\ell 1}A^*_{\ell 2} - V_{\ell 2}A^*_{\ell 1}) \; , \end{aligned} \tag{9}$$

where $f(r_\ell)$ is a smooth function of the order of unity [7]. These leading terms obviously vanish in the standard model where $V_{\ell i}/A_{\ell i} = 1$. From (3) and (9) we see that without such an accidental cancellation the natural scale of the width is $\Gamma \sim G_F^2 m_\ell^2 m_2^3$.

A similar situation may occur in models with composite leptons if the weak couplings between ground state leptons and excited leptons are of the Pauli magnetic moment type [8]:

$$L = (f/2\Lambda)(g\bar{\Psi}_{L*}\,\sigma^{\mu\nu}\,\frac{1}{2}\,\tau\,\Psi\,W_{\mu\nu} + g'\,\bar{\Psi}_{L*}\,\sigma^{\mu\nu}\,\frac{1}{2}\,Y\Psi\,B_{\mu\nu} + \text{H.c.}) \; , \tag{10}$$

where Ψ_{L*} and Ψ are left-handed weak doublets of the excited and ground state leptons, respectively, f is a parameter of O(1) and Λ is the composite scale. An excited neutrino $\nu_2 = \nu^*$ may then decay radiatively via loop diagrams, of which the dominant one looks like Fig.1b. There are no right-handed couplings, but the form of the interaction implies that the decay rate is again proportional to $m_{\nu^*}^3$.

Let us consider in some detail both types of models with right-handed interactions. In the left-right-symmetric model [5], based on the gauge group $SU(2)_L \otimes SU(2)_R \otimes U(1)$, a right-handed (1,2,-1) neutrino ν_R can decay radiatively to a left-handed (2,1,-1) neutrino ν_L (or vice versa), provided that the vector bosons $W_R^\pm$ and $W_L^\pm$ mix (see Fig.2). The lighter of the two vector-boson mass eigenstates, $W \equiv \cos\zeta\, W_L + \sin\zeta\, W_R$, couples to leptons according to

$$L = -\,W^\mu\,(g/\sqrt{2})(\cos\zeta\;\bar{\nu}_L\gamma_\mu\,\ell_L + \sin\zeta\;\bar{\nu}_R\gamma_\mu\,\ell_R) + \text{H.c.}, \tag{11}$$

where ζ is the bosonic mixing angle. (We neglect above for clarity the possible family mixing and the mixing between ν_L and ν_R, which are inessential to our argument.) The contribution of the diagrams mediated by the heavier vector boson W_2 can be neglected since $M_{W_2} \gtrsim 20 M_{W_1} \simeq 1.6$ TeV from the K_L-K_S mass difference and other low-energy constraints [9].

To apply (9) we note that now

$$V_{\ell 1} = A_{\ell 1} = \cos\zeta,\; V_{\ell 2} = -\,A_{\ell 2} = \sin\zeta. \tag{12}$$

The decay width for $\nu_R \to \nu_L\gamma$ (or $\nu_L \to \nu_R\gamma$) is then

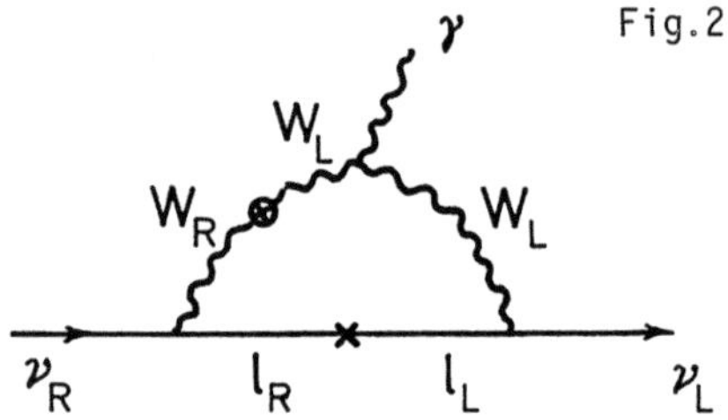

Fig.2: The typical diagram for the decay $\nu_R \to \nu_L \gamma$ in the left-right symmetric model.

$$\tau_{LR}^{-1} \approx (1.3\cdot 10^{18}s)^{-1} \left(\frac{m_R}{1\ eV}\right)^3 \sin^2 2\zeta , \tag{13}$$

where we have taken the charged lepton to be the τ and the cosmological density of the ν_R to be the same as the ν_L. The line LR in Fig.4 corresponds to the limit $\zeta = 0.004$ [9]. Family mixing and LR-mixing allows the region to the right of this line.

Some authors [10] have considered the version of $SU(2)_L \otimes SU(2)_R \otimes U(1)$ in which neutrinos aquire masses due to the "see saw" mechanism. In this model, in contrast to our assumption [4], the predominantly right-handed neutrinos are very heavy, and they therefore essentially decouple from low energy physics. The predominantly left-handed neutrinos (in each family) aquire a tiny right-handed piece during mass generation which enables them to couple to W_R. In the radiative decay of such a predominantly left-handed neutrino to another similar one of a different family, via family mixing, the right-handed pieces do not radically change the conclusions of the standard model, where no right-handed pices participate.

In the mirror fermion model [6] the situation is favorable in the following two respects. Firstly, the cosmological density of mirror neutrinos N_R is never suppressed since their neutral-current interactions have the same strength as those of the standard left-handed neutrinos. Secondly, charged mirror leptons, L, must be heavy, $25 < m_L < 310$ GeV [3] and therefore the radiative width increases. The relevant part of the weak Lagrangian is given by

$$\mathcal{L} = - W^\mu (g/\sqrt{2})(-\sin\Theta_\ell \bar{\nu}_L \gamma^\mu L_L + \cos\Theta_\ell \bar{N}_R \gamma^\mu L_R) + \text{H.c.}, \tag{14}$$

where Θ_ℓ is the ℓL mixing angle parametrizing the mixing between the weak eigenstates of the charged lepton ℓ and its mirror partner L. One typical diagram contributing to the decay $N_R \to \nu_L \gamma$ is shown in Fig.3.

From (14) one can read off the couplings

$$V_{\ell 1} = A_{\ell 1} = -\sin\Theta_\ell,\ V_{\ell 2} = -A_{\ell 2} = \cos\Theta_\ell . \tag{15}$$

The radiative decay rate $N_R \to \nu_L \gamma$ (or $\nu_L \to N_R \gamma$) is then ($\ell = e$)

$$\tau_M^{-1} \approx K(3.6\cdot 10^{15}s)^{-1} \left(\frac{m_N}{1\ eV}\right)^3 \cdot \sin^2 2\Theta_e . \tag{16}$$

This limit is drawn for Dirac neutrinos (K=1), m_L=310 GeV, and $\sin^2 2\Theta_e = 1$ as the line M in Fig.4.

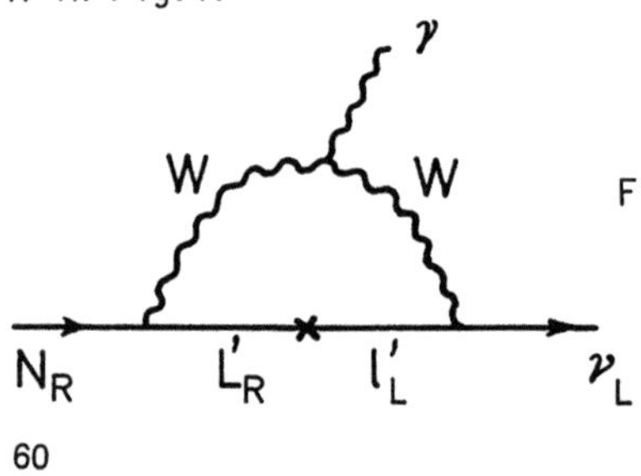

Fig.3: The typical diagram for the decay $N_R \to \nu_L \gamma$ in the mirror model. The primes indicate weak eigenstates of fermions.

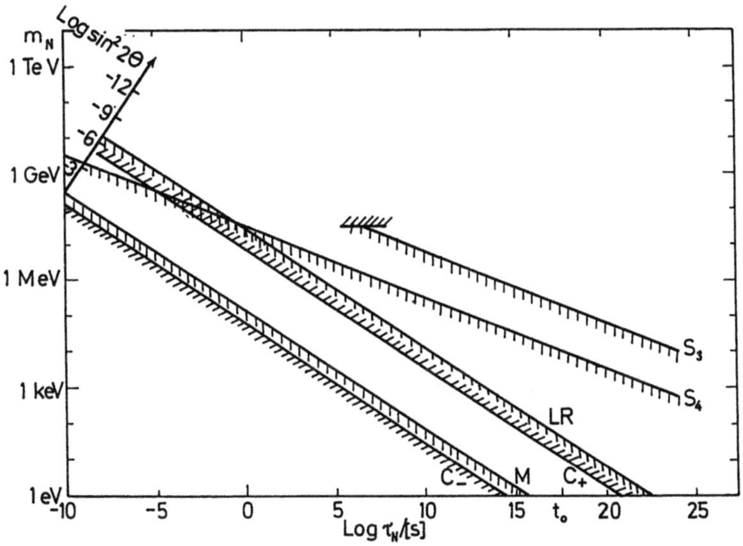

Fig. 4: Theoretical limits for neutrino mass m_N versus radiative lifetime τ_N. The notation is described in the text.

Let me finally quote [8] the radiative width of an excited neutrino ν^*:

$$\tau_C^{-1} \approx \frac{\alpha^3}{10\pi^2} \left(\frac{f}{\Lambda}\right)^2 \left[\ln\left(\frac{\Lambda^2}{M^2}\right)\right]^2 m_{\nu^*}^3, \tag{17}$$

where M is the mass of whichever particle is heavier, W or the charged lepton ν^*. Taking

$$200 \text{ GeV} \lesssim \Lambda/f \lesssim 250 \text{ TeV} ,$$

one obtains [8]

$$\tau_C \approx (3\cdot10^{14} - 4\cdot10^{20}) \text{ s} \left(\frac{1 \text{ eV}}{m_{\nu^*}}\right)^3 . \tag{18}$$

In Fig.4 these limits correspond to the lines C_- and C_+, respectively.

3. Limits on Neutrino Mass and Decay Rate

All relevant limits on neutrino mass, radiative decay rate and mixing angle from terrestrial particle physics experiments, from astrophysical observations as well as from cosmology are exhibited in Figs.4-6. The coordinates are neutrino mass m_N and inverse decay rate or radiative lifetime τ_N. However, one can equally well read off mass and mixing angle, since the mixing angle acts as a delay factor on the lifetime. Thus the corresponding mixing angle is $\sin^2 2\Theta = 1$ on each of the base lines S_3, S_4, LR and M in Fig.4, and smaller mixing angles allow the region to the right of the base line. In Fig.4 the $\sin^2 2\Theta$ scale for the mirror model M is shown.

Fig.4 also contains some model-specific limits: the upper and lower limits, C_+ and C_-, to the composite model are such. The standard model with 3 families has no heavier neutrino than ν_τ, for which the laboratory limit obtained by the ARGUS detector [12], $m_{\nu_\tau} < 35$ eV, is shown. The limit $\zeta < 0.004$ [9] in the LR-model was already incorporated. Note that the LR-line has been drawn for the third family: $\ell = \tau$. The LR-line corresponding to the second family would be shifted to longer lifetimes by the factor $(m_\tau/m_\mu)^2 = 285$.

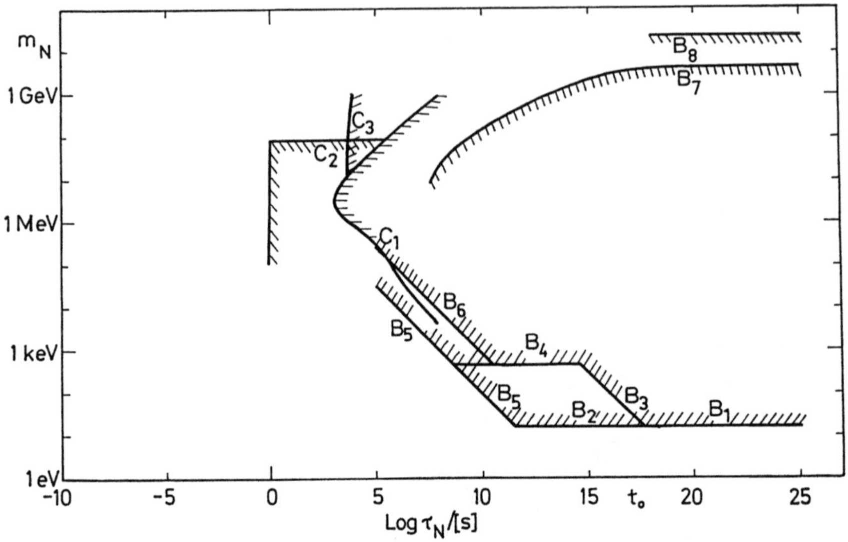

Fig.5: Limits from cosmology (B) and from primordial nucleosynthesis (C). For details, see the text.

Fig.5 contains cosmological limits (denoted B) and limits from primordial nucleosynthesis (C).

From the beginning of the matter-dominated era at time 10^{-6} t_0 up to now, at $t_0 \approx 10$ Gy ≈ 3.2 · 10^{17}s, neither a massive neutrino N nor its decay products are allowed to contribute excessively to the present mass density, or else, the age of the Universe gets too small. This sets the limit B_1, m_N (B_1) $\lesssim$ 24 eV [13].

At earlier times, during the radiation-dominated era, neither the N nor its decay products must significantly affect the dynamical evolution of the Universe. This sets the limits B_2 and B_5 [13].

Alternative cosmological scenarios set weaker limits. At limit B_3 [13], the present Universe is still radiation-dominated and has critical density $\Omega_0 = 1$. Perturbations caused by N-decay will cause large-scale structures of the right size. If the structures were too small, the Universe would be too young.

The same argument sets the limits B_4, B_6 [13], however in a scenario which assumes an early N-matter dominated era, followed by a radiation-dominated era when the N have decayed, followed by the usual matter-dominated era when the N decay products have become non-relativistic.

As the N gets heavier, their number density goes down so that their contribution to the total mass density of the Universe also decreases. This sets the lower limits B_7 for Dirac neutrinos [14] and B_8 for Majorana neutrinos [15].

At nucleosynthesis time, the abundance of ^{4}He and ^{2}H set the well-known limit C_1, most recently recalculated in Ref.[16]. This is slightly improved by the limit C_3 on photofission of ^{2}H [17]. Moreover, the total N energy density should not exceed the energy density stored in radiation at time later then 100 s. An early matter-dominated universe would speed up the expansion and affect nucleosynthesis. This yields [15] limit C_2.

Fig.6 contains astrophysical limits (denoted D) and limits from terrestrial experiments (E).

The limit D_1 refers to the extragalactic photon flux in the UV, visible, and IR regions of the spectrum, compiled in Ref.[18]. Preliminary results [19] from the last Apollo flight in 1985 add a small corner excluding the 13.5 - 17.5 eV region for decay rates >(4.5 - 5.8 · 10^{23}s)$^{-1}$. This is of special interest since it rules out previous claims and rumors for a radiatively decaying neutrino of mass 14.9 eV with a lifetime of 0.6 - 5.0 · 10^{23}s.

The limit D_2 refers to ionizing extragalactic photons [18]. The limits D_3 and D_4 are from measurements in the Coma cluster and Virgo cluster directions, respectively

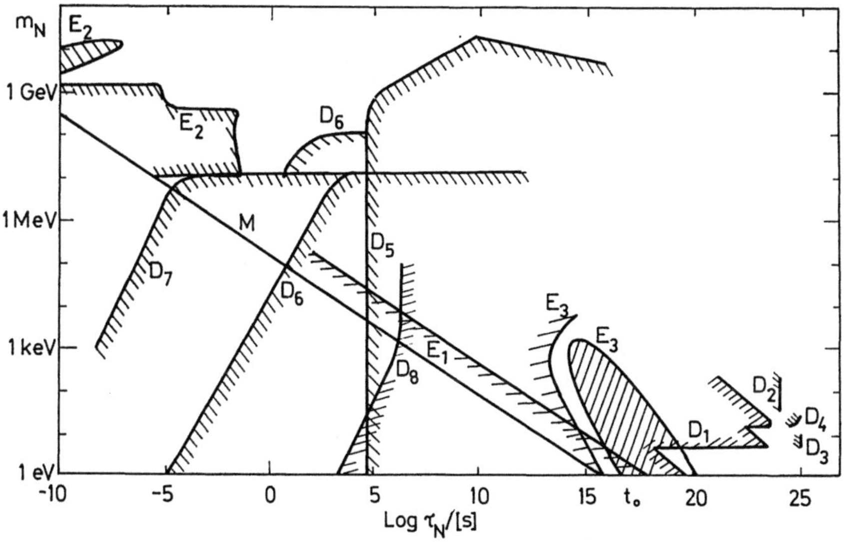

Fig.6: Limits from astrophysics (D) and from terrestrial particle physics experiments (E). For details, see the text.

[18]. Note that the flux from galaxies is about 50 photons/cm^2·s·sterad·Å, implying an ultimate limit of observability of extragalactic decay rates at $(2.6\cdot10^{24}s)^-$. In the direction of heavy masses, the absence of distorsions in the cosmic background radiation sets the well-known limit D_5, here plotted for a chemical potential $\mu < 0.01$ and $\Omega_B = 0.1$ [20]. Another calculation of the maximal distortion gives the limit [21] D_8.

The γ-spectrum from supernovae including SN1987A sets the limit [14,22] D_6, and supernova energetics set the limit [14] D_7.

Detection of electron neutrinos from the supernova SN1987A sets limits [23] on the maximal amount of ν_e mixing with any other species; this yields the limit E_3. If the decaying neutrino is the ν_μ or the ν_τ as is possible in the LR-model, the limit E_3 moves to the right by 1/2 - 1 order of magnitude.

A very large number of charged current and neutral current data can be used [7,24] to set limits on the mixing angle Θ_e (Θ_μ) in (14). These data include [24] the muon decay parameters, neutrino cross-sections, asymmetries in e^+e^--annihilation and in polarized lepton scattering, optical rotation in atoms, π/K decay ratio, neutrino oscillations into inert states, and the vector boson masses. In Fig.6 these data yield the limit E_1.

All other low-energy limits on mixing angle are summarized in the contours E_2. These include [24] data on pseudoscalar meson decays and on searches for heavy lepton decays into specific channels.

We also show the limit E_4 on $\nu_\mu \to \nu_\tau$ oscillations from accelerator experiments [25]. This is only relevant if the decaying neutrino is the ν_μ or the ν_τ, as may be the case in the LR-model.

4. Conclusions

By superimposing the experimental limits in Figs.5,6 on the theoretical lines in Fig.4 one may draw several conclusions. For clarity, Figs.7,8 contain the main outlines of the experimentally forbidden region and the theoretical lines M and LR, respectively.

One notes that heavy neutrinos, above 14 GeV, are tolerated at all times. The present laboratory limit of m > 25 GeV is not visible within the scale of Figs.5-8. Clearly the standard model cannot accommodate any heavy neutrinos, neither can the composite model (C). In the LR-model (Fig.8) and the standard model with 4 families avoiding GIM-suppression (S_4) such neutrinos are natural above a few GeV. In the

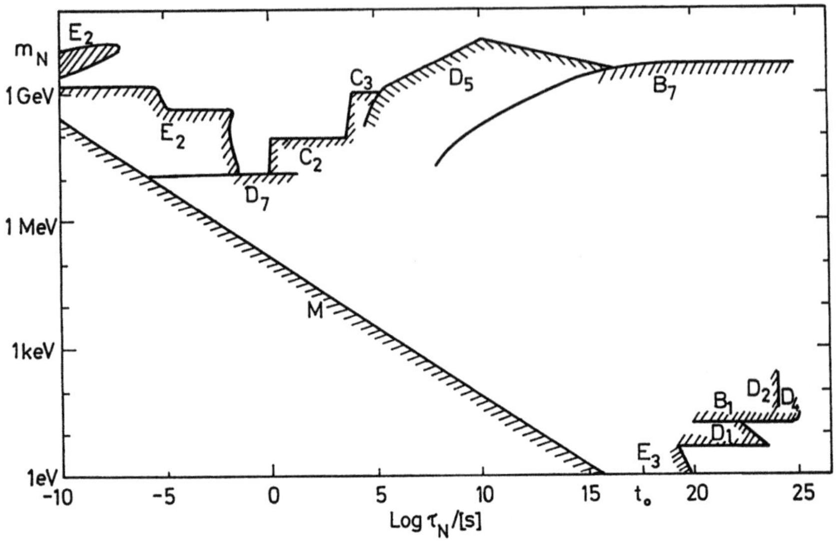

Fig.7: Limits on the mirror fermion model (M) summarized.

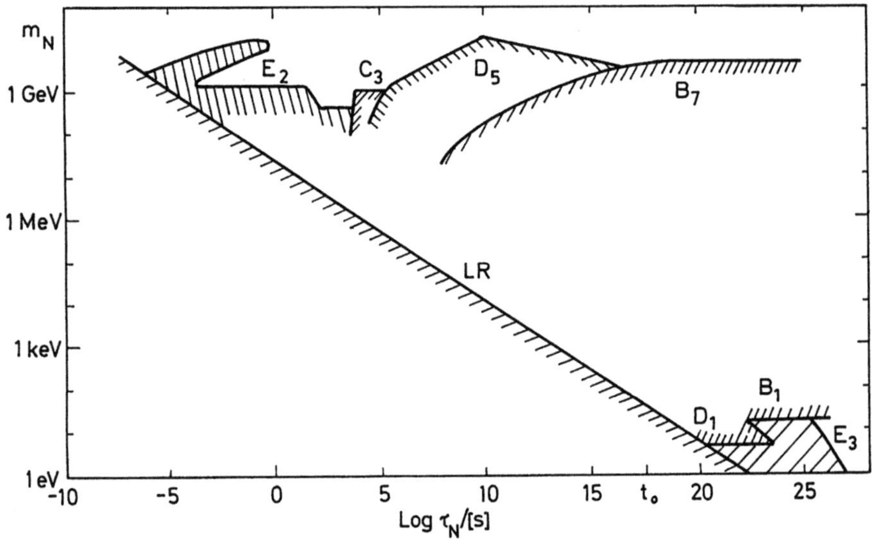

Fig.8: Limits on the LR-model summarized.

mirror fermion model (Fig.7) lower masses are allowed, even as low as 10 MeV in a window at τ = 0.35 - 1s. At all times small mixing angles are needed.

Turning now to the lower right-hand corner of Figs.7,8 where a small allowed window remains, some care must be exercised. The limit B_1 is indeed a limit on neutrino mass. On the other hand, the limit D_1 is a limit on twice the photon energy

$$2E_0 = (m_2^2 - m_1^2)/m_2 \tag{19}$$

plotted under the assumption $m_1 = 0$. The limits E_3, E_4, then again, are limits on $\Delta m^2 = m_2^2 - m_1^2$, here plotted assuming $m_1 = 0$. The limit E_4 applies only to the LR-model.

One then concludes the following:

- No light radiatively decaying neutrinos are possible in the standard models (S_3, S_4).
- No neutrinos heavier then 24 eV are allowed by the cosmological limit B_1 in any model. Expressed in photon wave lengths, the allowed range is

$$\lambda \gtrsim 1000\ \text{Å}. \qquad (20)$$

- Neutrinos lighter than 24 eV may decay radiatively into UV, visible, or IR photons. The composite model (C) can only accommodate very soft IR radiation, due either to neutrinos in the 1 meV - 1eV range, or to very degenerate neutrino pairs, as indicated by (19).
- The mirror model can accommodate any radiative decays in the range of (20), because of the flexibility allowed by the mixing angle.
- The LR-model is excluded at all times up to $\tau = 10^{26}$ s at least.
- If the problem of missing matter in the Universe can be solved by neutrinos at all, then they have masses in the 10-24 eV range and are relatively stable against radiative decay (as specified above).

Let me finally note that attempts have been made to explain both the SN1987A neutrino burst and the solar neutrino deficiency problem by resonant oscillations [23]. This would imply radiative stability at the level of $\tau > 10^{29}$s in the LR-model and $\tau > 4 \cdot 10^{22}$s in the mirror fermion model.

Acknowledgements

It is a pleasure to acknowledge help and useful discussions with J. Maalampi, Helsinki. Part of this work was done at the University of Mainz where I enjoyed the hospitality of K. Kleinknecht and F. Scheck, and the helpfulness of their staff. I am indebted to the Deutsche Akademische Austauschdienst (DAAD), to the Finish Ministry of Education, and to H.V. Klapdor and the Max-Planck-Institut für Kernphysik, Heidelberg, for financial support.

References

1. W. Marciano and A. Sanda, Phys.Lett. 67B,303(1977);
   S. Petkov, Yad.Fiz. 25,641(1977) and 25,1336(1977), [Sov.J.Nucl.Phys. 25, 340 (1977) and 25,698(1977)].
2. P.B. Pal and L. Wolfenstein, Phys.Rev. D25,766(1982) and references therein.
3. U. Amaldi et al., Phys.Rev. D36,1385(1987).
4. J. Maalampi, K. Mursula and M. Roos, Phys.Rev.Lett. 56,1031(1986).
5. C. Pati and A. Salam, Phys.Rev. D10,275(1974);
   R.N. Mohapatra and J.C. Pati, Phys.Rev. D11,566(1975) and 11,2558(1975);
   G. Senjanovic and R.N. Mohapatra, Phys.Rev. D12,1502(1975).
6. See, e.g. J.C. Pati and A. Salam, Phys.Lett. 58B,333(1975);
   F. Wilczek and A. Zee, Phys.Rev. D25,553(1982).
7. K. Enqvist, K. Mursula and M. Roos, Nucl.Phys. B226,121(1983).
8. J.A. Grifols, A. Mendez and M. Ruiz-Altaba, Phys.Lett. 171B,303(1986).
9. See, e.g., L. Wolfenstein, Phys.Rev. D29,2130(1984).
10. U. Chattopadhyay and P.B. Pal, Phys.Rev. D34,3444(1986);
    J. Liu, Carnegie Mellon Univ. preprint CMU-HEP87-10.
11. J. Maalampi, K. Mursula and M. Roos, Nucl.Phys. B207,233(1982).
12. K. Schubert, these Proceedings.

13. G. Steigmann and M.S. Turner, Nucl.Phys. B253,375(1985) and references therein.
14. A.D. Dolgov and Ya.B. Zeldovich, Rev.Mod.Phys. 53,1(1981) and references therein.
15. D. Fargion and M. Roos, Phys.Lett. 147B,34(1984).
16. N. Terasawa et al., Univ. of Tokyo preprint (1986).
17. M. Kawasaki et al., Phys.Lett. 178B,71(1986).
18. F.W. Stecker and R.W. Brown, Astrophys.J. 257,1(1982) and references therein.
19. S. Bowyer, private communications (1987).
20. P. Salati, Phys.Lett. 163B,236(1985).
21. M. Kawasaki and K. Sato, Phys.Lett. 169B,280(1986).
22. M. Takahara and K.Sato, Phys.Lett.174B,373(1986) and Mod.Phys.Lett.A2,293(1987).
23. J. Arafune et al., Phys.Lett. 194B,477(1987) and Phys.Rev.Lett. 59,1864(1987); S.P. Mikheyev and A.Yu. Smirnov, JETP Lett.(1987) and these Proceedings; D. Nötzold, Phys.Lett. 196B,315(1987).
24. J. Maalampi and M. Roos, in preparation.
25. V. Zacek, these Proceedings.

Part II

Neutrino Reactions and Properties

Neutrino Reactions and the Structure of the Neutral Weak Current

K. Winter

CERN, CH-1211 Geneva 23, Switzerland

The role of neutrino reactions in the investigation of the neutral weak current is reviewed from its discovery in 1973 to recent high precision tests of the Standard Model.

1. INTRODUCTION

In 1973 muon-less neutrino reactions were discovered at CERN in the elastic scattering of anti-muon-neutrinos on electrons [1],

$$\bar{\nu}_\mu e \rightarrow \bar{\nu}_\mu e;$$

the neutral neutrino current $\bar{\nu}_\mu \nu_\mu$ and the neutral electron current $\bar{e}e$ couple by exchanging a neutral intermediate boson. In the reaction [2]

$$\nu_\mu N \rightarrow \nu_\mu \pi^+\pi^- \pi^0 N$$

muon-neutrinos scatter inelastically on nucleons and transfer part of their energy to the neutral quark currents $\bar{u}u$ and $\bar{d}d$.

Observation of an asymmetry of the cross-section of the scattering of left-handed and right-handed electrons on deuterons at SLAC in 1978 [3] demonstrated that the neutral weak $\bar{e}e$ current is violating parity.

The discovery of parity violation of the weak $\bar{e}e$ current was then confirmed by the observation that the polarization plane of a laser beam passing through bismuth vapour is rotated [4].

The existence of $\bar{\mu}\mu$ and $\bar{\tau}\tau$ neutral currents was deduced from the observation of a weak forward-backward charge asymmetry in the annihilation of electrons and positrons at the PETRA Collider at DESY [5] in 1982. The existence of other neutral quark currents, e.g. $\bar{s}s$, was deduced from the analysis of the inelasticity distribution of deep inelastic neutral current neutrino scattering on nucleons [6]. Also the neutral current $\bar{\nu}_e \nu_e$ has been observed in neutrino experiments at CERN [7].

So far only diagonal neutral currents have been observed which do not change the flavour of the particles involved. With three families of leptons and quarks, the standard theory [8] predicts the existence of six diagonal neutral lepton currents and six diagonal neutral quark currents. The neutral current has a more complex helicity structure than the charged current. Experiments, e.g. observation of coherent π^0 production on nuclei (Chapter 3) has demonstrated the vector nature of the neutral current interaction and measurements of the inelasticity distribution in deep inelastic neutral current neutrino scattering [9] have shown that both left-handed and right-handed currents exist (Chapter 4). This observation shows directly the existence of a unified, electro-weak force. Its existence was also demonstrated by the observation of interference of amplitudes with γ and Z^0 interchange, e.g. in the helicity asymmetry of electron-deuteron scattering and in the optical activity of bismuth vapour.

The structure of the current, as deduced from these experiments implies that left-handed fermions transform as a doublet under a weak isospin rotation group and right-handed fermions transform as a singlet. The Standard Model of the electroweak gauge theory predicted this structure of neutral currents and it successfully describes a large amount of experimental data [10], which are all consistent with universal strength of the forces, g_2 and g_1, associated with the SU(2) and U(1) symmetry groups, respectively.

The fundamental quantities of the Standard Model, the coupling constants g_1 and g_2, and the masses of the weak bosons, m_W and m_Z, are related to the angle Θ_W that describes the mixing of the two local symmetries by the relations

$$e = \frac{g_1\, g_2}{\sqrt{g_1^{\,2} + g_2^{\,2}}} = g_2 \sin \Theta_W$$

$$\sin^2\Theta_W = 1 - \frac{m_W^{\,2}}{m_Z^{\,2}} . \qquad (1)$$

The value of the mixing angle is not predicted by the Standard Model. Grand Unified Theories predict a value of $\sin^2\Theta_W = 3/8$ at the unification mass.

The existence of a non-zero mixing angle defines both the structure and the strength of the neutral currents. The left-handed currents of the "up" particles of the weak isospin doublets couple with a coefficient ($\frac{1}{2} - Q \sin^2\Theta_W$), while the left-handed currents of the "down" particles of the doublets couple with a coefficient ($-\frac{1}{2} - Q \sin^2\Theta_W$). Q is the electric charge of the particle. The coupling coefficients of the right-handed currents are the same for "up" and "down" fermions, $- Q \sin^2\Theta_W$. Hence, the value of $\sin^2\Theta_W$ can be deduced from all neutral-current induced processes. In Born approximation, one finds the values $\sin^2\Theta_W^{unc}$ obtained before 1985, summarized in Table 1; they are in reasonable agreement with a common value of $\sin^2\Theta_W$. The values after radiative correction, also given in Table 1, are not in significantly better agreement, owing to their relatively large errors.

2. HIGHER ORDER CORRECTIONS

At a higher level of precision we expect that electroweak radiative corrections change this simple picture. The effective values of $\sin^2\Theta$ derived from the measurements will be

TABLE 1 Values of $\sin^2\Theta_W$ obtained from various processes and the values after electroweak radiative corrections have been applied

Experiment	$\sin^2\Theta_W^{unc}$	Condition	$\sin^2\Theta_W^{corr}$
Parity violation in Cs atoms [4]	0.221 ± 0.027		0.228 ± 0.027
eD scattering asymmetry [3]	0.224 ± 0.020	$\rho = 1$	0.218 ± 0.020
$\nu_\mu(\bar{\nu}_\mu)$e scattering [11]	0.215 ± 0.034	independent of ρ	0.215 ± 0.034
$\nu_\mu(\bar{\nu}_\mu)$e scattering [12]	0.209 ± 0.032	independent of ρ	0.209 ± 0.032
$\nu_\mu(\bar{\nu}_\mu)$p scattering [13]	0.220 ± 0.016	M_A = 1.05 GeV	0.220 ± 0.016
ν_μN deep inel. scattering [14]	0.232 ± 0.014	$\rho = 1$	0.223 ± 0.014
ν_μN deep inel. scattering [15]	0.239 ± 0.012	$\rho = 1$	0.226 ± 0.012

modified differently in each process. It is therefore of great interest to make measurements with improved precision to test whether the theory can correctly predict these radiative shifts. The shifts are in close analogy to the famous Lambshift of atomic energy levels and to (g-2) of the electron and the muon. The agreement of the high precision measurements and the higher order calculations of the shifts by the theory have given us confidence that Quantum-Electro-Dynamics is a successful theory. The same tests can now be performed for the Electroweak Interaction. While the QED shifts are small, those of the electroweak interaction are large, as we shall see.

A more advanced test of the theory, at the level of the radiative corrections of order α, can be performed by a precise determination of $\sin^2\Theta_W$ from semileptonic and from leptonic neutrino scattering and from the direct measurements of m_W and m_Z. The Fermi interaction, mediated by the weak bosons, leads to a relation between the fine-structure constant α, the Fermi coupling constant G_F, the boson masses, and the weak mixing parameter $\sin^2\Theta_W$.

$$m_W^2 = m_Z^2 \cos^2\Theta_W = \pi\alpha/\sqrt{2}G_F \sin^2\Theta_W. \tag{2}$$

Using $1/\alpha$ = 137.035963(15), as measured by the Josephson effect, and G_F^μ = 1.16637(2) × 10^{-5} GeV^{-2} as determined from the muon lifetime, leads in lowest order to the relations:

$$\sin^2\Theta_W^{unc} = (37.281 \text{ GeV}/m_W)^2$$
$$\sin^2 2\Theta_W^{unc} = (74.562 \text{ GeV}/m_Z)^2. \tag{3}$$

Because of the definition of $\sin^2\Theta_W$ in eq. (2) this relation must be radiatively corrected by using values of α and G_F at the mass of m_W. The change of the value of $G_F(m_W^2)$ is negligible, whereas α is changed to $\alpha(m_W) = \alpha(0)/(1 - \Delta r)$ with

$$\Delta r = 0.0713 \pm 0.0013, \tag{4}$$

assuming m_{Higgs} = 100 GeV and a top quark mass of m_t = 45 GeV [10]. The predicted corresponding radiative shift of the W mass is about 3 GeV. The radiative correction can be experimentally determined from the measurements of m_W and m_Z and of $\sin^2 \Theta_W$ through the relation

$$\Delta r = 1 - (\pi\alpha / m_W^2 \sqrt{2} G_F \sin^2 \Theta_W) \tag{5}$$

and the corresponding relation involving m_Z. A comparison with the calculated radiative shift thus tests the predictive power of the Standard Model at the level of vacuum polarization corrections. The electroweak radiative shift is approximately 50 times larger than (g - 2) of the muon, owing to the larger mass scale involved. The weak bosons $W^{\pm}$ and Z^0 which mediate the charged and the neutral weak currents were discovered at the CERN $p\bar{p}$ Collider in 1983 [11]. The measurements of m_W and m_Z are summarized in Table 2. The predictions of the Standard Model, using $\sin^2\Theta_W$ determined from deep inelastic neutrino scattering are given for comparison. The successful prediction of the new mass scale of the intermediate vector bosons has been one of the triumphs of the Standard Model.

TABLE 2 The W and Z masses. The first error is statistical and the second systematic, mainly due to energy calibration.

Group	M_W	M_Z	Reference
UA2 (CERN)	80.2 ± 0.8 ± 1.3	91.5 ± 1.2 ± 1.7	R. Ansari et al., Phys. Lett. 186B (1987) 440
UA1 (CERN)	$83.5^{+1.1}_{-1.0}$ ± 2.7	93.0 ± 1.4 ± 3.0	G. Arnison et al., Phys. Lett. 166B (1986) 484
UA1 + UA2 combined	80.9 ± 1.4	92.1 ± 1.8	
Prediction from deep inelastic ν scattering (with radiative corrections; $\sin^2\Theta_W$ = 0.233 ± 0.006).	80.2 ± 1.1	91.6 ± 0.9	

3. COHERENT π^0 PRODUCTION BY NEUTRAL CURRENT AND CHARGED CURRENT NEUTRINO INTERACTIONS

Is the Lorentz structure of the weak neutral current interaction of the helicity conserving vector type (V, A) or of the helicity-changing S, P or T type? Elucidating the same question for the weak charged current took nearly 25 years of experimental investigation. In neutrino-induced reactions, the answer is obscured by the so-called confusion theorem due to B. Kayser et al [12]. The process of coherent π^0 production by weak neutral-current (anti)neutrino interactions on a nucleus with mass number A probes directly the helicity structure of the interaction, owing to the intrinsic quantum numbers (0^-) of the neutral pion.

In the limit of zero momentum transfer, the helicity of the incident and of the outgoing neutrino is depicted in Figure 1a for V, A interaction and in Figure 1b for S, P or T interaction. In the latter case, the cross-section must vanish because of angular momentum conservation. Hence, the π^0 spectra are distinctly different for V, A and S, P, T interactions and experiments have shown that the interaction must be of the V, A type. Calculations of this process

$$\overset{(-)}{\nu}_\mu + A \rightarrow \overset{(-)}{\nu}_\mu + A + \pi^0 \tag{6}$$

were carried out by Nachtmann and Lackner [13,14] and later by Rein and Sehgal [15]. Measurements at mean energies of 2 GeV at the CERN-PS were published by the Aachen-Padoua group [16] and by the Gargamelle Bubble Chamber group [17], and at a mean neutrino energy of 31 GeV and antineutrino energy of 24 GeV by the CHARM Collaboration [18]. The corresponding charged-current reaction $\bar{\nu}_\mu + A \rightarrow \mu^+ + A + \pi^-$ was studied in BEBC [19].

The coherent π^0-production is characterized by a constructive interference of the neutrino interactions on neutrons and protons within a single nucleus. The predicted

cross-section for coherent π^0 production is proportional to the factor

$$\sigma_{\pi^0} \propto (u_a - d_a)^2 \rho^2 \tag{7}$$

where u_a and d_a are the axialvector coupling constants of u and d quarks, respectively, and ρ the ratio of the neutral and charged-current coupling constants. The standard model predicts $u_a = -d_a = 1/2$ and $\rho = 1$ so that a value of 1 is expected for expression (7).

In this experiment one determines the absolute value of the difference of u_a and d_a, i.e. the isovector axialvector coupling $|\beta| = |u_a - d_a|$, by comparing the measured cross-section with the calculated value, asumping $\rho = 1$ as suggested by other measurements [23]. Because of the pure axialvector nature of the process, the cross-sections of the neutrino and antineutrino-induced reactions are expected to be equal. The measurement of $|\beta|$ allows a determination of the relative sign of u_a and d_a. Previously this sign has been deduced from an analysis of elastic νp scattering and Δ production [24].

The method used by the CHARM Collaboration [18] to extract the coherent π^0 candidates followed closely the procedure used in the previously published analysis of $\overset{(-)}{\nu}_\mu e$-scattering [20,21,22].

The theoretical values in Table 3 were evaluated according to ref. [15], assuming $\beta = 1$ and $\rho = 1$. Their errors are due to uncertainties in the nuclear radii. Comparing the

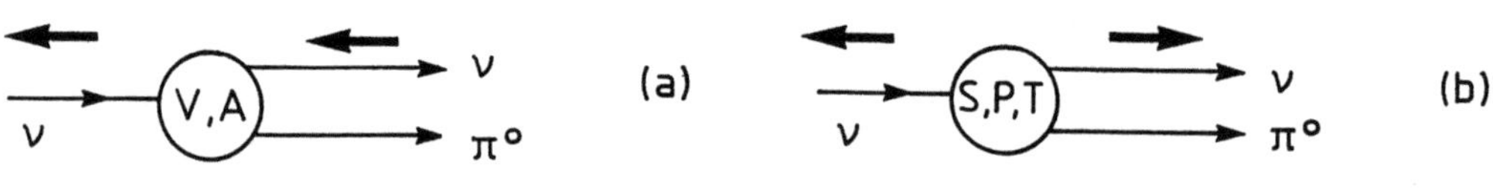

Figure 1 Helicity of incident and outgoing neutrino (denoted by a black arrow) in the reaction $\nu_\mu A \rightarrow \nu_\mu \pi^0 A$ for (a) V, A interaction (b) S, P or T interaction in the limit of $Q^2 \rightarrow o$

TABLE 3 Summary of experimental cross-sections for coherent π^0 production per average marble nucleus in units of $10^{-40}cm^2$

	Experiment	Theory
ν_μ	$\sigma = 96 \pm 42$	$\sigma = 74 \pm 38$
$\bar{\nu}_\mu$	$\sigma = 79 \pm 26$	$\sigma = 71 \pm 34$

TABLE 4 Summary of experimental determinations of $|\beta| = |u_a - d_a|$ from coherent π^0 production

Experiment	$\|\beta\|$
Aachen-Padova [16]	0.93 ± 0.16
CHARM [18]	1.08 ± 0.24
Skat [27]	0.99 ± 0.20
FNAL-15' [27]	0.98 ± 0.24
World average	0.99 ± 0.10

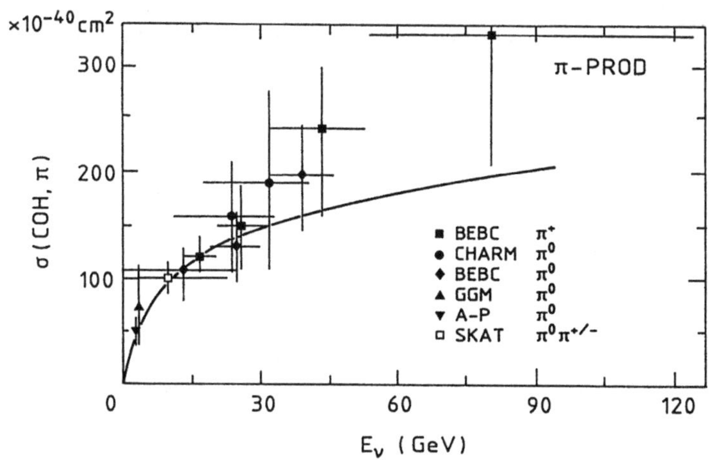

Figure 2 Cross-sections for coherent π^0 production as a function of neutrino energy showing a comparison of the experimental data with theoretical models. The horizontal error-bars indicate the range of neutrino-energies used for the particular points. For the CHARM results 68% of the events have a neutrino energy within the error bars.

experimental and theoretical values of Table 3, averaged for neutrinos and antineutrinos, the isovector neutral current coupling constant was determined:

$$|\beta| = |u_a - d_a| = 1.08 \pm 0.24$$

assuming $\rho = 1$. Statistical error (0.10) and systematic error (0.21) are quadratically combined. Figure 2 shows this result together with the theoretical predictions and experimental results from previous experiments [16,25,26]. A summary of the values of $|\beta|$ published before Summer 1987 is given in Table 4. The world average is

$$|\beta| = 0.99 \pm 0.10 \tag{8}$$

in close agreement with the value of 1.0 predicted by the standard model and hence with a negative relative sign of the axialvector coupling constants of u and d quarks.

4. DETERMINATION OF THE HELICITY STRUCTURE OF THE WEAK NEUTRAL CURRENT

Values of the coupling constants g_L^2 and g_R^2 of the left-handed and the right-handed u and d quarks have been determined from the total inclusive cross-sections of weak neutral-current (NC) and charged-current (CC) interactions induced by neutrinos and anti-neutrinos on isoscalar targets [27]. It should be noted that these scalar quantities do not allow a direct demonstration of the existence of a right-handed current; one needs a nut if one is looking for a screw. In analogy to the famous analysis of the kinematical correlation of the recoil proton in neutron beta-decay which leads to convincing direct evidence of the left-handedness (V - A) of the weak charged-current interaction, the investigation of the differential cross-section $d\sigma/dy$ in the inelasticity y can provide the nut and hence direct evidence of the conjectured contribution of a right-handed (V + A) current to the electroweak interaction. The differential cross-sections for both NC and CC events are directly related to the space-time structure of the weak-current involved. They give the admixture of V - A and V + A couplings and may indicate by a term in y^2 the presence of any scalar (S) or pseudo-scalar (P) terms. In addition, they yield information on the antiquark content, both strange and non-strange, of the nucleon and on the $s\bar{s}$ neutral-current.

For CC muon-neutrino interactions, all the initial energy E_ν appears as detectable energy in the final state and the inelasticity value of each event is unambiguously defined by

$$y = E_h/(E_h + E_\mu) = E_h/E_\nu \tag{9}$$

where E_h is the hadron energy and E_μ the muon energy. For NC interactions E_μ is replaced by $E_{\nu'}$, the energy of the outgoing neutrino which can obviously not be observed. One must resort to a narrow-band neutrino beam produced by a sign and momentum-selected nearly parallel beam of pions and kaons [28].

Within the context of the quark model, and under the assumption (see Chapter 3) that the weak currents contain only V and A terms, the differential cross-sections for deep inelastic scattering on isoscalar nuclei can be written as [28].

$$\begin{aligned} \frac{d\sigma}{dy}\,(\overset{(-)}{\nu} \to \overset{(-)}{\nu}) = A\Big[& g^2_{L(R)}\left(Q + \bar{Q}(1-y)^2\right) + g^2_{R(L)}\left(\bar{Q} + Q(1-y)^2\right) \\ & + g_s^2\, Q_s\left(1 + (1-y)^2\right) - F_L\Big] \end{aligned} \tag{10}$$

where A is a normalisation constant. The quark structure of the nucleon is described by Q, $\bar{Q}$ and Q_s ; for example $Q = \int \left(u(x) + d(x)\right) dx$ is the momentum-weighted valence quark content of the nucleon. $\alpha = \bar{Q}/(Q + \bar{Q})$ is the fractional momentum-weighted sea quark content and $\beta = Q_s/(Q + \bar{Q})$ the fractional strange quark content. The constants g_L^2 and g_R^2 are the left-handed and right-handed couplings of the weak neutral current to "up" and "down" quarks, while g_s^2 is the sum of the right and left-handed couplings to strange quarks. By fitting simultaneously these expressions to the four unfolded differential cross-sections the couplings of the weak neutral current have been determined; the resulting values are given in Table 5. The values of α, β and g_L^2, g_R^2 are determined by neglecting the term in g_s (fit A). Note that g_R^2 is determined with a significance of 5.6 standard deviations. The form of the NC y-distribution in eq. (10) precludes the possibility of varying all five parameters α, β, g_L^2, g_R^2 and g_s simultaneously in a fit. In particular, g_s is strongly correlated with the value of β. In a fit B the parameters α, β, $\sin^2\Theta_W$ and g_s are determined, and $g_s/g_d = 1.06 \pm 0.14$. Thus, one finds that the total coupling strength of the weak neutral current to the strange quark is consistent with being equal to that of the non-strange down quark, an assumption implied by the GIM mechanism [29].

TABLE 5 Results of simultaneous fits to CC and NC y distributions using an unfolding method

Parameter	Fit A	Fit B	Fit C
α	0.129 ± 0.013	0.129 ± 0.011	0.129 ± 0.011
β	0.134 ± 0.034	0.118 ± 0.037	0.124 ± 0.030
g_L^2	0.315 ± 0.004		
g_R^2	0.056 ± 0.010		
g_s^2		0.204 ± 0.054	
$\sin^2\Theta_W$		0.241 ± 0.016	0.236 ± 0.008

Motivated by this consistent result, one can go on to assume that the couplings of the u, d and s quarks can all be described in terms of the Glashow-Salam-Weinberg standard model. The neutral current sector is then described by a single parameter, $\sin^2\Theta_W$. This fit (C) gives $\sin^2\Theta_W = 0.236 \pm 0.008$ in full agreement with the result obtained from the ratio of NC and CC cross-sections for deep inelastic scattering using the same data set [23].

5. MEASUREMENTS OF $\mathrm{SIN}^2\Theta_W$

5.1 High Precision Measurements of $\sin^2\Theta_W$

Experimental tests have now, for the first time, reached the precision required to probe the theory at the level of the radiative corrections. The present error on $\sin^2\Theta_W$ of ± 0.004 corresponds to an error in m_W of 0.2 GeV, whereas the total radiative shift is predicted to be 3.3 GeV.

Future improvements of the luminosity of the CERN Sp$\bar{p}$S Collider using a new anti-proton collector ring (ACOL) are aimed at obtaining a data set with 10^4 nb^{-1} instead of the present integrated luminosity of 10^2 nb^{-1}. These data will then give statistical errors of the masses of

$$\Delta m_W, \Delta m_Z \sim \pm 0.15 \text{ GeV}.$$

The uncertainty in the experimental mass scale will then be the dominant error. Assuming an absolute calibration of the giant calorimeters of the detectors to ± 1% one derives a limiting accuracy of

$$\Delta \sin^2 \Theta \sim \pm 0.004.$$

The scale errors cancel in the ratio of the masses where we expect the full improvement due to the large statistics, corresponding to $\Delta\sin^2\Theta \sim \pm 0.004$.

Among the low energy experiments, those which use neutrino scattering on nucleons have already been improved to similar precision in $\sin^2\Theta_W$. A new experiment on neutrino-electron scattering, aiming at a higher precision (CHARM II) is presently being performed at CERN. These experiments and their limitations are described in some detail in sections 5.2 and 5.3. The small theoretical error on the radiative shift requires a measurement of m_Z to ± 50 MeV to be matched. This precision can be achieved at LEP.

5.2 Measurement of $\sin^2\Theta_W$ in semileptonic neutrino reactions

The cross-sections of neutrino scattering on isoscalar targets by the neutral and by the charged-current weak interaction are related by the following expression derived by Llewellyn-Smith [30]:

$$\frac{d^2\sigma_{NC}^{\nu,\bar{\nu}}}{dxdy} = (\tfrac{1}{2} - \sin^2\Theta + \tfrac{5}{9}\sin^4\Theta)\,\frac{d^2\sigma_{CC}^{\nu,\bar{\nu}}}{dxdy} + \tfrac{5}{9}\sin^4\Theta\,\frac{d^2\sigma_{CC}^{\nu,\bar{\nu}}}{dxdy}. \qquad (11)$$

The simplest measurements are those of the total cross-section ratios

$$R_\nu = \sigma^\nu(NC)/\sigma^\nu(CC), \quad r = \sigma^{\bar{\nu}}(CC)/\sigma^\nu(CC)$$

which give the relation

$$R_\nu = \tfrac{1}{2} - \sin^2\Theta + \tfrac{5}{9}\sin^4\Theta + r(\tfrac{5}{9}\sin^4\Theta). \qquad (12)$$

Eqs. (11) and (12) are valid if neutrino interactions with quarks and anti-quarks other than u and d can be neglected and if the Cabibbo angle is set to zero. Higher twist effects have also been neglected; they have been estimated [30] to give uncertainties smaller than $\Delta\sin^2\Theta \sim \pm 0.005$. Of course, weak isospin symmetry is implied by eq. (11). We know that it is broken by flavour-changing processes which have been observed to contribute to charged-current induced reactions but not to neutral-current reactions. The energy threshold of the flavour transition $s(d) + W^+ \rightarrow c$ crosses the peak energy of the neutrino beam used for the measurement. To correct the measured cross-sections we require knowledge of the mass of the charm quark to describe the threshold behaviour. At much higher energy, for instance at HERA, a measurement of the ratio R_ν does not require a threshold correction. The corrections are applied with the help of the quark model of the nucleon. Using the best knowledge of the charm quark mass, $m_c = 1.5 \pm 0.3$ introduces an uncertainty of $\Delta(\sin^2\Theta_W) = \pm 0.004$. Fixing the mass at $m_c = 1.5$ GeV, the remaining theoretical uncertainty is $\Delta(\sin^2\Theta_W) = \pm 0.003$.

The total theoretical uncertainty is $\Delta(\sin^2 \Theta_W) = \pm 0.005$. A high precision measurement has recently been performed at CERN by the CHARM Collaboration [31] and by the CDHS Collaboration [32]. Ten years after the discovery of the neutral-current interaction by the Gargamelle team at CERN with a signal to background ratio of one to six, events can now be classified as neutral-current (NC) or charged-current (CC) by direct recognition of the muon in the fine-grain calorimeter of the CHARM detector with less than 0.2% ambiguity. This progress is due to some important new features of this electronic detector:

- Fast timing is used and events occurring upstream are vetoed, thus eliminating the so-called associated neutron background which plagued the Gargamelle experiment;

- The lateral and longitudinal dimensions of the target-calorimeter are more than 10 times larger than the interaction length of hadrons, thus giving clear signatures to neutrino interactions and muon tracks;

- Detector elements of small lateral dimensions (fine-grain) and frequent segmentation of the target plates allow detection of hadron showers with high efficiency and good energy resolution ($\sigma(E_H)/E_H = 0.47/\sqrt{E_H \text{ GeV}}$) and the recognition of muons with momenta as low as 1 GeV/c.

- Nearly equal response of the calorimeter ($E_e/E_H = 1.17$) to electromagnetic and hadronic showers, allowing the definition of an effectively equal energy threshold in NC and CC events which have different π^0 content.

The feasibility of a precision measurement was discussed by the author at the 1982 Javea Workshop on Weak Interaction [33] and was demonstrated in detail by the CHARM Collaboration at a Physics Workshop at CERN [34].

A photograph of the CHARM detector is shown in Figure 3. A detailed description of this can be found in reference [35]. Figures 4 and 5 show schematic views of a CC and NC neutrino event, respectively. Scintillation counters and proportional drift tubes which are hit by the event are shown, and the range of a 1 GeV muon is indicated. Its track can also be recognized close to the hadron shower. Charged-current (CC) events for which the primary muon cannot be identified are classified as neutral-current (NC). Some of these lost CC events have a muon with an energy less than 1 GeV, or a muon that left the detector at the sides before depositing 1 GeV, or a muon that was obscured by the hadronic shower. A correction is required for these CC losses.

Table 6 shows the radiative and the various quark model corrections which have to be applied to determine $\sin^2\Theta_W$ in the definition of Sirlin and Marciano. The final result is

$$\sin^2 \Theta_W = 0.236 + 0.012\,(m_c - 1.5) \pm 0.005\ (\text{exp}) \pm 0.003\ (\text{theor}).$$

This result, obtained by the CHARM Collaboration, is compared with other recent high statistics results from semi-leptonic neutrino scattering experiments in Table 6, assuming a charm mass of $m_c = 1.5$ GeV/c^2. The agreement between the experiments is good and significant in view of the fact that different experimental methods have been used, as indicated in Table 7. A combined value for $m_c = 1.5$ GeV/c^2 is

Figure 3 Photograph of the CHARM detector

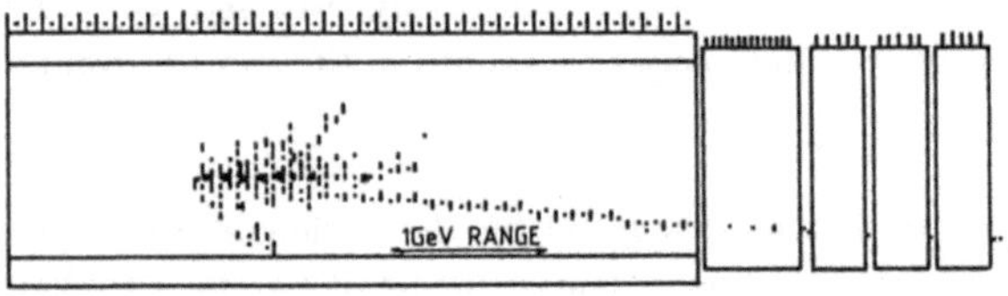

Figure 4

Schematic view of a CC neutrino event recorded by the CHARM I detector

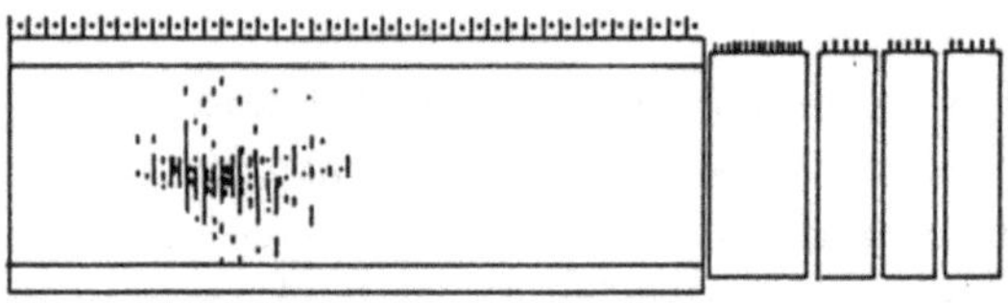

Figure 5

Schematic view of a NC neutrino event recorded by the CHARM I detector

TABLE 6 Corrections to $\sin^2\Theta_w$ ($E_h > 4$ GeV)

Source	$\Delta\sin^2\Theta_w$	Theoretical uncertainty
Muon mass	+ 0.0011	± 0.0001
W^2 thresholds, F_L	+ 0.0005	± 0.0005
K-M mixing matrix		± 0.0010
Strange sea for $m_c = 0$	- 0.0074	± 0.0010
Charm sea for $m_c = 0$	+ 0.0015	± 0.0010
Radiative corrections	- 0.0092	± 0.0020
Total uncertainty (fixed m_c)		± 0.0030
Charm mass ($m_c = 1.5$ GeV/c^2)	+ 0.0140	
Total ($m_c = 1.5$ GeV/c^2)	+ 0.0005	± 0.0030

TABLE 7 Values of $\sin^2\Theta_W$, derived from semi-leptonic neutrino scattering experiments. The common theoretical error is ± 0.003.

Experiment	$\sin^2\Theta_W$	Method	Ref.
FMMF	0.246 ± 0.016	event-by-event	[36]
CCFR	0.239 ± 0.010	event length	[37]
CDHS	0.225 ± 0.005	event length	[32]
CHARM	0.236 ± 0.005	event-by-event	[31]

TABLE 8 Comparison of values for $\sin^2\Theta_W$, measured in semi-leptonic neutrino scattering and from the masses of the intermediate bosons, with and without radiative correction

Experiment	Corrected $\sin^2\Theta_W$	Uncorrected $\sin^2\Theta_W$
νN	0.233 ± 0.004 ± 0.005	0.242 ± 0.004 ± 0.005
m_W, m_Z	0.226 ± 0.008	0.210 ± 0.008
Difference	0.007 ± 0.010	0.032 ± 0.010

$$\sin^2\Theta_W = 0.233 \pm 0.004 \pm 0.003.$$

Table 8 shows the averaged value of $\sin^2\Theta_W$ from the neutrino experiments and from the m_W and m_Z measurement, with and without the radiative corrections applied. The difference between the values from the two types of experiments shows good agreement for the corrected results, but a difference at the level of three standard deviations for the uncorrected values (the error in the difference is calculated by combining the statistical and theoretical/systematic errors quadratically). The radiative correction determined by the experiments

$$\Delta r\ (\text{exp}) = 0.077 \pm 0.025\ (\text{exp}) \pm 0.038\ (\text{syst})$$

agrees with the calculated value of Δr (theor) = 0.0713 ± 0.0013 assuming m_t = 45 GeV and m_H = 100 GeV [10].

Assuming therefore the validity of the radiative corrections, a precise value of the ρ parameter can be obtained by combining our measured value of R^ν with the value of $\sin^2\Theta_W$, as determined in the collider experiments from the W mass. Propagating the errors quadratically, we find

$$\rho = 0.990 - 0.013\ (m_c - 1.5) \pm 0.009 \pm 0.003$$

in good agreement with the minimal Standard Model.

New Physics Appendages to the Standard Model or a different input value for m_t would modify the prediction of Δr. We can therefore use the good agreement with experiment as a constraint. For example, if m_t were 250 GeV/c^2 Δr would become zero, whereas $\sin^2\Theta_W$ hardly changes [38]. From the present experimental value of Δr, Marciano and Sirlin [39] find the constraint

$$m_t \lesssim 180 \text{ GeV}.$$

In the case of a heavy fourth-generation charged lepton L with a massless neutrino companion the corresponding bound is [38],

$$m_L \lesssim 300 \text{ GeV}.$$

Future high-precision experiments should determine m_W and m_Z to ± 0.1 GeV and ± 0.02 GeV, respectively and, combined with the result of a determination of $\sin^2\Theta_W$ from $\nu_\mu e$ and $\bar{\nu}_\mu e$ scattering in the CHARM II experiment to ± 0.005 without theoretical uncertainty, should thus provide even better constraints or perhaps a hint of new physics.

A final comment is concerned with the comparison of $\sin^2\Theta_W$ predicted by Grand Unified Theories. The experimental result presented here contradicts the prediction of the minimal SU(5) model [10], $\sin^2\Theta_W = 0.214 \pm 0.004$ [SU(5)], which has already been ruled out by the experimental lower limit of proton decay [40]. It can be used to constrain the additional mass scales of non-minimal Grand Unified Theories. For example, in the case of supersymmetric theories (SUSY), the experimental result is in good agreement with the mass constraint $M_{SUSY} \gtrsim 1$ TeV [10].

5.3 Measurement of $\sin^2\Theta_W$ in leptonic neutrino reactions

The first experimental observation of a weak neutral-current phenomenon was one event (reproduced in Figure 6) in which an $\bar{\nu}_\mu$ scattered on an electron, found in 1972 at CERN in the Gargamelle bubble chamber [1]. Now, 15 years later, massive electronic detectors have achieved remarkable progress in this field. The CHARM Collaboration has collected about 83 events of $\nu_\mu e$ and 112 events of $\bar{\nu}_\mu e$ scatterings [41]. A US-Japan Collaboration working at Brookhaven has collected 107 events of $\nu_\mu e$ and 45 events of $\bar{\nu}_\mu e$ scattering [42, 43]. An event observed in the fine-grain calorimeter of the CHARM detector is reproduced in Figure 7. Compared to the bubble chamber event in Figure 6 it looks quite coarse. In spite of this, the events can be separated from the background and the cross-section determined.

In contrast to semileptonic neutrino scattering there are no theoretical uncertainties to extract the weak neutral-current coupling constants of the electron, g_V^e and g_A^e, and $\sin^2\Theta_W$ from the cross-sections of neutrino electron scattering:

$$g_V^2 + g_A^2 = (3\pi/4G_F^2\, m_e)\, (\sigma^\nu/E + \sigma^{\bar{\nu}}/E)$$
$$g_V^e \times g_A^e = (3\pi/4G_F^2\, m_e)\, (\sigma^\nu/E - \sigma^{\bar{\nu}}/E). \qquad (13)$$

The most precise determination of the value of $\sin^2\Theta$ in the leptonic sector has been obtained by the CHARM Collaboration [41], making use of the direct relation between the ratio of $\sigma(\nu_\mu e)$ amd $\sigma(\bar{\nu}_\mu e)$ and $\sin^2\Theta_W$:

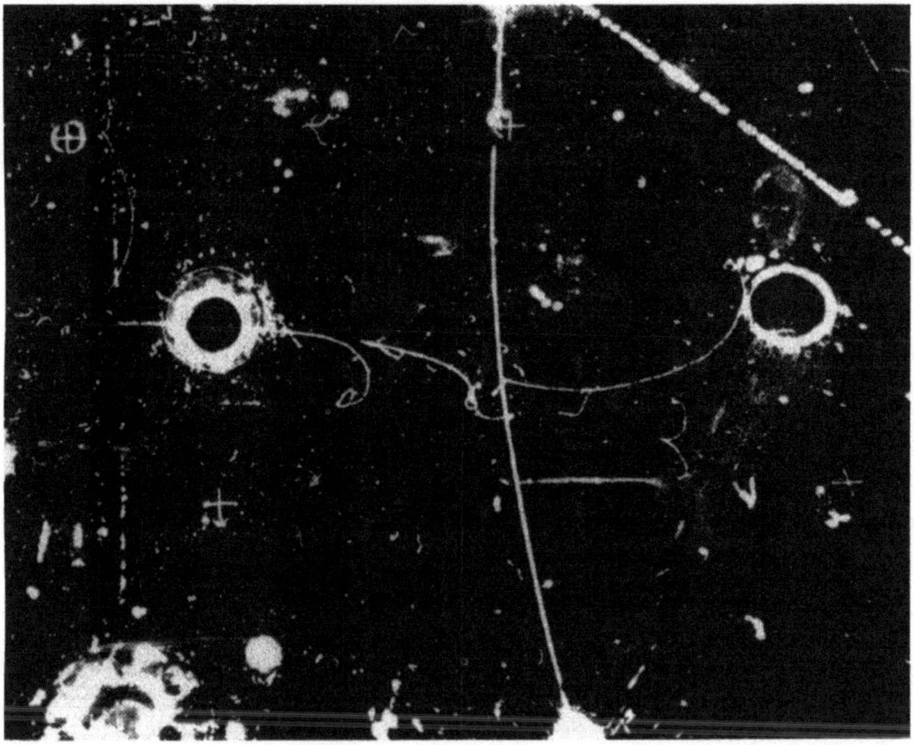

Figure 6 First event of $\bar{\nu}_\mu e$ scattering observed in Gargamelle [1]

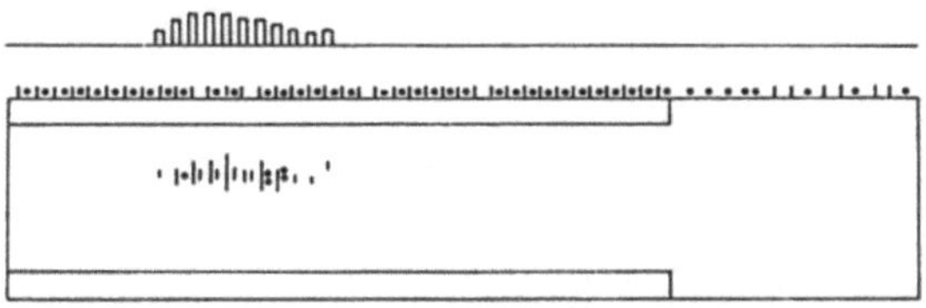

Figure 7

Neutrino-electron scattering events observed in the CHARM I detector

$$R = \frac{\sigma(\nu_\mu e)}{\sigma(\bar{\nu}_\mu e)} = 3\,\frac{1 - 4\sin^2\Theta_W + (\frac{16}{3})\sin^4\Theta_W}{1 - 4\sin^2\Theta_W + 16\sin^4\Theta_W} \qquad (14)$$

In the vicinity of $\sin^2\Theta_W = 1/4$ this relation gives $\Delta\sin^2\Theta_W \sim 1/8\ (\Delta R/R)$ and, hence, a very precise determination of the mixing angle. The detection efficiency cancels in the ratio, many systematic uncertainties are reduced, and no absolute neutrino flux measurement is required. Radiative corrections also cancel to a large extent in the ratio and the value of $\sin^2\Theta_W$ determined by eq. (14) is therefore practically equal to the value determined by the boson masses, as defined by Sirlin and Marciano [39]:

$$\sin^2\Theta_W = 1 - M_W^2/M_Z^2 .$$

Several experimental problems have to be solved [44], namely:

1) EVENT RATE requiring a large fiducial tonnage and high selection efficiency over a wide window of recoil electron energies;

2) BACKGROUND, dominantly due to quasi-elastic scattering of electron-neutrinos and to coherent π production has to be reduced by a factor of about 10 by precise measurements of the shower direction and efficient e/π discrimination;

3) MONITORING of the relative flux of the different beam components ν_μ, $\bar{\nu}_\mu$, ν_e and $\bar{\nu}_e$ is required to determine the cross-section ratio $\sigma(\nu_\mu e)/\sigma(\bar{\nu}_\mu e)$.

It has been demonstrated by the CHARM Collaboration [41] that these problems can be solved. In two different exposures of the fine-grain CHARM detector to the horn-focused wide-band neutrino beam of the CERN 400 GeV SPS, 83 events of $\nu_\mu e$ scattering and 112 events of $\bar{\nu}_\mu e$ scattering have been recorded.

Criteria based on shower properties in low Z material (marble) were used to select these events. The ratio of the normalized numbers of $\nu_\mu e$ and $\bar{\nu}_\mu e$ events is found to be

$$R_{exp} = N(\nu_\mu e)/N(\bar{\nu}_\mu e) \times F = \sigma(\nu_\mu e)/\sigma(\bar{\nu}_\mu e) = 1.20 \begin{array}{l} +0.41 \\ -0.28 \end{array}$$

The relation between R_{exp} and $\sin^2\Theta_W$ given by eq. (14) is shown in Figure 8. The measured value of R_{exp} agrees with the predictions of the standard model for

$$\sin^2\Theta_W = 0.211 \pm 0.035\ (stat) \pm 0.011\ (syst) \tag{15}$$

Four pairs of values of the neutral-current coupling constants, g_A^e and g_V^e, can be obtained from these measurements. The limits from other measurements in the lepton sector, those from the forward-backward asymmetry in the reaction $e^+e^- \rightarrow \ell^+\ell^-$ at PETRA and PEP [45] and from the $\bar{\nu}_e e$ scattering cross-section [46], shown in Figure 9 select a unique

$$g_A^e = -0.57 \pm 0.04\ (stat) \pm 0.06\ (syst)$$
$$g_V^e = -0.06 \pm 0.07\ (stat) \pm 0.02\ (syst). \tag{16}$$

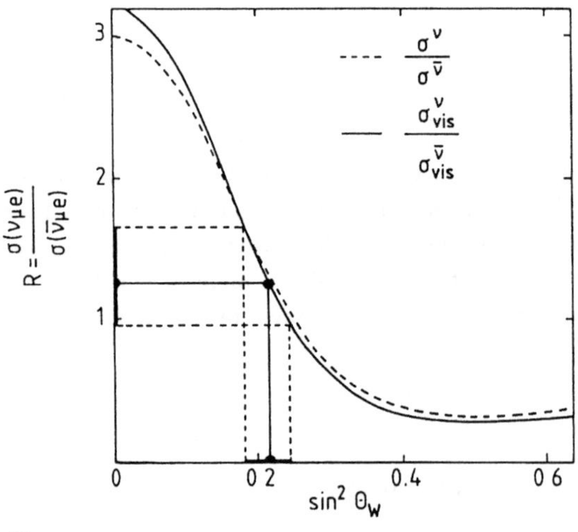

Figure 8

Relation between $R = \sigma(\nu_\mu e)/\sigma(\bar{\nu}_\mu e)$ and $\sin^2\Theta$ with the measured value of R and the corresponding value of $\sin^2\Theta$

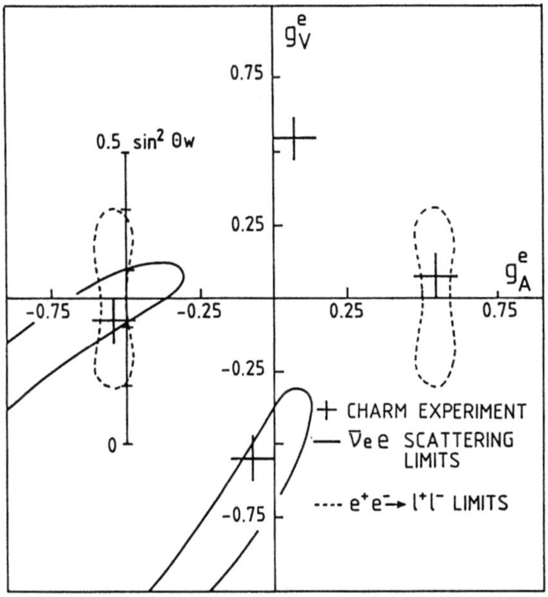

Figure 9

Values of g_A^e and g_V^e, CHARM Collaboration [41]. The limits from forward-backward asymmetry in $e^+e^- \to \ell^+\ell^-$ [49] and from $\sigma(\bar{\nu}_e e)$ measurements [46] select a unique solution.

A value of $g_A^e = -\frac{1}{2}$ is predicted by the Standard Model [8], in agreement with the experiment.

The relative strength of the neutral and charged-current coupling constants is found to be

$$\rho = 1.14 \pm 0.07 \text{ (stat)} \pm 0.12 \text{ (syst)}, \tag{17}$$

in agreement with the prediction of the Standard Model, $\rho = 1$, provided the Higgs fields form an isospin doublet. At a higher level of precision small deviations of ρ from one are expected, due to fermion doublets with large mass differences between the up and down components.

Can the present technique used for studies of neutrino-electron scattering be sufficiently improved to match the aim of $\Delta\sin^2\Theta = \pm 0.005$? A new detector, dedicated to this task (CHARM II) [47], has been built and is presently taking data at CERN. The technique of low Z material calorimetry is used again together with fine-grain detection systems.

The limiting accuracy of shower direction measurements is given by the Z number of the target material

$$\sigma(\Theta) \sim \text{const. } Z/\sqrt{E}. \tag{18}$$

The mean Z of the present marble target is 13, with glass (SiO_2) a value of ~ 11 is achieved. The accuracy depends further upon the sampling frequency (plate thickness), the grain size of the detector and the detection method.

The structure of the new, dedicated detector is sketched in Figure 10. It consists of 420 modules of 3.7 × 3.7 m^2 surface area, each composed of a 4.8 cm thick target plate (glass) and of a plane of streamer tubes with 1 cm wire spacing, read out by the wires and by crossed (90^0) cathode strips of 2 cm width. Using pulse height measurements the centroid

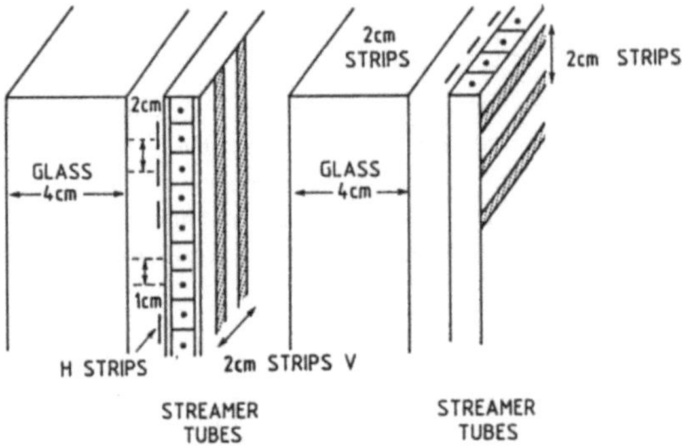

Figure 10

Sketch of the structure of the CHARM II detector [48]

Figure 11 Photograph of the CHARM II detector

position of a track can be reconstructed with ± 3 mm accuracy from the cathode strips, whereas the wires are read out digitally to obtain unambiguous information about the track multiplicity near the vertex. A simulation of this detector by Monte Carlo methods gives an electron shower angular resolution of $\sigma(\Theta) 18$ mrad/$\sqrt{E/\mathrm{GeV}}$, about equal to the natural angular spread of recoil electrons. A photograph of this new detector is shown in Figure 11. Two typical events, one due to charged-current neutrino interaction (a), the other due to $\nu_\mu e \rightarrow \nu_\mu e$ scattering (b) are shown in Figure 12. An exposure leading to 2000 $\nu_\mu e$ and 2000 $\bar{\nu}_\mu e$ events is planned.

Figure 13 shows the measurements of elastic neutrino scattering at low Q^2 [48]. In the forward direction, at Q^2 values less than 0.03 GeV^2 the ν_μ and $\bar{\nu}_\mu$ distributions differ

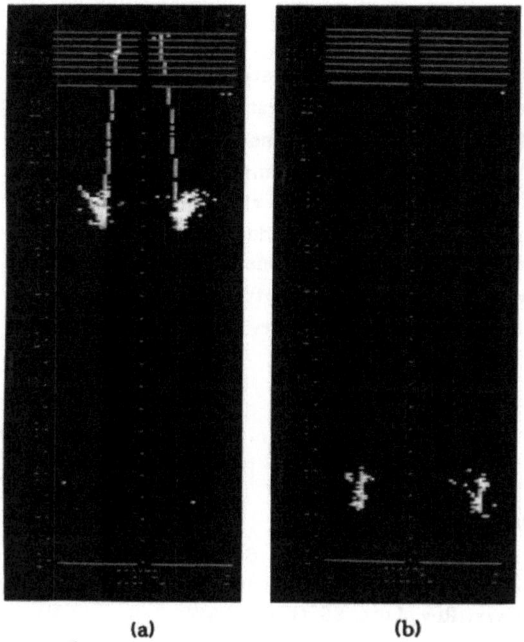

Figure 12

Typical neutrino events in the CHARM II detector, (a) ν_μ charged-current interaction (b) a candidate of $\nu_\mu e \rightarrow \nu_\mu e$ scattering

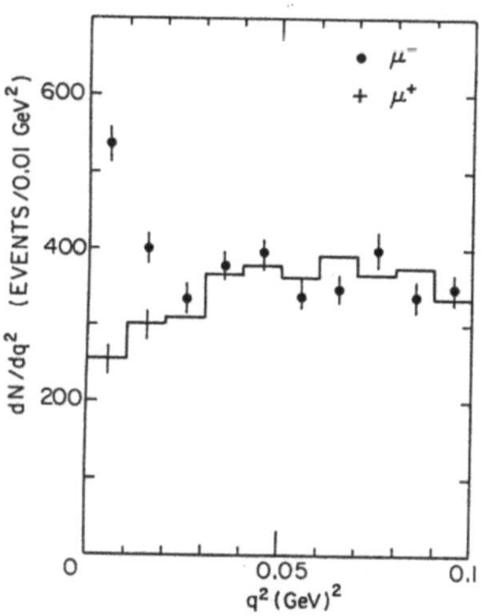

Figure 13

Observed Q^2 dependence of quasi-elastic $\nu_\mu(\mu^-)$ and $\bar{\nu}_\mu(\mu^+)$ events

because of contributions from the reaction $\nu_\mu e \rightarrow \mu^- \nu_e$. At larger values of Q^2 the cross sections differ by the V, A interference term proportional to Q^2/E, which can be corrected for. The detection efficiencies and background contributions from delta resonance production differ by small amounts. A total flux ratio error of ± 2% is estimated, leading to a total error of ± 4% on R, and a corresponding experimental error of $\Delta\sin^2\Theta = \pm\ 0.005$. There is no theoretical uncertainty in extracting this result from eq. (14).

CONCLUDING REMARKS

With the recent advances in experiments determining the Lorentz structure and measuring $\sin^2\Theta_w$ in semi-leptonic and leptonic neutrino scattering, our understanding of the neutral weak current has progressed. Improvements from further measurements of $\sin^2\Theta_w$ in semi-leptonic neutrino processes are likely to be limited by theoretical uncertainties. Neutrino-electron scattering is a field that is expected to progress, following the demonstration that experiments at high energy using low Z fine-grain calorimeters can detect these rare events and separate them reliably from the dominant semi-leptonic reactions. Future results from the CERN $p\bar{p}$ collider on m_w and m_Z and from neutrino electron scattering are expected to match each other in precision of $\sin^2\Theta_w$ and, hence, to test the theory of electroweak interaction at the level of vacuum polarisation.

REFERENCES

[1] F.J. Hasert et al., Phys. Lett. 46B (1973) 121

[2] F. J. Hasert et al., Phys. Lett. 46B (1973) 138

[3] C.Y. Prescott et al., Phys. Lett. 77B (1978) 347 and Phys. Lett. 84B (1979) 524

[4] M.A. Bouchiat and C.A. Piketty, J. Phys. (France) 46 (1985) 1897 S. Gilbert, N. Noecker, R. Watts and C. Wiemann, Phys. Rev. Lett. 55 (1985) 2680

[5] See e.g. S.L. Wu, Phys. Reports 107 (1984) 59

[6] M. Jonker et al., CHARM Collab., Phys. Lett. 102B (1981) 67

[7] J. Dorenbosch et al., CHARM Collab., Phys. Lett. 180B (1986) 303

[8] S.L. Glashow, Nucl. Phys. 22 (1961) 579
A. Salam and J. Ward, Phys. Lett. 13 (1963) 168
S. Weinberg, Phys. Rev. Lett. 19 (1967) 1264

[9] See ref. [6] and I. Abt, PhD Thesis, Univ. Hamburg 1986

[10] U. Amaldi et al., CERN-EP/87-93 and Phys. Rev. D (to be published)
G. Costa et al., CERN-TH/4675-87

[11] G. Arnison et al., UA1 Collab., Phys. Lett. 122B (1983) 103 and 126B (1983) 398
M. Banner et al., UA2 Collab., Phys. Lett. 122B (1983) 476 and 129B (1983) 130

[12] B. Kayser et al., Phys. Lett. 52B (1974) 385

[13] O. Nachtmann, Nucl. Phys. B22 (1970) 385

[14] K.S. Lackner, Nucl. Phys. B153 (1979) 526
and Doctoral Thesis, Heidelberg 1978, unpublished

[15] D. Rein and L.M. Seghal, Nucl. Phys. B223 (1983) 29

[16] H. Faissner et al., Phys. Lett. 125B (1983) 230

[17] E. Isiksal, D. Rein and J.G. Morfin, Phys. Rev. Lett. 52 (1984) 1096

[18] F. Bergsma et al., CHARM Collab., Phys. Lett. 157B (1985) 469

[19] P. Marage et al., Phys. Lett. 140B (1984) 137

[20] M. Jonker et al., CHARM Collab., Phys. Lett. 105B (1981) 185

[21] F. Bergsma et al., CHARM Collab., Phys. Lett. 117B (1982) 272

[22] F. Bergsma et al., CHARM Collab., Phys. Lett. 147B (1984) 481

[23] J.V. Allaby et al., Z. f. Phys. C36 (1978) 611

[24] P.Q. Hung and J.J. Sakurai, Ann. Rev. Part. Sci. 31 (1981) 375

[25] P. Marage, WA59 Collab., Doctoral Thesis, Brussels 1984, unpublished

[26] J. Panman, Electroweak Neutral Current Interactions, Proc. of the 1987 Int. Symposium on High-Energy Lepton and Photon induced Processes, Hamburg, (to be published)

[27] See e.g. M. Jonker et al., CHARM Collab., Phys. Lett. 99B (1981) 265
[28] M. Jonker et al., CHARM Collab., Phys. Lett. 102B (1981) 67
F. Bergsma et al., CHARM Collab., Phys. Lett. 177B (1986) 446
and I. Abt, PhD Thesis, Hamburg University (FRG) 1986
[29] S.L. Glashow, J. Iliopoulos and L. Maiani, Phys. Rev. D2 (1970) 1285
[30] C.H. Llewellyn-Smith, Nucl. Phys. B228 (1983) 205
[31] J.V. Allaby et al., CHARM Collab., Phys. Lett. 177B (1986) 446
and Z. f. Phys. C36 (1987) 611
[32] H. Abramowicz et al, CDHS Collab., Phys. Rev. Lett. 57 (1986) 298
[33] K. Winter, Weak Interaction Workshop, Javea (Spain), 1982
[34] J. Panman, Workshop on SPS Fixed-Target Physics in the years 1984-1989, CERN 83-02, Vol. 11 p. 146
[35] A.N. Diddens et al, CHARM Collab., Nucl. Instr. Methods 178 (1980) 27, 200 (1982) 183, 215 (1983) 361 and A253 (1987) 203
[36] D. Bogert et al, FMMF Collab., Phys. Rev. Lett. 55 (1985) 1969
[37] P. Reutens et al, CCFRR Collab., Phys. Lett. 152B (1985) 404
[38] W.J. Marciano, Proc. 23rd Int. Conf. on High Energy Physics, Berkeley 1986 (World Scientific, Singapore, 1987) Vol. 2 p. 299
[39] W.J. Marciano and A. Sirlin, Phys. Rev. D29 (1984) 945
[40] M. Goldhaber, Proc. 23rd Int. Conf. on High Energy Physics, Berkeley 1986 (World Scientific, Singapore, 1987) Vol. 1 p. 248
[41] F. Bergsma et al., CHARM Collab., Phys. Lett. 147B (1984) 481
and Z. f. Phys. C (to be published)
[42] L.A. Ahrens et al., Phys. Rev. Lett. 54 (1985) 18
[43] K. Abe et al., Phys. Rev. Lett. 56 (1986) 1107
and Phys. Rev. Lett. 58 (1987) 636
[44] K. Winter, Invited Talk on $\nu_\mu e$ scattering, Europhysics Study Conference on Electro-Weak Effects at High Energy, Erice, Sicily 1983 (Plenum Press, New York, 1985) p. 141
[45] See e.g. S.L. Wu, Rapporteur Talk, Proc. Int. Symp. on Lepton and Photon Interactions at High Energies, Hamburg 1987 (to be published)
[46] F. Reines, H.S. Gurr and H.W. Sobel, Phys. Rev. Lett. 37 (1976) 315
[47] J.P. De Wulf et al, CHARM II Collab., Nucl. Instr. Methods A252 (1986) 443
[48] F. Bergsma et al, CHARM Collab., Phys. Lett. 122B (1983) 465

Analysis of the Lepton Mixing Matrix from Neutrino Oscillation Experiments

K. Kleinknecht

Institut für Physik der Universität Mainz, D-6500 Mainz,
Fed. Rep. of Germany

Dedicated to Ulrich Bonse on the occasion of his 60th birthday

Abstract

An analysis of experimental results in the framework of simultaneous oscillations of three neutrino flavours is used, in a combined fit, to constrain the elements of the 3 x 3 unitary lepton mixing matrix and to obtain limits on the three lepton mixing angles $\alpha_{e\mu}$, $\alpha_{e\tau}$ and $\alpha_{\mu\tau}$. No evidence for mixing is found.

If the lepton number is not conserved, neutrinos of different flavour may mix. In analogy to quark mixing, this lepton mixing can be described by a 3 x 3 unitary matrix. In fact, the weak eigenstates ν_e, ν_μ and ν_τ are related to the mass eigenstates ν_1, ν_2 and ν_3 by a unitary matrix U_{ik} /1/:

$$|\nu_\ell\rangle = \Sigma_k U_{\ell k}|\nu_k\rangle \qquad (\ell = e,\mu,\tau;\ k = 1,2,3) \qquad (1)$$

and this matrix $U_{\ell k}$ can be parametrized in analogy to the quark mixing matrix. We use the parametrization of Maiani /2/, given in table 1.

Table 1
Maiani parametrization of lepton mixing matrix

$$U = \begin{pmatrix} C_\beta C_\Theta & C_\beta S_\Theta & S_\beta \\ -S_\gamma C_\Theta S_\beta e^{i\delta'} - S_\Theta C_\gamma & C_\gamma C_\Theta - S_\gamma S_\beta S_\Theta e^{i\delta'} & S_\gamma C_\beta e^{i\delta'} \\ -S_\beta C_\gamma C_\Theta + S_\gamma S_\Theta e^{i\delta'} & -C_\gamma S_\beta S_\Theta - S_\gamma C_\Theta e^{-i\delta'} & C_\gamma C_\beta \end{pmatrix}$$

Up to now, the scarcity of data and the complexity of the problem have prevented a general analysis of experimental data without restrictive assumptions, i.e. with three free angles and neutrino masses. Instead, experimentalists searching for neutrino oscillations have usually made the ad hoc assumption that only one angle and one neutrino mass difference is different from zero, and have given their results in terms of these two quantities.

We present here an update of an analysis of available experimental data with five free parameters, i.e. three angles and two neutrino mass differences /3/.

The time evolution of an initially pure ν_ℓ state can be expressed in terms of mass differences and the neutrino energy E. If this energy is large compared to the neutrino mass, the probability to find a $\nu_{\ell'}$ after a distance L(m) from the production point of an initially pure ν_ℓ state of energy E(MeV) is /4/:

$$P_{\ell\ell'} = \sum_{k,k'=1} U_{\ell k}\, U_{\ell k'}\, U_{\ell' k}\, U_{\ell' k'} \cos\,(2.54\;\Delta m^2_{k'k}\; L/E) \qquad (2)$$

where $\Delta m^2_{k'k} = |m^2(\nu_{k'}) - m^2(\nu_k)|$ is measured in (eV^2).

As a simplification, the elements $U_{\ell k}$ are assumed to be real. This leaves us with five free parameters to be determined, because for three neutrino masses only two Δm^2 values are independent.

For the Maiani parametrization, in the limiting case of only one angle being finite, this angle can be directly related to oscillations between two flavours of neutrinos. The Maiani angles Θ, β and γ are thus uniquely related to the oscillations channels $\nu_e \leftrightarrow \nu_\mu$, $\nu_e \leftrightarrow \nu_\tau$ and $\nu_\mu \leftrightarrow \nu_\tau$ respectively. We call these angles $\Theta = {}_{e\mu}$, $\beta = \alpha_{e\tau}$ and $\gamma = \alpha_{\mu\tau}$. We note that this decoupling of angles does not occur in the Kobayashi-Maskawa notation.

The experimental data on neutrino oscillations can be divided into two classes: 1. Disappearance experiments, where the flux of neutrinos of one flavour is measured at two distances from their production point. 2. Appearance experiments, where neutrinos of flavour ℓ are searched in a beam of neutrinos of flavour $k \neq \ell$.

For the simultaneous fit of oscillation experiments of different kinds, we have used all data from reactor and accelerator experiments available at this time. They are listed in Table 2 /4/. For each experiment, the original data were used, such that for a given neutrino energy each experiment gives a contribution to the global χ^2 in the fit.

Table 2

Oscillation experiments included in the fit /4/. The last column is the χ^2 contribution of the experiment for the hypothesis of no oscillation.

experiment	measured quantity	Δm^2 range $[eV^2]$	$\chi^2_{no\ osc.}$	Data points	χ^2_{best}
(1) CCFRR	$p_{\mu\mu}$ ratio	30. - 1000.	15.5	15	15.4
(2) CDHS	$p_{\mu\mu}$ ratio	0.24 - 90.	15.3	15	15.4
(3) CHARM	$p_{\mu\mu}$ ratio	0.60 - 20.	1.8	4	1.7
	$p_{e\mu}$ difference	1. - 10.	0	1	0.03
(4) GOESGEN	p_{ee} absolute	> 0.01	15.5	16	11.7
(5) GOESGEN	p_{ee} ratio	0.03 - 3.	7.4	16	7.3
(6) BUGEY	p_{ee} ratio	0.02 - 5.	20.9	9	18.1
(7) ν_τ Exp.'s	$p_{\mu\tau}/p_{\mu\mu}$	> 0.2	6.3	6	6.2
(8) BNL	$p_{\mu e}$ absolute	> 0.43	2.4	7	2.4
(9) FNAL E531	$p_{e\tau}/p_{\mu\mu}$, $p_{\mu\tau}/p_{\mu\mu}$	> 9	0.0	2	0.
(10) PS 191	$p_{\mu e}/p_{\mu\mu}$	5 - 10	8.5	1	8.5
(11) IMB	$p_{\mu\mu}$ ratio	10^{-5} - 10^{-4}	0.0	1	0.
(12) UCI	p_{ee} ratio	0.06 - 6.	2.7	1	3.9
(13) BEBC	$p_{\mu e}$ abs	> 0.1	0.04	1	0.
all data (D.F. = 90)			96.34	95.	90.63
data excluding PS 191 (D.F. = 89)			87.9	94.	82.6

For each experiment we checked that, making the restrictive assumptions of the authors, the procedure used here gave back the quoted result of the authors in the two-parameter model with only one Θ and one Δm^2.

A total chisquare function was constructed from the measured quantities Q^m (ratios of oscillation probabilities at different L/E or absolute probabilities from a comparison to calculated initial fluxes) and the corresponding computed values Q^C. The data points of each experiment i are allowed to vary together by a scale factor N_i within the quoted normalization uncertainty $\sigma(N_i)$. The expression

$$\chi^2 = \sum_{i=1} \left[\sum_j \left(\frac{N_i Q^m_{ij} - Q^c_{ij}}{\sigma(Q^m_{ij})} \right)^2 + \left(\frac{N_i - 1}{\sigma(N_i)} \right)^2 \right] \qquad (3)$$

is minimized by fitting the mixing angles and mass differences, where the index i labels the experiments and the index j gives the bin in L/E for a specific experiment.

The results of the fitting procedure are listed in Table 3. Limits on the oscillation parameters can be derived from eq. (3) in three parameter planes ($\sin^2 2\alpha_{e\mu}$, Δm^2_{12}), ($\sin^2 2\alpha_{e\tau}$, Δm^2_{13}), and ($\sin^2 2\alpha_{\mu\tau}$, Δm^2_{23}). The curves in Fig. 1 give the largest allowed $\sin^2 2\alpha$ values (for 90% C.L.) as a function of the corresponding Δm^2. Each such point has been obtained under the condition that in the other two planes all combinations ($0. < \sin^2 2\alpha < 1.$, $0.01 < \Delta m^2 < 1000\ eV^2$) are possible. There is one best fit solution with a $\chi_{best} = 90.6$ for 90 D.F.. The parameters of this point in parameter space are $\Delta m^2_{12} = 0.2\ eV^2$, $\Delta m^2_{13} = 1.1\ eV^2$ and $\Delta m^2_{23} = 0.9\ eV^2$, with $\sin^2 2\alpha_{e\mu} \sim 0.06$, $\sin^2 2\alpha_{e\tau} \sim 0.05$ and $0 < \sin^2 2\alpha_{\mu\tau} < 1$. The limits on the mixing angles derived from this fit are $\sin^2 2\alpha_{e\mu} < 0.15$ and $\sin^2 2\alpha_{e\tau} < 0.15$ in the range $\Delta m^2_{12} > 0.06\ eV^2$, $\Delta m^2_{13} > 0.04\ eV^2$ and $\Delta m^2_{23} > 2 eV^2$, at 90% confidence level. These limits correspond to the ranges for the elements of the lepton mixing matrix as given in Table 3:

Table 3

90% C.L. limits on mixing matrix elements for best fit values $\Delta m^2_{12} = 0.2\ eV^2$ and $\Delta m^2_{13} = 1.1\ eV^2$

$$U = \begin{bmatrix} 0.97 - 1.00 & 0. - 0.17 & 0. - 0.18 \\ -0.25 - 0. & 0.67 - 1.00 & 0. - 0.70 \\ -0.17 - 0. & -0.72 - 0. & 0.70 - 1.0 \end{bmatrix}$$

$\sin^2 2\alpha_{e\mu} < 0.15$; $\sin^2 2\alpha_{e\tau} < 0.09$; $0 < \sin^2 2\alpha_{\mu\tau} < 1.0$

For any specific mass difference $\Delta m_{kk'}$, the 90% C.L. limit and the 3 standard deviation limit for the mixing parameters can be obtained from Fig. 1 (In this figure $\Theta_{12} = \alpha_{e\mu}$, $\Theta_{13} = \alpha_{e\tau}$ and $\Theta_{23} = \alpha_{\mu\tau}$). It becomes clear that these limits are less significant in those regions of Δm^2 where the Bugey and PS 191 experiments found indications for oscillation. This is also evident in table 2 where these two experiments are the only ones at variance with the hypothesis of no oscillation, as indicated by their χ^2 contributions larger than expected.

Still, even including these data, the hypothesis of no oscillation is consistent with the data, with $\chi^2/D.F. = 96.3/90$.

In conclusion, we have shown that an analysis of existing data on oscillations of three neutrino flavours yields limits on the mixing

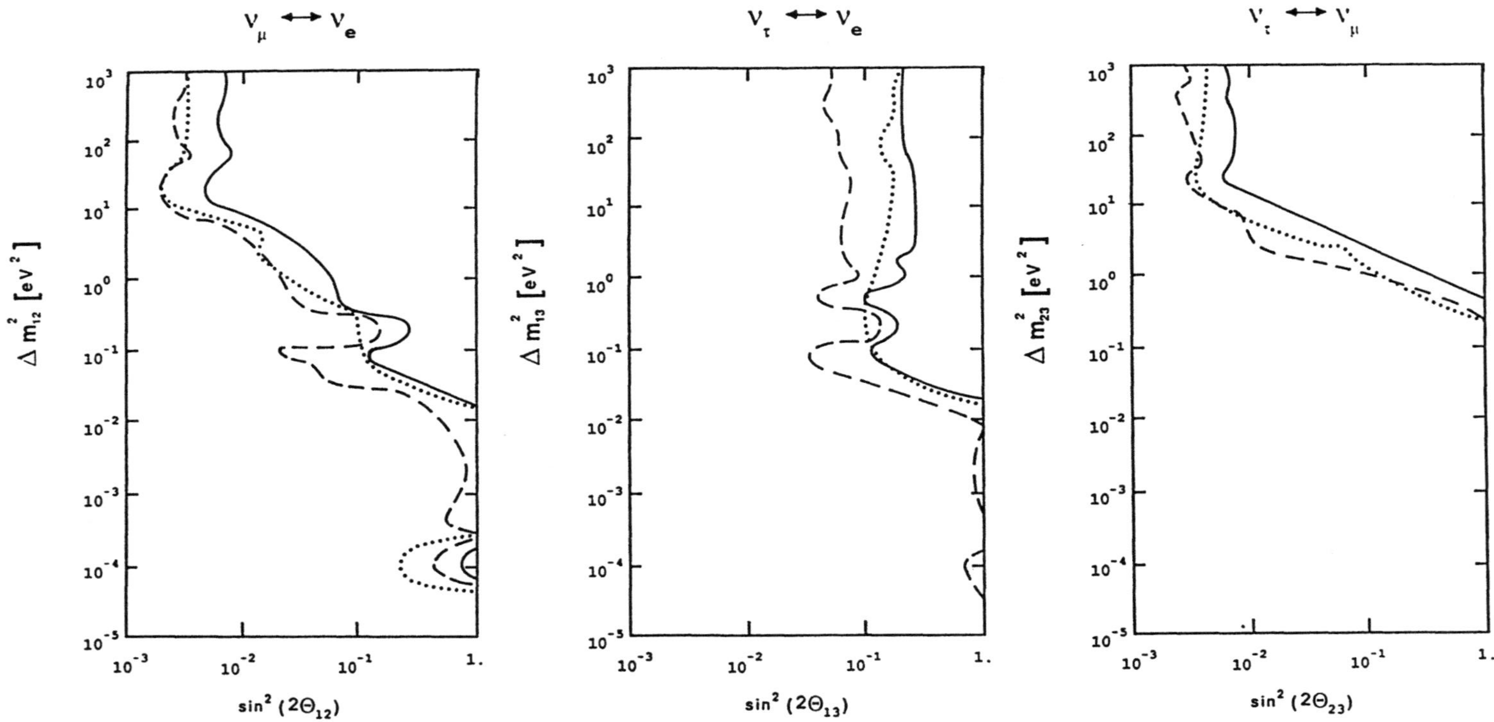

Fig. 1: Combined limits for three neutrino - oscillation parameters. The curves indicate different levels of significance: Solid line for 3 sigma, dashed line for 90 % C.L.
The dotted curves show for comparison the 2 neutrino limits from the most sensitive experiments

matrix and on the three mixing angles each corresponding to one channel of oscillations. While this method does not in general allow stringent constraints on the mixing parameters to be extracted from one experiment, the combination of the different experimental data gives limits on angles and the mixing matrix elements which are about as restrictive as earlier Two-neutrino analyses on separate oscillation channels. The ensemble of data is consistent with no oscillation occurring.

If neutrino masses are finite /5/, then mixing angles of the order of $\sqrt{(m_e/m_\mu)}$ or $\sqrt{(m_\mu/m_\tau)}$ are expected in some models invoking family symmetries /6/. Experiments have nearly reached the level of sensitivity needed for testing such models. As in the case of quark mixing, the problem of family mixing is still waiting for a solution.

References

/1/ S.M. Bilenky and B. Pontecorvo: Phys.Rep. 41C, 225 (1978)
A. de Rujula et al..: Nucl.Phys. B168, 54 (1980)
S.P. Rosen: Preprint Los Alamos LA-UR-84-1789 (1984)

/2/ L. Maiani: Proc.Intern.Symp. on Lepton and Photon Interactions at High Energies (Hamburg 1977), p. 877

/3/ H. Blümer and K. Kleinknecht: Phys.Lett. 161B, 407 (1985) and references quoted therein.

/4/ Experimental results are taken from the following references:
CCFRR (1) C. Haber et al., Phys.Rev.Lett. 52, 1384 (1984)
CDHS (2) F. Dydak et al., Phys.Lett. 134B, 281 (1984)
CHARM (3) F. Bergsma et al., Phys.Lett. 142B, 103 (1984)
Goesgen (4) J.L. Vuilleumier et al., Phys.Lett. 114B, 298 (1982)
Goesgen (5) K.Gabathuler et al., Phys.Lett. 138B, 449 (1984)
Bugey (6) J.F. Cavaignac et al., Phys.Lett. 148B, 387 (1984)
ν_τ search (7) N.J. Baker et al., Phys.Rev.Lett. 47, 1576 (1981)
N. Armenise et al., Phys.Lett. 100B, 182 (1981)
O. Errique et al., Phys.Lett. 102B, 73 (1981)
N. Ushida et al., Phys.Rev.Lett. 47, 1694 (1981)
G.N. Taylor et al., Phys.Rev. D28, 2705 (1983)
H.C. Ballagh et al., Phys.Rev. D30, 2271 (1984)
BNL (8) L.A. Ahrens et al., Phys.Rev. D31, 2732 (1985)
FNAL E531 (9) N. Ushida et al., Phys.Rev.Lett. 57, 2897 (1986)
PS 191 (10) G. Bernardi et al., Phys.Lett. 181B, 173 (1986)
IMB (11) J.M. Lo Secco et al., Phys.Rev.Lett. 54, 2299 (1985)
UCI (12) H.W. Sobel: 12th Int.Conf. on Neutrino Physics and Astrophysics (Sendai 1986), World Scientific, Singapore, p. 148

BEBC (13) M. Baldo-Ceolin: 12th Int.Conf. on Neutrino Physics and Astrophysics (Sendai 1986), World Scientific, Singapore, p. 170

/5/ V.A. Lubimov et al.: Phys.Lett. 94B, 266 (1980); Proc.EPS Conf. on High Energy Phys., Brighton (1984) p. 386; Phys.Lett. 159B, 217 (1985)
K. Bergkvist: Phys.Lett. 154B, 224 (1985)
J.J. Simpson: Phys.Rev.Lett. 54, 1891 (1985)
T. Altzitzoglou et al., Phys.Rev.Lett. 55, 799 (1985)
M. Fritschi et al., Phys.Lett. 173B, 485 (1986)
H. Kawakami et al., Proc. 12th Int.Conf. on Neutrino Physics and Astrophysics (Sendai 1986), World Scientific, Singapore, p. 36
R.G.H. Robertson et al., ibid. p. 49

/6/ Z.G. Berezhiani: Phys.Lett. 150B, 177 (1984)
H. Fritzsch and P.Minkowski: Phys.Rep. 73, 67 (1981)
H. Fritzsch: Nucl. Phys. B155, 189 (1979)
B. Stech: Phys.Lett. 130B, 89 (1983)
M. Shin: Phys.Lett. 145B, 285 (1984)

Search for $\nu_\mu \rightarrow \nu_\tau$ Oscillations Motivation and Feasibility

V. Zacek

CERN, CH-1211 Geneva 23, Switzerland

Theoretical prejudices derived from solar-ν matter oscillations and assumptions of neutrino mass hierarchies suggest. that ν-oscillations are observable in laboratory with mass parameters of $\Delta m^2 = 10^{-3} - 10^4\,eV^2$. In particular $\nu_\mu \rightarrow \nu_\tau$ appearance searches at accelerators and $\bar{\nu}_e \rightarrow \bar{\nu}_x$ disappearance experiments at reactors seem strongly motivated.

INTRODUCTION

The question whether the neutrino possesses a finite rest mass remains one of the most challenging problems in todays physics. Experimental evidences for a neutrino mass effect in Tritium decay and neutrino oscillations are still heavily disputed and contradicted by other experiments[1]. In addition there is little guidance from theory and much room for speculations. One of the common theoretical prejudices. namely. that neutrino masses scale similar to the quark or lepton masses, has interesting consequences, when combined with the MSW explanation of the solar neutrino puzzle. In particular one can speculate, that the only type of experiment to establish a finite neutrino mass effect in the laboratory will be an improved $\nu_e \rightarrow \nu_\tau$ or $\nu_\mu \rightarrow \nu_\tau$ oscillation search. In order to push the experimental limits on oscillation parameters further, new techniques and facilities are required.

NEUTRINO MASSES AND NEUTRINO OSCILLATIONS

Whereas in the minimal version of the SU(5) model there is no room for neutrino mass generating processes, extended models beyond SU(5) and superstring theories easily allow for neutrino masses in a vast range from 10^{-6} eV to 10 eV. In addition a mass hierarchy is expected and it is a common feature of most models, that neutrino masses for different flavors (m_k , k=1,2,3) scale like the masses of the charged leptons or quarks of the corresponding family:

$$m_1 : m_2 : m_3 \approx m^q_{u,e} : m^q_{c,\mu} : m^q_{t,\tau} \quad (1)$$

The exponent q is model dependent, but usually lies between 1 and 2. q= 2 corresponds to the popular See-Saw model[2] , which predicts light lefthanded Majorana neutrinos with mass e.g. like $m_{\nu_e} \approx m_u^2/M$. The mass scale is essentially fixed by a grand unification mass M in the region between 10^{11} GeV to 10^{14} GeV. Since such absolute scale predictions are too uncertain, it is safer to base speculations on the ratios of the masses alone. Taking lepton, as well as quark masses into account a conservative estimate for the range of the ratios of the squared masses gives:

$$m_1 : m_2 : m_3 \approx 1 : 10^{4,8} : 10^{8,16} \quad (2)$$

The mass values m_k need not necessarily coincide with the eigenstates of the weak interaction. If this is the case, lepton flavor is no more a good quantum number and neutrino mixing can occur, in close anology to the Kobayashi-Maskawa mixing in the quark sector:

$$\nu_l = \sum_k U_{l,k} \nu_k \qquad l = e, \mu, \tau, \ldots \quad k = 1, 2, 3, \ldots \quad (3)$$

U is a unitary matrix and for three flavors it is completely determined by three mixing angles and one CP violating phase. The weak interaction eigenstates ν_l are now coherent superpositions of the states ν_k with definite masses m_k. The latter determine the space - time behaviour and consequently the propagation of a neutrino wavepacket will exhibit flavor oscillations.

The probability for a neutrino with energy E_ν emitted with definit flavor e.g. ν_l to be found in the state $\nu_{l'}$ at a distance L from the source is given by:

$$P(\nu_l \rightarrow \nu_{l'}) = \delta_{ll'} - 2*\mathrm{Re} \sum_{k>k'} U_{l'k} U^*_{lk} U_{lk'} U^*_{lk'} \left[1 - \exp(i\, 2.53\, \Delta m^2_{kk'}\, L / E_\nu)\right] \quad (4)$$

where $\Delta m^2_{kk'} = |m^2_k - m^2_{k'}|$ is the mass parameter. If $\Delta m^2_{kk'}$ is measured in eV^2, L and E_ν are expressed in units of km and GeV, respectively. The frequency of the flavor modulation is defined by the ratio $E_\nu / \Delta m^2_{kk'}$, the amplitude is governed by the coefficients U_{lk}. If only two neutrino flavors are involved, (4) reduces to

$$P(\nu_l \rightarrow \nu_{l'}) = \delta_{ll'} - \sin^2 2\vartheta \sin(1.27\, \Delta m^2_{kk'}\, L / E_\nu) \quad (5)$$

Where ϑ is the mixing angle, which relates weak interaction and mass eigenstates. It is important to keep in mind, that being an interference experiment, the study of neutrino oscillations might be the only way to prove a finite neutrino mass, if the masses are too small to be detected in kinematic or double beta decay measurements.

MATTER OSCILLATIONS AND THE SOLAR NEUTRINO PUZZLE

A particularly interesting phenomenon can arise if the oscillating neutrinos travel through matter. As observed by Wolfenstein[3], the slightly differing weak interaction of electrons with ν_e (via W and Z - exchange) and with ν_μ, ν_τ (via Z - exchange only) leads to differing effective neutrino masses in matter. In addition Mikheyev and Smirnow[4] proved, that as a consequence neutrino oscillations can be resonance enhanced in matter. In fact, for a definite neutrino energy, full mixing, i.e. $\sin^2 2\vartheta_m = 1$, can occur in matter for a certain electron density, irrespective how small the mixing might be in vacuo (MSW - effect). Therefore large oscillations can be observed, if a neutrino beam exits a resonant layer of mattter.

In particular the solar neutrino puzzle gets an elegant explanation in the frame of the MSW-effect: matter enhanced neutrino oscillations of the type $\nu_e \rightarrow \nu_\mu$ or $\nu_e \rightarrow \nu_\tau$ are at the origin of the apparent solar neutrino flux deficit, as recorded over a period of more than 10 years by the ^{37}Cl-experiment. Given the parameters of the standard solar model, resonance enhancement happens for mass parameters Δm^2 in the range of $10^{-4}\,eV^2$ -$10^{-7} eV^2$ [5]. However the resulting enormously long oscillation lengths (10^7m- 10^{10} m for E_ν =1GeV) will render laboratory observations impossible. Nevertheless it turns out, that the situation is not hopeless, provided the following conditions are fullfilled:

1) $\nu_e \rightarrow \nu_\mu$ transitions dominate the oscillation effects in the sun. ($\nu_e \rightarrow \nu_\tau$ transtions would imply vanishing laboratory effects) .

2) The mixing occurs between at least three neutrino types.

3) The neutrino mass ratios behave according to the theoretical prejudices in equ.2.

4 The mixing angles are not too small, but at least comparable to those in the quark sector.

The mass parameters expected in this scenario are then Δm^2_{12} =10^{-4} -$10^{-7}eV^2$ and Δm^2_{13}= Δm^2_{23} = 10^{-3} - 10^4 eV^2. These have to be compared to the accessible ranges of $\Delta m^2 \geq 10^{-2}$ eV^2 at reactors and $\Delta m^2 \geq 1\,eV^2$ at accelerators and thus are largely covered by experiment.

What about mixing angles and transition probabilities? If mixing is small then the matrix can be approximated by

$$U_{jk} \approx \begin{pmatrix} 1 & \vartheta & \beta \\ -\vartheta & 1 & \gamma \\ -\beta & -\gamma & 1 \end{pmatrix} \qquad \vartheta, \beta, \gamma \ll 1 \qquad (6)$$

and the transition probabilities for the three experimentally interesting channels become :

$$P(\nu_e \rightarrow \nu_\tau) \approx 4\,\beta^2 \sin^2 (1.27\, \Delta m^2_{13} L/E_\nu) \qquad (7)$$

$$P(\nu_\mu \rightarrow \nu_\tau) \approx 4\,\gamma^2 \sin^2 (1.27\, \Delta m^2_{13} L/E_\nu) \qquad (8)$$

$$P(\nu_\mu \rightarrow \nu_e) \approx 4\,\beta\,\gamma^2 \sin^2 (1.27\, \Delta m^2_{13} L/E_\nu) \qquad (9)$$

Thus $\nu_\mu \to \nu_e$ transitions are expected to be less likely than $\nu_\mu \to \nu_\tau$ appearance oscillations. Furthermore long baseline reactor experiments, which probe (7) in the ν_e- disappearance mode could be promising in a search for very small $\Delta m^2 \approx 10^{-3}$ eV^2, as long as β is sufficiently large.

To be more quantitative, if one assumes similar mixing as in the quark sector, modulation amplitudes of $P(\nu_e \to \nu_e) \approx 6 * 10^{-4}$ result for ν_e- disappearence. On the other hand $P(\nu_\mu \to \nu_\tau) \approx 10^{-2}$ and $P(\nu_\mu \to \nu_e) \approx 10^{-4}$ will be expected for appearance oscillations in ν_μ- beams. A search for $\nu_\mu \to \nu_\tau$ transitions would thus be strongly favorable in such a scenario. Also a search for ν_e - appearance in ν_μ-beams looks attractive, although new techniques would be necessary to reduce the uncertainty in beam impurities, which limit present experiments at the permille level. In contrast to accelerator searches, the case of reactor experiments would be in this case rather hopeless, since their sensitivity to mixing effects is constrained by statistics and systematical uncertainties in the percent region where no further improvement can be expected.

ν_τ - OSCILLATION EXPERIMENTS

The present experimental situation is summarized in fig.1 a) and 1 b). Good limits on the mass parameters are obtained in disappearance experiments, where one looks simply for a flux reduction of the original flavor. These experiments are of the inclusive type, since they check several channels at once. Interpreted in terms of $\nu_\mu \to \nu_\tau$ oscillations, a limit of $\Delta m^2_{23} \geq 0.2$ eV^2 for full mixing can be infered from the CHARM[14] and CDHS[15] results. Similarly for the $\nu_e \to \nu_\tau$ channel reactor experiments give $\Delta m^2_{13} \geq 0.02$ eV^2 for full mixing. Since it is difficult to measure small flux changes precisely, these experiments show poor sensitivity to mixing angles.

High energy accelerators with neutrino beam energies above the τ-production threshold at 3.5 GeV allow studies of the exclusive type $\nu_{\mu,e} \to \nu_\tau$. They provided up to now the most sensitive tests of the mixing angles down to $\sin^2 2\vartheta = 10^{-3}$ for $\Delta m^2 \geq 1$eV^2. In practice the search for $\nu_\mu \to \nu_\tau$ oscillations concentrates on the detection of a few ν_τ charged current interactions among a majority of ν_μ induced charged current events. If limited by statistics only, the sensitivity on the oscillation parameters is given by

$$\Delta m^2 \cdot \sqrt{\sin^2 2\vartheta} \leq 0.8 \, (E_\nu / L)_{ave} \sqrt{\sigma_{\nu_\mu} N_\tau^{cc} / \sigma_{\nu_\tau} N_\mu^{cc}} \tag{10}$$

Here N_τ^{cc} and N_μ^{cc} are the numbers of recorded charged current events, induced by ν_τ and ν_μ respectively, σ_{ν_τ} and σ_{ν_μ} are the corresponding crossections, E_ν is the neutrino energy in GeV, L in km the distance between detector and neutrino source. The charged current detection reaction is

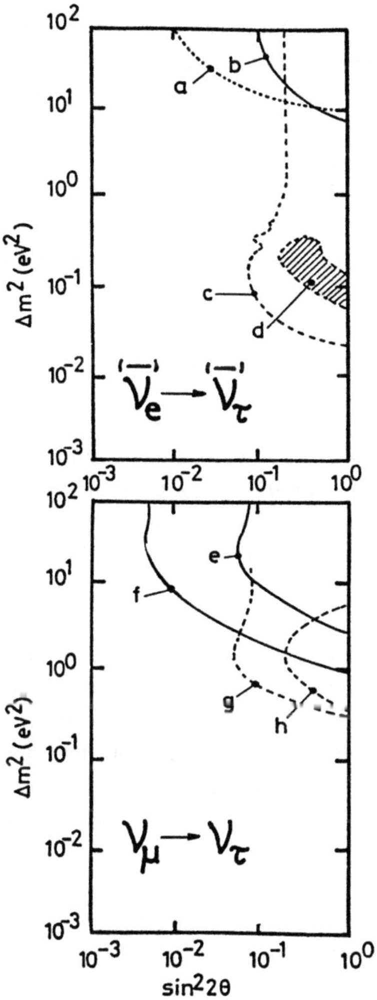

Fig.1) Summary of experimental limits from $\nu_{e,\mu} \rightarrow \nu_\tau$ oscillation searches. Broken lines belong to disappearance type experiments, solid lines to appearance searches. Excluded regions of the oscillation parameters Δm^2 and $\sin^2 2\vartheta$ are to the right of the curves.

a) BEBC SPS-WBB [13]
b) FNAL emulsion [12]
c) GOESGEN reactor [10]
d) BUGEY reactor [11]
(shaded area allowed)
e) BNL BC [16]
f) FNAL emulsion [12]
g) CDHS PS [15]
h) CHARM [14]

$$\nu_\tau + N \rightarrow \tau^- + X \qquad (11)$$

$$\tau^- \rightarrow \nu_\tau + \text{hadrons} \quad (BR = 66\%)$$
$$\tau^- \rightarrow \nu_\tau + e^- + \bar{\nu}_e \quad (BR = 17\%)$$
$$\tau^- \rightarrow \nu_\tau + \mu^- + \bar{\nu}_\mu \quad (BR = 17\%)$$

The τ will decay within $3.5*10^{-13}$ sec, i.e. with a decay length of $60\mu m/$ GeV. Even if detectors do not have the spatial resolution to observe the track of the intermediate τ-lepton, its decay into two neutrinos and an electron makes it possible, that any high energy experiment searching for $\nu_\mu \rightarrow \nu_e$ appearance will also be sensitive for $\nu_\mu \rightarrow \nu_\tau$ oscillations, with a sensitivity reduced by the small branching ratio. The best result obtained in this way comes from an experiment performed in the Fermilab wide-band beam with the 15 ft bubble chamber by the BNL-Columbia collaboration[16].

Better limits can be obtained in dedicated experiments, which try to identify the ν_τ - interaction unambiguously by observation of the τ-decay and its topology. This provides an ideal, virtually background free signature. Excellent spatial resolution below 50μm is required, which at present can be achieved only in nuclear emulsions or holographic bubble chambers. An example of this kind is the experiment of the Canada-FNAL-Japan-Korea collaboration [12], which exposed 23 l nuclear emulsion to the Fermilab horn-focused narrow-band beam. Since not a single interaction containing a τ in the final state was observed, stringent limits on Δm^2 and $\sin^2 2\vartheta$ could be derived according to equ. 10 .

FUTURE PROSPECTS

I . There is an interesting outlook at LAMPF, where a new low energy neutrino beam is installed (E_ν = 60 - 250 MeV). A very small ν_e-contamination is expected, since the proton energy is below K production threshold. Therefore $\nu_\mu \rightarrow \nu_e$ experiments can probe sensitively the term $\beta\gamma^2$ of equ. 9 to the 10^{-4} level and down to very small Δm^2 .

II. At CERN the CHARM II collaboration considers the project of a vertex detector based on scintillating optical fibres . Recent progres in fabrication of coherent fibre bundles, promises the feasibility of active target detectors with spatial resolution of several tens of μm down to several μm. A detector of this type would be the equivalent of a continuously acitive, triggerable bubble chamber and would allow to tag the τ-lepton in a $\nu_\mu \rightarrow \nu_\tau$ appearance search.

What size of target is needed? If this detector would be installed in front of the CHARM II detector in the CERN SPS wide-band neutrino beam (i.e $L_{ave} \approx$ 670m, $E_{ave} \approx$ 20GeV), a minimum mass of about 40 kg will be required at $5*10^{18}$ pot in order to start improving on the existing emulsion limits .

The detector consists of coherent fibre bundles oriented parallel to the neutrino beam. By recording the transverse projection of the event at the end-face of a bundle, advantage can be taken of the Lornentz invariant nature of the event picture characterised by decay length and impact parameter. In addition the light yield in the fibres is increased by roughly the Lorentz factor due to the small opening angles of the reaction products. The required spatial resolution in the transverse plane is given by the fibre diameter. Therefore thin fibres with crossections of about 20 μm will be used. The end-faces of individual target blocks are optically coupled to an imaging system, which consists of a combination of image intensifiers and CCD arrays (fig.2).

First tests were made with different fibre bundles[6], containing fibres of 12.5μm and 100μm diameter, each surrounded by a thin cladding with low

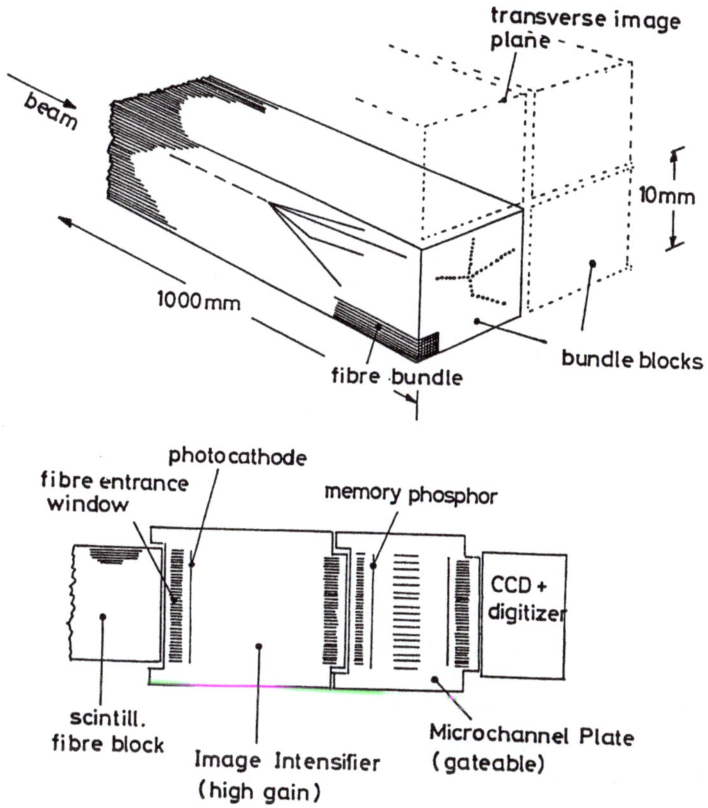

Fig.2) Scheme of high resolution scintillating fibre tracker and CCD-read out system.

refractive index followed by a layer of a black extramural absorber. The fibres were doped with Tb_2O_3 as scintillator.They showed well seperated absorption and emission lines at 380 nm and 550 nm resp., but too short attenuation lengths ($\lambda_{att} \approx 10$cm) , presumably due to rare earth impurities. To achieve adequate attenuation lengths of $\lambda_{att} \geq 1$m development work is in progres.

III. An improved oscillation search for $\nu_\mu \rightarrow \nu_e$, ν_τ with an already existing neutrino detector, like e.g. the CHARM II calorimeter can be achieved by adding a tagging detector downstream in the neutrino beamline. The tagger determines flavor and energy of the neutrino from a simultaneous measurement of the charged particles following π- , $K_{\mu 2}$- or K_{e3}- decays.

Several schemes were studied at CERN [7], SERPUKHOV [8] and BNL[9]. Usually a narrow-band beam is required, with a several seconds lasting slow extraction and an efficient cosmics reduction in the main detector. The installation of a neutrino tagger would be very efficient in reducing uncertainties in beam composition. In principle mixing down to $\sin^2 2\vartheta \sim 10^{-4}$ should be accessible in the $\nu_\mu \rightarrow \nu_e$ channel. The actual limitation will rather be given by the π^0/ν_e discrimination power of the used detector.

CONCLUSIONS

The study of neutrino oscillations at accelerators and reactors is still a fruitful and exciting enterprise. In view of common theoretical prejudices especially searches for $\nu_\mu \rightarrow \nu_\tau$ appearance oscillations at accelerators and $\bar{\nu}_e \rightarrow \bar{\nu}_x$ disappearance at reactors are promising. Challenging new experimental techniques and facilities will be required for the next round of experiments. In particular a scintillating fibre tracker for τ-tagging in ν_τ - appearance searches and long-baseline kton reactor experiments could lead to substantial improvements in sensitivity .

REFERENCES

1. J.L.Vuilleumier, Rep.Prog.Phys. **42** (1986) 1293
2. M.Gell-Mann, Supersymmetry, North-Holland,Amsterdam, p.317
3. L.Wolfenstein, Phys.Rev. **D17** (1978) 2369
4. S.P.Mikheyev and A.Yu.Smirnov, Sov.J.Nucl.Phys. **42** (1985) 913
5. J.M.Bouchez et al, Z.Phys. **C32** (1986) 499
6. by courtesy SCHOTT GLASWERKE/ MAINZ
7. CHARM II, internal report (1987)
8. A.A. Boikov et al, IHEP (1980) Preprint 80-158
9. F.Vannucci, Proc. of the BNL NEUTRINO WORKSHOP, (1987) 5
10. G.Zacek et al, Phys.Rev. **D34** (1986) 2621
11. J.F.Cavaignac, et al, Phys.Lett. **B148**, (1984) 387
12. N. Ushida et al, Phys.Rev.Lett. (1986) 2898
13. O.Erriquez et al, Phys. Lett. **102B** (1981) 73
14. F.Bergsma et al, Phys. Lett. **B142** (1984) 251
15. F.Dydak et al, Phys. Lett. **B134** (1984) 281
16. N.J.Baker et al, Phys. Rev. Lett. (1981) 1576

A Search for Neutrino Oscillations at LAMPF

S.J. Freedman

Argonne National Laboratory, Argonne, IL 60439, USA

Experiment E645 is a search for neutrino oscillations, employing the LAMPF beam stop as a neutrino source. Data collection is now underway and we describe the experiment, the quality of the preliminary data, and the sensitivity to neutrino oscillations that we expect to reach by the end of the 1987 running cycle.

Attempts to discover neutrino masses and mixings by detecting neutrino oscillations now constitutes a mature area of experimental physics which exploits a wide variety of neutrino sources at accelerators and reactors. Pion factories are copious sources of low energy neutrinos produced from the decay of stopped pions and muons, and, a neutrino oscillation search, experiment E645, is now collecting data at the Los Alamos Meson Physics Facility (LAMPF). The E645 experiment is a collaboration between Argonne National Laboratory, California Institute of Technology, Lawrence Berkeley Laboratory, LAMPF, Louisiana State University, and Ohio State University.[1]

After passing through production targets the remaining $\simeq$ 600 μA of the LAMPF proton beam degraded to $\simeq$ 750 MeV, is absorbed in a water cooled copper beam dump. Most of the pions produced in the beam dump quickly thermalise. The negative pions are captured, and the positive pions decay, producing three types of neutrinos: $\pi^+ \to \mu^+ + \nu_\mu$ and $\mu^+ \to e^+ + \nu_e + \bar{\nu}_\mu$. There are $\simeq$ 0.09 stopped π^+ per incident proton.[2] The resulting source intensity is $\simeq 3 \times 10^{14}$ ν/sec of each type and Fig. 1 displays the neutrino energy spectra. Contamination of $\bar{\nu}_e$'s from negative pions which manage to decay before being captured is well below 10^{-3}. E645 is a simple appearance experiment: we look for the conversion of either ν_μ, $\bar{\nu}_\mu$, or ν_e into $\bar{\nu}_e$ with a detector sensitive only to $\bar{\nu}_e$.

Since there are yet no clear uncontroversial indications that neutrinos oscillate, it is convenient to discuss the sensitivities of experiments in terms of a simple two component mixing scheme. In the most common senario, a $\bar{\nu}_e$ signal would appear from $\bar{\nu}_\mu$'s, the probability depending on the neutrino energy E_ν (in MeV) and the distance to the source L (in meters). We have,

$$P(\bar{\nu}_\mu \to \bar{\nu}_e) = \sin^2 2\theta \; (1.27 \; \Delta m^2 L/E_\nu) \; , \qquad (1)$$

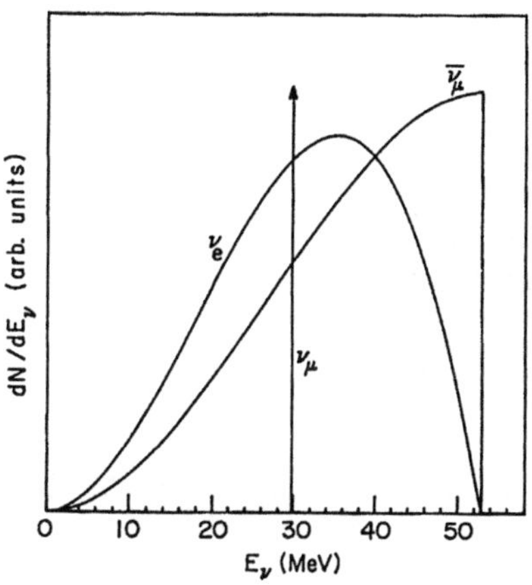

Fig. 1 The energy spectra of the three types of neutrinos produced by stopping π^+ decay

where $\Delta m^2 = |m_1^2 - m_2^2|$ is a measure of the neutrino eigenstate mass differences in eV^2 and θ is a fundamental mixing parameter. Of course the real physics might be much more complicated, involving more than two neutrino types and mixings between neutrinos and antineutrinos as well.

Appearance experiments with low backgrounds have good sensitivity for small θ and sensitivity at small Δm^2 comes from arranging for large L/E_ν. The E645 detector is about 24m from the LAMPF beam stop so $L/E_\nu \simeq 0.6$ and the overall sensitivity is similar to experiments at higher energy but larger distances at Brookhaven and CERN.

The experimental arrangement at LAMPF is shown schematically in Fig. 2. We detect $\bar{\nu}_e$'s via inverse beta decay, the $p(\bar{\nu}_e,e^+)n$ reaction, by observing the final state positron.

The LAMPF duty cycle, typically $\simeq$ 0.5 msec pulses at 120 Hz, is not at all ideal for neutrino experiments, making cosmic ray rejection a challenging problem. An unrecognized muon decay imitates the e^+ signal of $\bar{\nu}_e$-appearance and the muon veto must be nearly flawless. The active cosmic ray shield is a mostly continuous (there are actually three major isolated sections) 15cm thick cylindrical liquid scintillator tank surrounding the central detector. The liquid scintillator is viewed with 360 hemispherical phototubes each 13cm in diameter. A cosmic ray muon loses about 30 MeV in the shield and the scintillation light pulse is seen by several phototubes resulting in high rejection efficiency and low backgrounds from radioactivity.[4] Passive shielding consisting of 13cm thick lead and 5cm thick steel is put just inside the active shield to suppress spurious events from muon decay bremsstrahlung photons which would not be vetoed by the active shield. Neutral cosmic ray particles are suppressed by locating the entire detector in a tunnel under 2500 gm/cm^2 of overburden as shown in Fig. 2. The experimental trigger is coincident hits in three out of four consecutive central detector scintillation planes; the active shield is used on-line to suppress the trigger rate from cosmic rays.

Typically the detector is vetoed for 2 μsec for cosmic rays which interact with two isolated sections of the shield (passing muon veto) and 10 μsec when only one is hit (stopping muon veto). A potential 1.6 Hz trigger rate from cosmic rays is reduced to 0.5 Hz on-line.

The neutrino detector consists of 40 vertical modules, each consisting of a layer of 12 liquid scintillation counters and two layers of 45 proportional drift tubes (PDT), one oriented vertically and one horizontally. Scintillation counters are made from 0.4cm-wall extruded acrylic tanks 3cm x 30 cm x 3.7m with a photomultiplier at each end. Since the detector is incapable of distinguishing positrons from electrons materials which have large cross sections for ν_e interactions (odd neutron nuclei like ^{27}Al) are avoided. The PDT's are constructed with laminated paper and only a thin layer of aluminium. The PDT cross sectional area is 3.8cm x 7.6cm but they are capable of better than 1cm tracking resolution.[5] The scintillators measure both total energy and differential energy

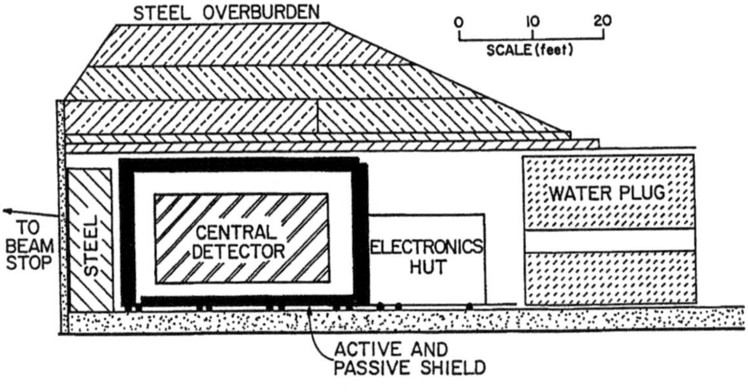

Fig. 2 The experimental arrangement. The neutrino detector and cosmic ray shield are housed in a tunnel under 2500 g/cm^2 of overburden 24m from the LAMPF beam stop. LAMPF itself is approximately 2200m above sea level

loss (dE/dx) which is important for finding positrons in a background of neutron induced recoil protons. A minimum ionizing particle at normal incidence loses about 9 MeV in each module. Each scintillator layer is covered with a plastic sheet coated with gadolinia (Gd_2O_3) which can be used to detect the neutron remnants of inverse beta decay. Some of these neutrons thermalize inside the fiducial volume, capture on gadolinium, and register as delayed γ-ray signals in the scintillators. The use of a delayed coincidence with neutrons decreases the overall detection efficiency by nearly a factor of 5, but it is a useful contingency in the event that a substantial beam excess of events is found. In all the detector weights ≃ 20 metric tons of which 2.3 tons is proton target.

Signals from phototubes and PDT's are shaped, amplified, and digitized in flash ADC's every 80 nsec for the detector and every 150 nsec for the shield. The results are stored in individual memories and a 150 μsec long "event history" is read out for each on-line trigger. Later, stringent off-line cuts are applied to the data in the event histories. The detector is calibrated and monitored with cosmic rays. Stopping muons are a particularly valuable since the decay electron characteristics approximates the $\bar{\nu}_e$-appearance signal (without a neutron left over of course). Figure 3 is an example of a stopping muon event. The data that is in time with the trigger shows a minimum ionizing track consistent with an electron with E < 50 MeV and there are no corresponding hits in the shield. The information from 3 μsec before the trigger, however, clearly indicates the electron signal is from a cosmic-ray muon which was seen by the shield and stopped in the detector.

The experiment operated with beam for a few weeks in the winter of 1986, and the data allowed us to find a hole in the beam stop shielding. With improved shielding the experiment began taking production data in June 1987; it has operated nearly without interruption since then. There are already several different preliminary analyses of the data and depending on the severity of the cuts the cosmic-ray background has been rejected to a level between 0.1 to 0.5 per LD (LD = the effective time the beam is on per day about 5600 sec). The trigger efficiency is ≃ 52% from a Monte Carlo analysis and the off-line analysis efficiencies vary from 30% to 50%, again depending on the cuts imposed. At some level we expect beam associated backgrounds from ν_e-nucleus scattering and from decaying μ's produced with very low momentum by energetic ν_μ's from π's decaying in flight. The bulk of the signal of ν_e-nucleus scattering (most of the scattering should come from $^{12}C(\nu_e,e^-)^{12}N$) would appear below 35 MeV. Scattering of ν_e's from ^{27}Al and ^{13}C is a potential background of higher energy electrons but at a much lower rate. The "decay-in-flight" neutrino backgrounds should be associated with other events in which the produced μ is observed before it decays, but there is yet no evidence for a significant rate. If we assume that no signal of neutrino oscillations is seen by the end of the 1987 LAMPF running cycle we can estimate our final sensitivity from our presently measured backgrounds and analysis efficiencies. Figure 4b shows our "expected" limits for 1987.

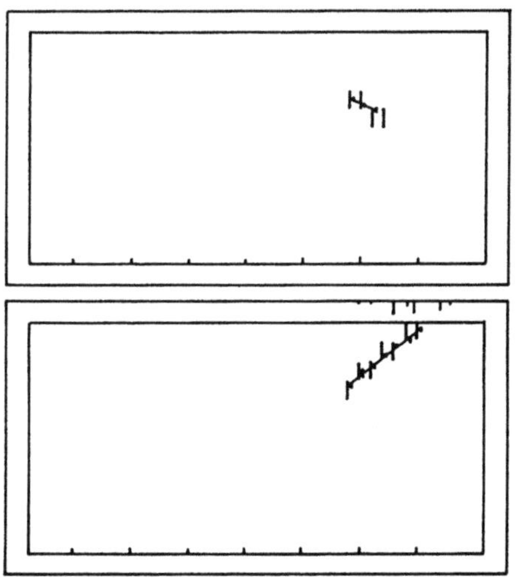

Fig. 3 Graphical display of a stopping muon event. The top panel shows the scintillator and PDT hits along the decay electron track. The lower panel is the detector 3 μsec earlier showing the track of the muon which entered the detector through the active shield

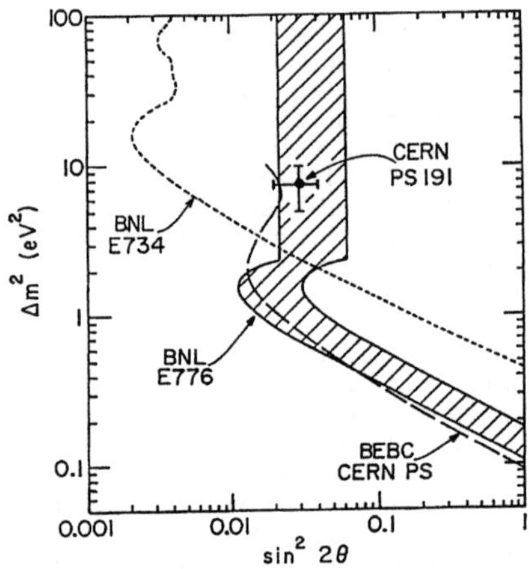

Fig. 4a Exclusion limits (90% cl) and possible positive indications (1σ) from $\nu_\mu \to \nu_e$ searches. (See References in the text, especially Refs. 7 and 8)

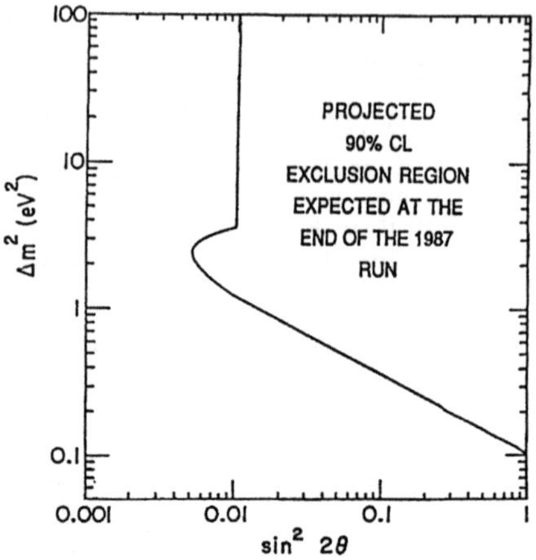

Fig. 4b Projected limits (90% cl) from the present E645 $\bar{\nu}_\mu \to \bar{\nu}_e$ appearance search. The limits indicate the expected sensitivity if no evidence for neutrino oscillations is seen in the rest of 1987 run

Figure 4a shows the most stringent existing limits from $\nu_\mu \to \nu_e$ appearance experiments at Brookhaven[5] and CERN.[6] Two preliminary observations of excess electron-like events, which could indicate that neutrinos oscillate from two other experiments are also shown.[7,8] The positive indications for neutrino oscillations are very preliminary and obviously needs further scrutiny; there are already indications of inconsistencies with limits from other experiments. We expect that E645 will reach adequate sensitivity to help settle the issue by the end of 1987.

This work supported in part by the U. S. Department of Energy, Nuclear Physics Division, under contract W-31-109-ENG-38.

REFERENCES

1. The E645 collaboration is: S. J. Freedman and J. Napolitano (Argonne National Laboratory); B. Fujikawa and R. D. McKeown (California Institute of Technology); K. T. Lesko and E. B. Norman (Lawrence Berkeley Laboratory); R. D. Carlini, J. B. Donahue, G. T. Garvey, and V. D. Sandberg (LAMPF); C. Choi, A. Fazely, R. Imlay, and W. Metcalf (Louisiana State University); L. S. Durkin, R. Harper, T. Y. Ling, J. Mitchell, T. A. Romanowski, E. Smith, and M. Timko (Ohio State University). For other descriptions of the E645 experiment see: E. Smith, Proceedings of the Sixth Moriond Workshop, Tignes France (Jan. 25-Feb. 1, 1986) p. 287-292; J. Napolitano, Proceedings of the Second Lake Louise Winter Institute (Feb. 15-21, 1987).
2. R. C. Allen et al., Phys. Rev. Lett. **55**, 2401 (1985) and H. H. Chen et al., Nucl. Inst. and Meth. **160**, 393 (1979); The character of the neutrino flux from the LAMPF beam stop is being studied in a new experiment, E866.
3. J. Fitch et al., Nucl. Inst. and Meth. **226**, 373 (1984)
4. S. J. Freedman et al., Nucl. Inst. and Meth. **215**, 17 (1983).
5. L. A. Ahrens et al. (E734), Phys. Rev. **D 31**, 2732 (1985).
6. C. Angelini et al. (BEBC), Phys. Lett. **B 181**, 307 (1986)
7. G. Bernardi et al. (PS191), Phys. Lett. **B 181**, 173 (1986); The point with errors in Fig. 4a is the best oscillation parameters reported in this reference; A second version of this experiment at BNL (E816) sees a consistent excess; P. Astier, Proceedings of the BNL Neutrino Workshop (Feb. 5-7, 1987), p. 25-38.
8. C. Chi et al., (E776) Proceedings of the BNL Neutrino Workshop (Feb. 5-7, 1987), p. 39-58. The region of possible oscillation parameters in Fig. 4a is calculated from the electron excess reported in this reference.

Generation Mixing and Heavy Neutrinos

J. Deutsch and R. Prieels

Institut de Physique Nucléaire, Université Catholique de Louvain,
Chemin du Cyclotron, 2, B-1348 Louvain-la-Neuve, Belgium

The astonishingly successful Standard Model has too many free parameters and ad-hoc features to be considered as really satisfactory. As a consequence, many experiments were undertaken to search for deviations from the predictions of this model in the entire energy-range accessible. These deviations would clearly provide important indications for more global unification-schemes of particles and interactions[1],[2],[3].

In this brief note we comment on a particular class : on searches for heavy neutrinos.

The Standard Model assumes neutrinos of zero mass having strictly conserved flavour quantum-numbers characteritic of each generation. This assumption corresponds to no known invariance-principle and introduces also an unpleasant distinction between quarks and leptons[3],[4].

Direct neutrino mass-measurements gave - up to now - only upper limits or controversial but highly stimulating non-vanishing values [5]. An other class of experiments can observe a neutrino of non-vanishing mass only if there is also a generation-mixing, i.e. a violation of the flavour quantum-number. These searches for phase-space anomalies in neutrino-emission, for neutrino-decay or for neutrino flavour-oscillation in flight did not provide up to now non-controversial evidence for heavy neutrinos neither[6],[7].

If the neutrino-mass is proportional to the square of the corresponding charged lepton mass, as predicted by various unification schemes[4], these methods, sensitive in a broad range of neutrino-mass, could prove very powerful indeed : e.g. the observation of a predominantly τ-neutrino of 0.1 - 70 MeV[1] would imply, in these

[1] 70 MeV is the actual upper limit of the ν_τ[8] ; cfr. however the 35 MeV/c^2 limit reported by the Argus-collaboration (H. Schubert) at this Workshop

models, a massive "electron-neutrino" of 10^{-2} - 10^{1} eV, quite below the possibilities of direct mass-measurement.

In this paragraph we describe briefly the novel upper limits we obtained on an eventual admixture to the electron-flavour of a heavy neutrino in the mass-range of 0.2 - 70 MeV.

The figure shows the upper limits of U^2_{ei} obtained from the absence of any anomaly in the beta-decay spectra of ^{35}S[9], of ^{64}Cu[10] and of the pion[11]. We show also (with dotted lines) the upper limits obtained from the $(\pi \rightarrow e\nu)/(\pi \rightarrow \mu\nu)$ branching ratio[12] and from the non-observation of neutrino-decay[13]. It should be noted, that to the contrary of the mentioned spectrum-measurements, these two later limits necessitate additional theoretical assumptions : the absence of pseudoscalar couplings as introduced by some extensions of the Standard Model[14] or the absence of "invisible" decays of the heavy neutrino into a Goldstone-boson[15].

We note the absence of any useful limit in a rather extended mass-range of 0.6 MeV to 20 MeV.

Two informations from <u>beta-decay</u> may provide additional limits on U^2_{ei} in this mass-range : the precision-measurements of the superallowed O^+ - O^+ ft-values and those of energetic beta-spectra[16].

The superallowed O^+ - O^+-transitions have 3-8 MeV energy-release[17] and the ft-values computed assuming the absence of any heavy-neutrino admixture in this mass-range should be <u>constant</u>. This feature could not be exploited before as the ft-values seemed to feature un <u>upward</u> trend with increasing energy-release, contrary to the effect of neutrino-admixture[18]. A new careful evaluation of the radiative corrections eliminated this anomaly[19] and so the data could be used for obtaining rather stringent upper limits to heavy-neutrino admixture into the electron-flavour[20]. The corresponding limits are also represented in the figure.

As a second approach, we exploit a precision-measurement of the ^{20}F beta-decay (5.9 MeV energy-release) performed in this laboratory[21].

Spectral shape factors of the form $C(E) = 1 + kE - 8.9 \times 10^{-5} E^2 - (4.0 \pm 0.1) \times 10^{-3}/E$ (E being expressed in electron mass units) were fitted to ten different high precision experiments performed in various experimental conditions. They all yielded slope-parameters

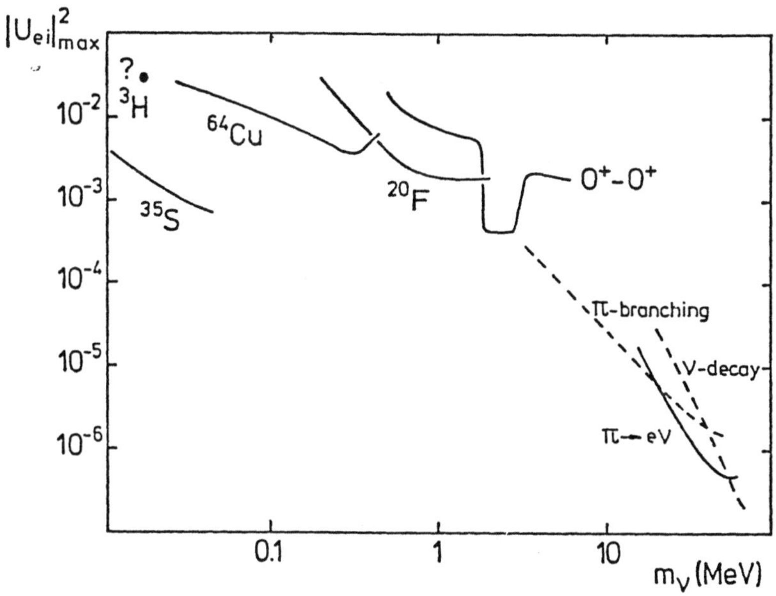

Upper limits on heavy-neutrino admixture amplitudes U_{ei} to the electron-flavour for $m_{\nu_3} < 70$ MeV/c^2.

k in good agreement with the prediction of the Conserved Vector Current theory : $k = (6.0 \pm 0.3) \times 10^{-3}$ [21],[22]. The normalized data were combined and searched for "kinks" indicative of an eventual heavy-neutrino admixture. No such kinks were found and the corresponding upper limits[22] are also shown in the figure.

The subsisting gap around 10 MeV neutrino-mass could be filled performing a dedicated high-resolution observation of the 70 MeV positron-peak in π-decay.

It is interesting to note that the upper limits obtained on the admixture of eventual heavy neutrinos from the searches shown in the figure can be confronted with lower limits deduced from the cosmic deuterium-abundance. If primordial heavy neutrinos exist below 70 MeV, they had to decay away fast enough so as to save deuterium from photofission induced by the electrons produced by their decay : $\nu_H \rightarrow \nu e^+ e^-$. The astrophysical "leg" of this confrontation is still somewhat controversial[23], but the confrontation begins to leave small or no room for heavy neutrinos in the mass-range we consider[24].

References

[1] M. Duffi : "Beyond the Standard Model", HEP 85, Bari, 18-24 July 1985, Proceedings p. 1017, Ed. Nitti and Preparata and refs. cited

[2] M.K. Gaillard : Future Prospects for Elementary Particle Physics: High and Low Energy Probes", LBL-20433, Oct. 1985

[3] R.N. Mohapatra : "Unification and Supersymmetry : the Frontiers of Quark-Lepton Physics", Springer Verlag, 1986

[4] J.D. Vergados, Phys. Reports 133 (1986) 1

[5] Review of Particles Properties, Phys. Lettes 170B (1986) 1

[6] F. Vanucci : "Baryon and Lepton Number Conservation and Neutrino Masses", HEP 85, Bari, 18-24 July 1985, Proceedings p. 1017, Ed. Nitti and Preparata

[7] J. Deutsch "Generation Mixing by Heavy Neutrinos", HEP 85, Bari, 18-24 July 1985, Proceedings p. 429, Ed. Nitti and Preparata

[8] H. Albrecht et al., Phys. Letters B163 (1985) 404

[9] T. Ohi et al., Phys. Letters 160B (1985) 322 ;
cfr. also : A.M. Apalikov et al., JETP Letters 42(5) (1986) 289

[10] K. Schreckenbach et al., Phys. Letters B129 (1983) 265

[11] N. De Leener-Rosier et al., Phys. Letters B177 (1986) 228

[12] D.A. Bryman et al., Phys. Rev. Letters 50 (1983) 1546

[13] G. Bernardi et al., Phys. Letters 166B (1986) 479

[14] D.A. Bryman et al., Phys. Rev. D33 (1986) 1211 and ref. 9 quoted

[15] B.G. Gelmini and J.W.F. Valle, Phys. Letters B142 (1984) 347

[**16**] F.P. Calaprice and D.J. Millener, Phys. Rev. C27 (1983) 1175

[**17**] I.S. Towner and J.C. Hardy, in Proc. of 7th Intern. Conf. on Atomic Masses, Ed. O. Klepper (GSI, Darmstadt, 1984), p. 564 and V.T. Koslowsky, ibidem, p. 572

[**18**] F. Calaprice, Princeton Report, unpublished
M. Lebrun, UCL Report, unpublished

[**19**] A. Sirlin and R. Zucchichi, Phys. Rev. Letters 57 (1986) 1994

[**20**] J. Deutsch, M. Lebrun and R. Prieels, to be published

[**21**] L. van Elmbt, J. Deutsch and R. Prieels, Nucl. Phys. A469 (1987) 531 and L.R. van Elmbt, Thesis, UCL, 1981, unpublished

[**22**] J. Deutsch, R. Prieels, to be published

[**23**] S. Sarkar and A.M. Cooper, Phys. Letters 148B (1984) 347 ; the upper bounds on the admixture obtained in this reference has however to be increased by a factor of 100 to account for photon-photon scattering as considered by D. Lindley, Astroph. Journal 294 (1985)1 (private communication from S. Sarkar)

[**24**] K. Sato, "Cosmological constraints on the lifetime of neutrino", Neutrino '86, Sendai, Japan, June 3-8, 1986, Proceedings p. 252, Ed. T. Kitagaki and H. Yuta (World Scientific) ;
R. Prieels "New constraints on Neutrino Masses and Electron Flavor Mixing from $\pi \to e\nu$ decay", Nuclear Beta Decays and Neutrino, Osaka, Japan, June 1986, Proceedings p. 403, Eds. T. Kotani, H. Ejiri and E. Takasugi (World Scientific)

Searches for Lepton-Flavor Violation

P. Depommier

University of Montreal, C.P. 6128, Station "A",
Montreal, Quebec, Canada H3C 3J7

1 The Muon Puzzle[1]

The problem arose soon after the discovery of the muon, as it was recognized that this particle is not the meson predicted by Yukawa. The existence of a new lepton, very similar to the electron but for its mass and instability, was very puzzling. One remembers the famous interrogation of Rabi "Who ordered that?". Then came the next question: how does the muon communicate with the electron? If the muon were some kind of heavy electron, it should easily decay via $\mu \to e\gamma$ or transform into an electron when captured by a nucleus (A,Z), giving rise to the process of *muon-to-electron conversion*, i.e. $\mu^-(A,Z) \to e^-(A,Z)$. Early experiments ruled out these two possibilities. Using cosmic rays, Hincks and Pontecorvo[2] established the first upper limit (of the order of one percent) on the $\mu \to e\gamma$ decay. Lagarrigue and Peyrou[3] showed that muon-to-electron conversion does not occur with a branching ratio greater than a few percent.

The electron spectrum from muon decay was analyzed according to the four-fermion theory developped by Michel[4]. Muon decay is a three-body decay with two neutrinos emitted. Assuming a conservation law for leptons, one considers these two particles as being neutrino and antineutrino. However, since the distinction between neutrino and antineutrino is still an open question, it is only possible to say that the two neutral particles emitted in muon decay are neutrinos of different helicities.

The possibility that these two neutrinos mutually annihilate leads back to the occurrence of the $\mu \to e\gamma$ decay. Assuming that the weak interaction is transmitted by a vector boson W, one has the following diagrams (Fig. 1):

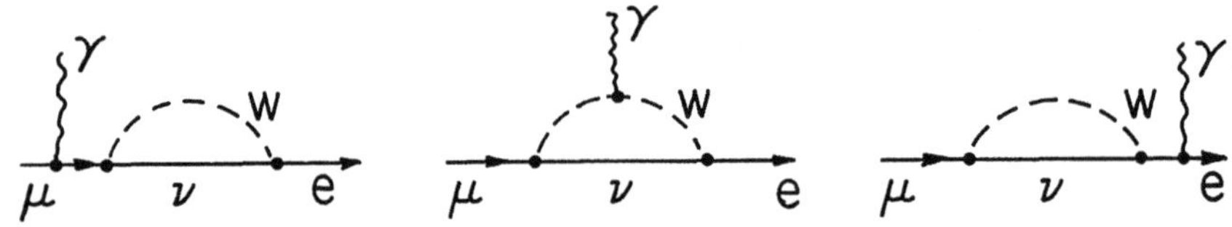

Fig. 1. Diagrams for the $\mu \to e\gamma$ decay: the neutrino emitted at the muon vertex is absorbed at the electron vertex. The photon is emitted from the charged current

In spite of computational difficulties and ambiguities (in the absence of a good theory of weak interactions) it became clear that the theoretical expectation for the branching ratio ($\sim 10^{-4}$) exceeded the experimental upper limit by three orders of magnitude. This led to the *two neutrino hypothesis* : the neutrino coupled to the electron and the one coupled to the muon are of different nature. Like the electron and the muon, they carry different lepton numbers, electronic and muonic, and these partial lepton numbers are separately conserved. This hypothesis was rapidly confirmed by experiment. As a result, one had two distinct lepton families, and no communication between them (Fig. 2).

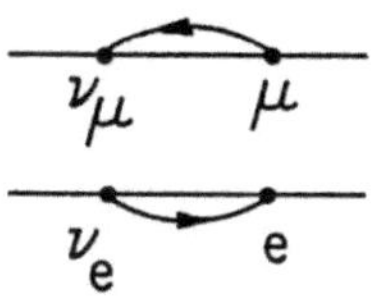

Fig. 2. In muon decay partial lepton numbers are conserved. There is no communication between the two lepton families

2 Three Generations of Quarks and Leptons

In the hadronic sector, one witnessed the discovery of strange particles, the advent of the SU(3)-based quark model, the discovery of charm. Quarks and leptons were recognized as being the building blocks of matter. Quarks differ from leptons in several respects (electric charges, confinement) and also because quarks mix via the Cabibbo matrix whereas leptons do not seem to mix. Later, this picture was extended with the discoveries of the τ lepton and b quark and the strong belief in the t quark. At present, we have evidence for three generations of quarks and leptons, but we do not know how many generations exist. Again we observe that quarks mix via the Cabibbo-Kobayashi-Maskawa matrix whereas no lepton mixing is evident. The experimental situation forces us to conclude that there is a conservation law for partial lepton numbers, electronic, muonic and tauonic. But, do we have an explanation for such a conservation?

3 The Standard Model and its Extensions

Nowadays, every particle physicist is familiar with the Standard Model of electroweak and strong interactions, based on the group structure SU(3) X SU(2) X U(1). This model is very successful[5] in explaining essentially all known experimental facts, with a minimal number of particles. In this model neutrinos are massless, therefore there is no lepton mixing. With only one Higgs doublet, no heavy neutral leptons, no other particles, the model does not possess the ingredients which could lead to lepton-flavor violation. Therefore, the conservation law for partial lepton numbers happens to be accidental in the Standard Model. The $\mu \rightarrow e\gamma$ decay, muon-to-electron conversion and similar processes should not exist, according to the Standard Model[6]. Since they are not observed, why should one worry?

The answer lies in the future of the gauge theories of elementary particles and their interactions. It is obvious that the Standard Model can not be the final theory of nature, although it could be the correct low-energy limit of a more fundamental theory. The Standard Model is too arbitrary, contains too many parameters and too many ad hoc assumptions. Many avenues are explored to build a more satisfactory theory: extension of the gauge group (additional fermions and gauge bosons), extension of the Higgs sector, left-right symmetric theories, grand unification, supersymmetry, supergravity, superstrings, composite quarks and leptons, etc... It is a general feature of these theories that the conservation of partial lepton numbers must be violated at some level. There exists an abundant literature on the subject and I find it convenient to refer to the review papers by Scheck[7], Vergados[8], Costa and Zwirner[9]. A strong theoretical activity was triggered in the beginning of 1977 by a rumor that the $\mu \rightarrow e\gamma$ had been observed at SIN. It was suddenly realized that many models could predict a $\mu \rightarrow e\gamma$ branching ratio of the order of 10^{-8} in a very natural way. The $\mu \rightarrow e\gamma$ and similar processes, $\mu \rightarrow e\gamma\gamma$, $\mu \rightarrow 3e$ and muon-to-electron conversion in a nucleus became the subject of intense experimental activities, mainly at the meson factories LAMPF, SIN and TRIUMF.

The importance of these processes is easily understood. They do not exist in the Standard Model, therefore they open a window on the physics beyond the Standard Model. Their discovery would be a major step forward and the beginning of a new era in theoretical and experimental physics. However, even if they are not found, upper limits on their branching ratios are extremely useful since they put stringent constraints on theoretical models. But why should one study all these processes and not just the $\mu \to e\gamma$ decay? It is because the various muon-number violating processes are strongly model dependent. As an example, a model has been developped[10], which predicts that the $\mu \to e\gamma\gamma$ decay is faster than the $\mu \to e\gamma$ decay, something which is a priori not obvious. The model dependence of the $\mu \to 3e$ to $\mu \to e\gamma$ ratio has been studied by several authors[11][12]. This ratio can vary by several orders of magnitude. Therefore, it is important to look for all lepton-number-violating processes because they can help in distinguishing between the numerous theoretical models.

One straightforward extension of the Standard Model consists in giving to the known neutrinos ν_e, ν_μ and ν_τ small masses (compatible with our present experimental knowledge). Mixing between lepton flavors can occur, giving rise to interesting phenomena like neutrino oscillations. But this is not sufficient to bring the rates of the lepton-number-violating decays to the level of observation. In fact, assuming two generations, the branching ratio of the $\mu \to e\gamma$ decay is given by[13]:

$$R(\mu \to e\gamma) = \frac{3\alpha}{32\pi} \sin^2\theta \cos^2\theta \frac{\left(m_1^2 - m_2^2\right)^2}{M_W^4}$$

where m_1 and m_2 are the neutrino masses, α the fine-structure constant, M_W the W-boson mass and θ the mixing angle in the lepton sector. Using the experimental upper limit on the heaviest neutrino ν_τ one calculates a branching ratio which is of the order of 10^{-18}. Of course, the situation would be different with heavy neutral leptons. Therefore the $\mu \to e\gamma$ and similar processes are sensitive to new particles not contained in the Standard Model and to a range of high masses.

4 Searches for Lepton-Flavor Violation at Meson Factories[14]

These studies were part of the mission of the meson factories, even if in the planning of these facilities the question of lepton-flavor violation was not so strongly emphasized. Experiments started in a rather modest way, using two sodium-iodide crystals to detect the electron and photon of the $\mu \to e\gamma$ decay. For a muon decaying from rest it is sufficient to measure the energies of both particles. One unavoidable source of background is the radiative muon decay $\mu \to e\nu\nu\gamma$, due to the finite energy resolution of the detectors. Another important source of background comes from accidental coincidences between ordinary muon decay and radiative muon decay. A measurement of the angle between electron and photon is redundant to define the kinematics but helps in reducing the background. At SIN, the sodium-iodide crystals were equipped with a wire chamber for the electron and an hodoscope for the photon. Such a device is limited to a sensitivity of $\sim 10^{-9}$ and therefore a second generation of experiments started with the construction of more complex detectors.

At LAMPF, the Crystal Box achieved on the $\mu \to e\gamma$[15] and $\mu \to e\gamma\gamma$[16] the best upper limits (in the following, all upper limits correspond to a 90% confidence level, unless otherwise indicated):

$BR(\mu \rightarrow e\gamma) \leq 4.9 \times 10^{-11}$

$BR(\mu \rightarrow e\gamma\gamma) \leq 7.2 \times 10^{-11}$

The Crystal Box is a big modular sodium-iodide assembly, providing a large solid angle and high detection efficiency for the γ rays. The positrons are tracked in a good-resolution wire chamber arrangement placed in a magnetic field. The apparatus allows a good measurement of the angle between positron and γ-ray.

At SIN, the detector SINDRUM1 has achieved the best upper limit on the $\mu \rightarrow eee$ decay[17]:

$BR(\mu \rightarrow eee) \leq 1.0 \times 10^{-12}$

SINDRUM1 consists of several cylindrical wire chambers placed in a magnetic field and surrounded by plastic scintillator hodoscopes for triggering. Good timing resolution is essential as well as a good reconstruction of the vertex.

At TRIUMF, the first operating *Time Projection Chamber* [18] was built to study muon-to-electron conversion in Titanium. One looks for the process in which the nucleus remains in its ground state. The signature is a monoenergetic electron with a kinetic energy $E_e = m_\mu - B - R$, where B is the binding energy of the muonic atom and R the nuclear recoil energy (negligible). To a good approximation $E_e = m_\mu \sim 106$ [MeV]. This process has some advantages over the muon decays. The nuclear quarks act coherently, leading to an increase of the transition probability. The electron energy of 106 [MeV] is far above the end-point energy of the muon decay spectrum. The background comes mainly from the muon decay in orbit: since the nucleus can participate in the momentum-energy sharing, the electron can (with a small probability however) take an energy up to 106 [MeV]. Another source of background is due to muon radiative capture. The photon can convert externally to give an asymmetric e^+-e^- pair, with an electron having an energy close to 106 [MeV]. Other background sources are cosmic rays and residual pions in the muon beam. The best result on muon-to-electron conversion has been obtained with the TPC at TRIUMF[19]:

$BR(\mu^- \rightarrow e^-) \leq 4.6 \times 10^{-12}$

The same experiment has given an upper limit for the branching ratio of muon-to-positron conversion $\mu^-\,Ti \rightarrow e^+\,Ca$[20]:

$BR(\mu^- \rightarrow e^+) \leq 9.0 \times 10^{-12}$ for $E_e \geq 96$ [MeV]

and

$BR(\mu^- \rightarrow e^+) \leq 1.7 \times 10^{-10}$ (integrated over all positron energies).

But this process is not coherent; there is some nuclear rearrangement and this leads to many excited states in the final nucleus. Therefore the previous integrated limit is based on the use of a nuclear model (giant-resonance-excitation model) and it is subject to some ambiguity.

Using a high-statistics muon-decay spectrum, a TRIUMF group has searched for the exotic decay $\mu \rightarrow eX$ where X is a particle with a mass in the range 0 to 45 [MeV]. They have also analyzed other experiments and they present an upper limit for the decay $\mu \rightarrow eX$ in the mass range 0 to 104 [MeV][21].

So far these forbidden processes have not been seen, they are still searched for, but how long can one still go? Obviously there is a hard limit which is not very far. Several on-going or proposed experiments are aiming at improvements in sensitivity by factors of 10 or even 100. The MEGA (MuEGAmma) experiment at LAMPF (a large US collaboration) will search for the $\mu \rightarrow e\gamma$ decay at a level of $1\text{-}2 \times 10^{-13}$ for the branching ratio[22]. The detector must combine extremely good energy resolution, high detection efficiency and the capability of handling high rates (the muon stopping rate will be 3×10^7 [s^{-1}] at a duty cycle of 6%). The good energy resolution will be achieved by tracking the positrons and the electron-positron pairs from photon conversion in a large magnetic spectrometer. Four concentric conversion layers will allow a good detection efficiency for the photons. The very powerful multimicroprocessor system developped at Fermilab (Advanced Computer Program) will be used for data acquisition. Engineering runs should take place in the fall of 87 and the summer of 88. Data taking is expected in 89-90[23]. At SIN, there is a letter of intention[24] for the $\mu \rightarrow e\gamma$ decay, but the priority has been given to a new search for muon-to-electron conversion[25], using an upgraded version of the SINDRUM detector, SINDRUM2. The experimental difficulties are enormous. For instance, the pion contamination in the beam must be as low as 10^{-9}. At high rates it is not possible to play the usual tricks to veto the beam pions, and one has to get rid of the pions in the muon beam itself.

Until now there are not very good data on muonium-to-antimuonium conversion, but two experiments are aiming at a better sensitivity. At SIN, a Heidelberg-Yale collaboration has proposed to look for muonium-antimuonium conversion in vacuum[26]. Assuming that the process goes via a four-fermion interaction with a coupling constant G_m they hope to be sensitive to a ratio G_m/G_F of the order of one (G_F is the usual Fermi constant as obtained from normal muon decay). At TRIUMF, an experiment uses a radiochemical technique and aims at the same sensitivity[27]. Positive muons are stopped in SiO_2 powder where muonium is formed. It diffuses into vacuum, converts to antimuonium, producing a negative muon which is allowed to capture on Tungsten. This gives ^{184}Ta, which has an 8.7 [h] half-life. The resulting activity is counted off-line in a low-background environment. Muonium-to-antimuonium conversion is forbidden by the additive conservation law of partial lepton numbers but is allowed by the multiplicative law. There are experimental predictions with G_m/G_F larger than one[28][29], claiming for new experimental results.

5 Searches in Other Sectors

Tau decays have been extensively studied at the DORIS II storage ring (DESY) with the ARGUS detector. These experiments have not yet reached the levels of sensitivity achieved at the meson factories but they are already providing useful constraints on theoretical models. They are interesting because they are complementary to the searches at meson factories: they test lepton-flavor violation and lepton-number violation in a different sector (involving the third generation). Also, because of the higher mass of the τ lepton, they offer a larger variety of decays and they can test more violation mechanisms. For instance, in the $\tau^- \rightarrow \mu^- K^{*o}$ decay the generation number does not change since the decrease of -1 in the lepton sector is compensated by the increase of +1 in the quark sector (Fig. 3).

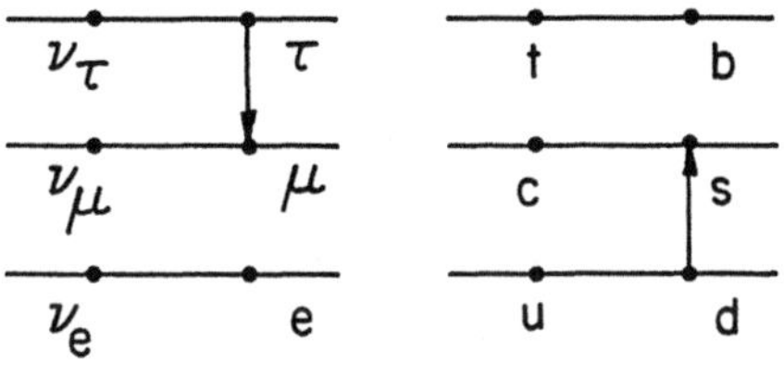

Fig. 3. The $\tau^- \to \mu^- K^{*o}$ decay

Using e^+e^- annihilation at center of mass energies near 10 [GeV], the ARGUS collaboration has obtained the following results[30], which are compared to the MARK II results [31] (Table 1).

Table 1 Upper limits (90% C.L.) on forbidden τ decays[32]

decay mode	ARGUS	MARK II
$\tau^- \to e^-e^+e^-$	3.8×10^{-5}	4.0×10^{-4}
$\tau^- \to e^-\mu^+\mu^-$	3.3×10^{-5}	3.3×10^{-4}
$\tau^- \to \mu^-e^+e^-$	3.3×10^{-5}	4.4×10^{-4}
$\tau^- \to \mu^-\mu^+\mu^-$	2.9×10^{-5}	4.9×10^{-4}
$\tau^- \to \ell^-+\ell^+-\ell^-$	3.8×10^{-5}	
$\tau^- \to e^-\pi^+\pi^-$	4.2×10^{-5}	
$\tau^- \to \mu^-\pi^+\pi^-$	4.0×10^{-5}	
$\tau^- \to e^-\rho^o$	3.9×10^{-5}	3.7×10^{-4}
$\tau^- \to \mu^-\rho^o$	3.8×10^{-5}	4.4×10^{-4}
$\tau^- \to \ell^-+\pi^+-\pi^-$	6.3×10^{-5}	
$\tau^- \to e^-\pi^+K^-$	4.2×10^{-5}	
$\tau^- \to \mu^-\pi^+K^-$	1.2×10^{-4}	
$\tau^- \to e^-K^{*o}$	5.4×10^{-5}	(e^-K^o) 1.3×10^{-3}
$\tau^- \to \mu^-K^{*o}$	5.9×10^{-5}	(μ^-K^o) 1.0×10^{-3}
$\tau^- \to \ell^-+\pi^+-K^-$	1.2×10^{-4}	

The MARK III Collaboration has obtained upper limits on the decays $\tau \to eX^o$ and $\tau \to \mu X^o$, where X^o is a light weakly interacting particle[33]. If X^o is a massless Goldstone boson G, one gets the following limits (95% C.L.):

$$B(\tau \to \mu G)/B(\tau \to \mu\nu\nu) \leq 12.5\% \quad \text{and} \quad B(\tau \to eG)/B(\tau \to e\nu\nu) \leq 4.0\%$$

from which one deduces a lower limit on the mass scale of the corresponding symmetry breaking:

$$M \geq 3 \times 10^6 \text{ [GeV]}.$$

Several groups have searched for the forbidden reaction $D^o \to \mu e$. The MARK II group has published an upper limit of 2.1 x 10^{-3} for the branching ratio[34]. The ACCMOR group at CERN, using a 200 [MeV] negative pion beam incident on a fixed Be target and very good microstrip silicon detectors, has obtained an upper limit of 9 x 10^{-4}[35]. An even better (and model-independent) upper limit has been obtained at SLAC by the MARK III group, 1.5 x 10^{-4}[36]. They used e^+-e^- collisions at the ψ (3770 [MeV]) resonance. These D^o decays imply a change in family number both in the quark and lepton sectors .

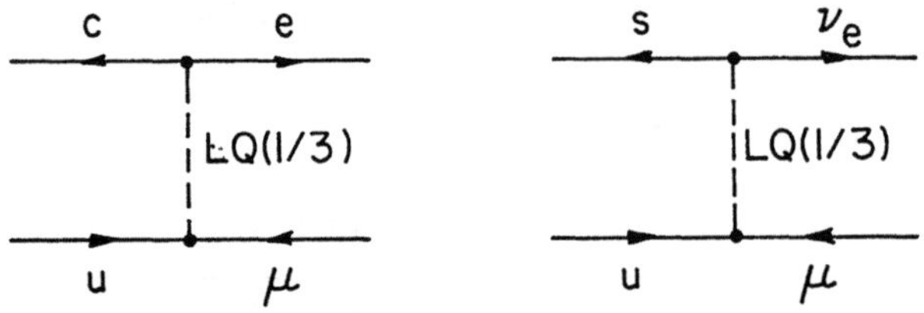

Fig. 4 The $D^o \to \mu e$ and $K^o \to \mu e$ decays via leptoquark exchange

They could be induced by massive leptoquarks (Fig. 4) which are predicted in a variety of unification models. In certain models, the leptoquarks couple "up-type" quarks (u, c, t) to charged leptons and "down-type" quarks (d, s, b) to neutral leptons. Then the $D^o \to \mu e$ decay should be favored over the $K^o \to \mu e$ and $B^o \to \mu e$ decays. In any case, it is difficult to suppress all these decays simultaneously. If Higgs bosons were responsible for lepton-flavor violation, they should be more strongly coupled to the charmed quarks than to the strange quarks. Therefore the forbidden D^o decays could be more sensitive than the K^o decays. There is also an upper limit on the $B^o \to \mu e$ decay, obtained at CLEO [37]:

$$BR(B^o \to \mu e) \leq 0.9 \times 10^{-4}$$

The kaon sector is presently being investigated very actively. Several experiments, at Brookhaven and KEK, aim at sensitivities of 10^{-10}[38], 10^{-11}[39] and 10^{-12}[40] for the $K_L \to \mu e$ decay. The present situation is somewhat confused. An experiment[41] gave an upper limit of 1.6 x 10^{-9} on the $K_L \to \mu e$ branching ratio. But it failed to observe the $K_L \to \mu^+\mu^-$ decay and gave for its branching ratio a upper limit of 1.8 x 10^{-9}, a factor of 5 below the measured branching ratio which is 9.1 x 10^{-9}. The upper limit of 6 x 10^{-6} quoted in the Table of Particle Properties[42] is a conservative estimate, but some experts think that the experiment should have seen the effect at the 10^{-8} level[43]. On the $K \to \pi\mu e$ mode we still have the upper limits of Diamant-Berger et al.[44]:

$$B(K^+ \to \pi^+\mu^+e^-) \leq 5 \times 10^{-9} \quad \text{and} \quad B(K^+ \to \pi^{+-}\mu^{-+}e^+) \leq 7 \times 10^{-9}$$

A proposed BNL experiment aims at a sensitivity of 10^{-11}[45]. The $K^+ \to \mu^+\nu_e$ decay has been searched for[46] by looking for the ν_e interactions in a neon/hydrogen bubble chamber. The upper limit on the branching ratio is 0.4%.

6 A Very Arbitrary Selection of Theoretical Topics

It is impossible, in a short paper, to review the entire theoretical literature. Until 1986 the field has been well covered by the review papers mentioned above[47]. In fact, lepton-flavor-violating processes have been discussed abundantly and many mechanisms have been considered. Here I will only review some recent developments.

A straightforward generalization of the Standard Model introduces heavy neutral leptons. One keeps the general structure of the model, allowing for massive neutrinos[48] and mixing between different lepton flavors. Additional generations of quarks and leptons could provide these heavy neutrinos. Using the data on $\mu \to e\gamma$, etc... and muon-to-electron conversion, it is possible to put limits on heavy neutrino masses and mixing matrix elements[49]. Another possibility lies in theories with right-handed currents, where the right-handed charged lepton is put in a doublet with a heavy neutral partner. Heavy neutral leptons with masses in the multi-[GeV] range could bring the $\mu \to e\gamma$ decay to the level of observation[50]. Some models have also introduced heavy doubly-charged leptons[51].

It has been shown a long time ago that with two Higgs doublets lepton-flavor violation becomes possible[52]. One can also introduce other Higgs multiplets, as in the *Majoron Models* [53][54]. In fact, there are several good reasons to enlarge the Higgs sector. For instance, it is an economical way to give Majorana masses to the neutrinos[55]. It also happens naturally in left-right symmetric theories and there are consequences for the various lepton-flavor violating decays[56] (Fig. 5).

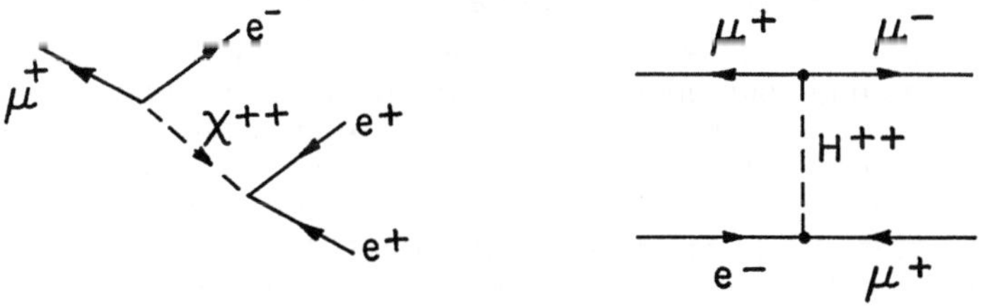

Fig. 5. The decay $\mu \to eee$ and muonium-antimuonium conversion via doubly-charged Higgs exchange

These theories have obvious advantages: they can account for parity violation in the low-energy domain, they naturally give rise to neutrino masses. They are more aesthetic and less arbitrary than the left-handed theories. Bilenky and Petcov have discussed lepton-flavor violation in the Triplet Majoron Model[57]. One interesting feature is the occurrence of the $\mu \to 3e$ decay with a branching ratio larger than that of the $\mu \to e\gamma$ and possibly close to the present upper limit. Santamaria et al.[58] have studied the influence of the scalar particles of the model on the parameters of muon decay and have also considered the exotic decays $\mu \to e\theta\theta$, $\mu \to e\rho_L\rho_L$ and $\mu \to e\rho_L\theta$, where θ is the massless Goldstone boson associated with the breaking of lepton number, the Majoron, and ρ_L a light scalar whose mass is proportional to the triplet vacuum expectation value. A model with three Higgs doublets has been studied by Cheng and Sher[59].

In several unification schemes one introduces leptoquarks which can also contribute to lepton-flavor violation (Fig. 6). Limits on leptoquark masses and couplings have been discussed by several authors[60][61].

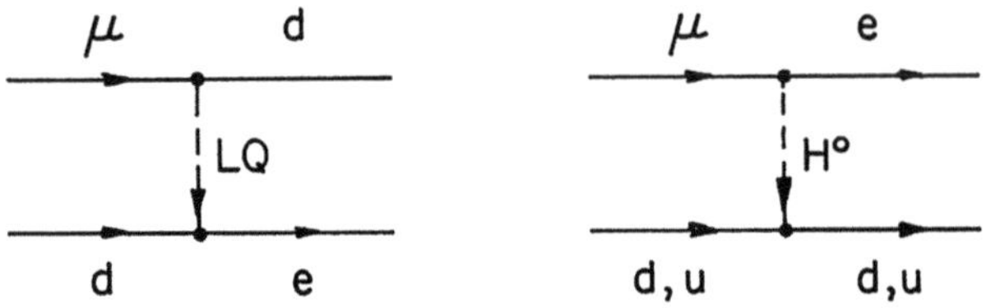

Fig. 6. Muon-to-electron conversion via leptoquark or Higgs exchange

The recurrence of several quark and lepton generations strongly suggests that these particles are made of more elementary constituents[62]. In this context also the lepton-number-violating processes are important[63][64]. It is conceivable that e, μ and τ are radial excitations ($1s_{1/2}$, $2s_{1/2}$ and $3s_{1/2}$) of a system of bound preons. In such a model the muon could decay by double E_1 transition to the electron rather than by a single M_1 transition (Fig. 7). The two-gamma transition involves a virtual P state. With the $\mu \rightarrow e\gamma\gamma$ it is possible to put stringent limits on the mass of the P states in a potential model of composite leptons.

The existence of several quark and lepton families could be related to the presence of a new symmetry, called family symmetry. This symmetry is of course broken and if it is a global symmetry there should appear massless Goldstone bosons. They have received the name of *familons* [65]. These particles could be important for cosmology and also could serve as axions to solve the "Strong CP Problem"[66]. Familons are responsible for $s \rightarrow df$ and $\mu \rightarrow ef$ transitions and can be searched for in $K \rightarrow \pi f$ and $\mu \rightarrow ef$ decays. They would manifest themselves in the $\mu \rightarrow ef$ decay as a peak in the electron energy spectrum, at the end-point energy. The corresponding electrons would also have a longitudinal polarization less than one[67]. A nice experimental possibility results from the fact that with positive muons, at the end-point energy, no positrons are emitted at 180° with respect to the muon spin if one assumes V-A coupling (for helicity reasons). An experiment was done at TRIUMF to search for right-handed currents[68]. It has also been analyzed to extract an upper limit for familon emission. The result can be used to extract a lower limit for the scale of family-symmetry breaking, $F \geq 9.9 \times 10^9$ [GeV]. Another limit has been obtained by a LAMPF group[69]. They have analysed their $\mu \rightarrow e\gamma$ data[70] to search for the $\mu \rightarrow e\gamma f$ decay. They obtain for F a lower limit of 2.9×10^9 [GeV]. A similar analysis has been performed by Bélanger and Ng[71]. In a more general way the role of various scalar particles in the $\mu \rightarrow e\gamma$ decay and other observables has been studied by Lopes and Martins Simoes[72].

Horizontal gauge groups have been introduced by several authors to connect the generations of quarks and leptons. Rare decays provide constraints on *horizontal gauge bosons* and their interactions[73]. Recently, horizontal gauge interactions have been discussed in the context of CP violation. Predictions for lepton-flavor violating decays of muons, kaons, D and B mesons have been made[74] and they show interesting rates.

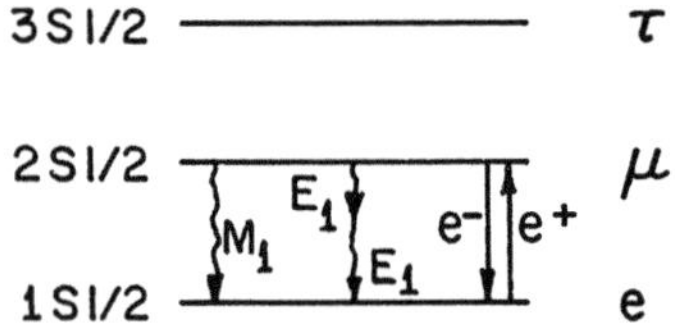

Fig. 7. The decays $\mu \rightarrow e\gamma$, $\mu \rightarrow e\gamma\gamma$ and $\mu \rightarrow eee$ in a composite model of leptons

Following the discovery of e^+-e^- peaks in heavy ion collisions at GSI[75], the search for axions has been revived. Rare muon decays can be used for that purpose. The $\mu \to e\gamma\gamma$ data had been analyzed to search for $\mu \to ea$, followed by $a \to \gamma\gamma$[76]. More recently, at SIN, a measurement of the $\mu \to eee$ decay has been analyzed to put a limit on the decay $\mu \to ea$ followed by $a \to e^+e^-$. Combining this result with the study of the $\pi^+ \to e^+\nu_e e^+e^-$ decay, the SIN group can rule out the 1.7 [MeV] axion which had been postulated to explain the GSI results[77].

When discussing possible deviations from the Standard Model it is convenient to parametrize these deviations with a scale parameter Λ which is simply related to the distance within which new interactions come into play[78][79]. Rates for the various forbidden muon decays can be expressed as a function of Λ and useful lower limits on this parameter can be obtained. For instance, using the work of Buchmüller and Wyler one calculates the following lower bounds on Λ (in [TeV]): 14000, 175 and 340 from the $\mu \to e\gamma$, $\mu \to 3e$ decays and muon-to-electron conversion, respectively.

Majoron emission in μ and τ decays has been studied by Santamaria and Valle[80] in the context of spontaneous R-parity breaking in supergravity. Their main motivation was the solar neutrino problem and the generation of neutrino masses, but their model is also consistent with a branching ratio for the $\mu \to ef$ decay which is close to the present experimental upper limit.

Supersymmetry has been introduced to cure several problems in unification theories and also comes naturally from superstrings. Supersymmetric partners of ordinary particles can induce lepton-flavor violation (Fig. 8). Rare processes can be used to put useful limits on flavor-violating interactions induced by supersymmetry[81]. Ellis and Nanopoulos have used the $\mu \to e\gamma$ decay to put constraints on slepton mass differences[82].

Supersymmetry and compositeness can be combined to yield supercomposite models, which are characterized by two different energy scales: M_S, where supersymmetry breaking occurs, and M_C, the scale of compositeness. Campbell and Peterson have studied the limits which can be placed on these energy scales by using. experimental results on μ-e universality (the $\pi \to e\nu/\pi \to \mu\nu$ ratio) and muon-to electron conversion[83]. Using for the latter the TRIUMF result on Ti (upper limit of 4.5×10^{-12}) one obtains the constraint:

$$\frac{C}{4\pi^2}\frac{M_Z^2}{M_C M_S} \leq 1.7 \times 10^{-7}$$

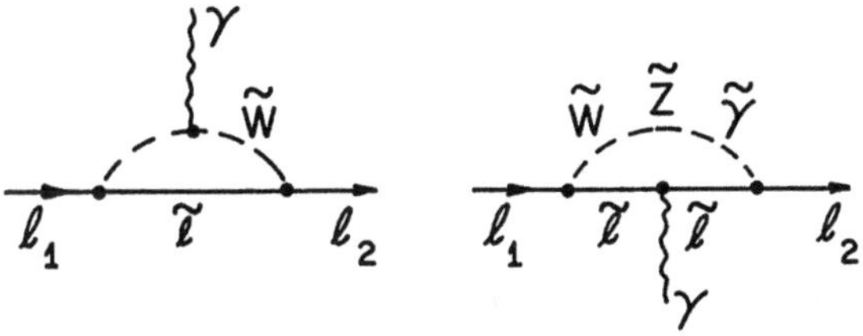

Fig. 8 Some diagrams for $\ell_1 \to \ell_2\gamma$ decay and and Z^0 decay via supersymmetric particles

where C is calculable in the model (0.143) and M_Z is the Z^o mass. Assuming that M_S is not very different from the Fermi scale, one obtains for M_C the bound:

$$M_C > 2.3 \times 10^6 \text{ [GeV]}$$

A slightly lower value is deduced from μ-e universality. The authors conclude that under the assumption that supersymmetry breaking occurs at the Fermi scale the supercompositeness scale must exceed 100 TeV. In another paper, rare decays have been discussed in the context of an effective-Lagrangian study of three generations of supersymmetric composite quarks and leptons[84].

There is presently, among theorists, a great deal of enthusiasm for *superstring theories*, which look very promising and could, according to the optimists, lead to the "Theory of Everything". There is, however, a lot of skepticism about the possibility to test these theories in the low-energy domain, the one which is presently accessible to experiments, or will be in the near future. But one should recognize that some efforts have been made to derive low-energy phenomenology from the superstrings. There is still a considerable confusion on how to do that and there is no unique and universally accepted way to come down from the Planck mass to the laboratory energies. The superstring-inspired E_6 unification group has been largely exploited[85] (in fact it has always been an attractive possibility). It introduces a sector of particles associated with the 27 representation. In addition to the fermions of the Standard Model one has two Higgs doublets H and $\overline{H}$, two colour field triplets D and D^c and two neutral fermions: ν^c is a right-handed neutrino already present in SO(10) and N is an SO(10) singlet. New interactions appear, which can contribute to lepton-flavor violation. Campbell et al.[86] have made a complete analysis of the lepton-flavor-violating decays $K_L \to \mu e$, $K^+ \to \pi^+\mu e$, $D^o \to \mu e$, $D^o \to \mu e X$, $B^o \to \mu e$, $\mu^-(A,Z) \to e^-(A,Z)$, $\mu \to 3e$, $\mu \to e\gamma$, together with other rare decays and other fundamental observables: mixing in the neutral K, D and B sectors, neutron dipole moment, etc... Whether evidence for superstrings can be obtained from lepton-flavor-violating processes represents a formidable challenge for theorists and experimentalists. The paper by Campbell et al. shows that the rare processes, especially the $\mu \to e\gamma$ decay and muon-to-electron conversion, provide powerful constraints on masses and coupling constants. The role of the E_6 exotic particles has also been discussed by Masiero et al.[87]. They are active in such processes as the rare kaon decays, the $\mu \to e\gamma$ decay and muon-to-electron conversion, and the $\pi^o \to \mu e$ decay. Borzumati and Masiero[88] have studied renormalization effects in the scalar-lepton mass matrix in spontaneously broken N = 1 supergravity theories. They show that photino-lepton-slepton couplings can lead to appreciable rates for $\mu \to e\gamma$ and muon-to-electron conversion. In these two cases the present upper limits on rare muon processes can be used to put useful limits on the parameters of the model (masses and coupling constants). Lepton-number nonconservation in E_6 superstring models has been discussed by Ma[89]. Various scenarios are considered together with their phenomenological implications.

Interesting connections can be made between lepton-flavor violation and other fundamental observables. For instance, in a wide class of models[90], the $\mu \to e\gamma$ decay rate can be related to the magnitude of the electric dipole moment of the electron. The corresponding CP violation originates in the phases between several flavor-changing neutral currents[91]. The predicted upper limits on the electric dipole moments are on the 10^{-26} [e-cm] scale, still a factor of ~ 100 below the present experimental limit[92].

An interesting question is: will it be possible to observe lepton-flavor violation at the forthcoming e^+-e^-colliders? Eilam and Rizzo[93] have considered the lepton-flavor-violating decay $Z^o \rightarrow \mu\tau$. Such a decay could result from the mixing of the new E_6 fermions with the ordinary quarks and leptons. The conclusion is that the $Z^o \rightarrow e\mu$ has a too small branching ratio to be seen ($\leq 7.2 \times 10^{-13}$) as constrained by the $\mu \rightarrow eee$ decay) whereas there is still hope to see the $Z^o \rightarrow e\tau$ decay ($\leq 3.4 \times 10^{-4}$) and the $Z^o \rightarrow \mu\tau$ decay ($\leq 4.2 \times 10^{-4}$), where the constraints come from the upper limits on the $\tau \rightarrow 3\mu$, $3e$, $2\mu e$ and $2e\mu$ decays.

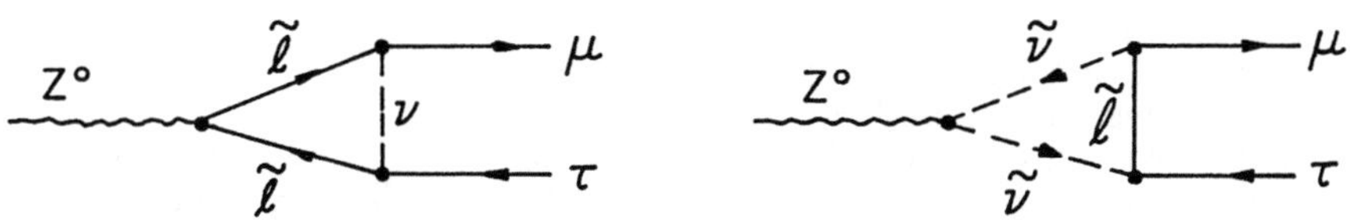

Fig. 9. Some diagrams for the $Z^o \rightarrow \mu\tau$ decay via sleptons

The $Z^o \rightarrow \mu\tau$ decay via slepton mixing has been considered by Levine[94]. He uses softly-broken supersymmetry with flavor mixing in the supersymmetric sector (Fig. 9). The calculated branching ratio for the forbidden $Z^o \rightarrow \tau^+\mu^-$ decay is of the order of 10^{-6}-10^{-7}. With this mechanism alone there is no hope to observe this process. Bernabéu and Santamaria have studied lepton-flavor-violating decays of the Z^o in the Scalar Triplet Model[95]. There is no GIM supression in their model and they find for the $Z^o \rightarrow \mu\tau$ and $Z^o \rightarrow e\tau$ decays the following branching ratios:

$$R(Z^o \rightarrow \mu\tau) \leq 5.9 \times 10^{-6}$$

$$R(Z^o \rightarrow e\tau) \leq 7.9 \times 10^{-6},$$

still difficult to detect. Using a superstring-inspired model with new neutral fermions, Bernabéu et al.[96] have predicted branching ratios in the 10^{-5} range, accessible to experiments at SLC or LEP. But it should be noted that these predictions rest on the present experimental upper limits for the $\tau \rightarrow \mu\gamma$ and $\tau \rightarrow e\gamma$ decays, therefore one must exert some caution.

7 Conclusion

The search for lepton-flavor-violating processes and the corresponding theoretical activity have played a useful role in the quest for "new physics", i.e. physics which lies beyond the Standard Model. These processes are sensitive to mass scales which can not be reached by the present and even future accelerators. The meson factories will soon complete they mission to search for rare muon and pion decays. The more distant future belongs to the kaon, charm and beauty factories, and to the large colliders. In view of the present experimental situation there is still some hope that lepton-flavor violation will be observed but it is also possible that it will escape all human efforts to detect it. Whatever the outcome is, lepton-flavor violation will remain as a beautiful chapter in the history of physics.

References

1. See for instance S. Frankel, *"Rare and Ultrarare Muon Decays"* , in "Muon Physics", V.W. Hughes and C.S. Wu, Editors, Academic Press, New York (1975)
2. E.P. Hincks and B. Pontecorvo, Phys. Rev. 73, 257 (1948)

3. A. Lagarrigue and C. Peyrou, C.R. Heb. Acad. Sci. Paris 234, 873 (1952)
4. L. Michel, Proc. Phys. Soc. A63, 514 (1950); see also C. Bouchiat and L. Michel, Phys. Rev. 106, 170 (1957)
5. P. Langacker, these proceedings
6. See reference 9, page 12
7. F. Scheck, Phys. Rep. 44, 187 (1978)
8. J.D. Vergados, Phys. Rep. 133, 1 (1986)
9. G. Costa and F. Zwirner, Nuovo Cimento Rivista 9, 1 (1986)
10. J.D. Bowman et al., Phys. Rev. Lett. 41, 442 (1978)
11. S.M. Barr and S. Wandzura, Phys. Rev. D16, 707 (1977)
12. W.J. Marciano and A.I. Sanda, Phys. Letters B67, 303 (1977)
13. S.M. Bilenky et al., Phys. Letters B67, 309 (1977)
14. R. Engfer and H.K. Walter, Ann. Rev. Nucl. Part. Sci. 36 , 327 (1986); see also the contributions of R. Engfer, R. Mischke and P. Depommier in "Proceedings of the International Symposium on Weak and Electromagnetic Interactions", Heidelberg, July 1986, Editor H.V. Klapdor, Springer Verlag; see also J. Schacher, Comments Nucl. Part. Phys., 8, 97 (1978)
15. R.D. Bolton et al., Phys. Rev. Lett. 56, 2461 (1986)
16. D. Grosnick et al., Phys. Rev. Lett. 57, 3241 (1986)
17. U. Bellgardt et al., SIN preprint, (September 4, 1987)
18. D. Bryman et al., in "The Time Projection Chamber", J.A. Macdonald, editor, AIP Conf. Proc. Number 108, American Institute of Physics, (1983)
19. S. Ahmad et al., Phys. Rev. Lett. 59, 970 (1987)
20. See reference 19
21. D.A. Bryman and E.T.H. Clifford, Phys. Rev. D57, 2787 (1986)
22. M.D. Cooper, Spokesman, LAMPF Experiment N°. 969, (1985)
23. M.D. Cooper, private communication
24. W. Bertl et al., *"Search for the decay $\mu \to e\gamma$"*, Letter of intent for an experiment at SIN, (February 1985)
25. A. Badertscher et al., *"Search for $\mu^- \to e^-$ conversion"*, Proposal for an experiment at SIN, (November 1986)
26. K.P. Arnold et al., *"Search for Muonium-Antimuonium Conversion in Vacuum"*, Proposal for an experiment at SIN, (January 1985)
27. A. Olin et al., TRIUMF- Research Proposal E304, (April 1987)
28. E. Derman, Phys. Rev. D19, 317 (1979)
29. A. Halprin, Phys. Rev. Lett. 48, 1313 (1982)
30. H. Albrecht et al., Phys. Letters B185, 228 (1987)
31. K.G. Hayes et al., Phys. Rev. D25, 2869 (1982)
32. See reference 30
33. R.M. Baltrusaitis et al., Phys. Rev. Lett. 55, 1842 (1985)
34. K. Riles et al., Phys. Rev. D35, 2914 (1987)
35. H. Palka et al., Phys. Letters B189, 238 (1987)
36. J.J. Becker et al., Phys. Letters B193, 147 (1987)
37. P. Avery et al., Phys. Letters B183, 429 (1987)
38. M.P. Schmidt, W.M. Morse et al., *"A search for flavor-changing neutral currents $K_L \to \mu e$ and $K_L \to e^+ e^-$"*, BNL experiment AGS-780
39. T. Inagaki et al., KEK proposal 85-1 (1985)
40. S.G. Wojcicki et al., *"Study of very rare K_L decays"*, BNL experiment AGS-791
41. A.R. Clark et al., Phys. Rev. Lett. 26, 1667 (1971)
42. Particle Data Group, Phys. Letters B170, 1 (1986)
43. D. Bryman, in *"New Frontiers in Particle Physics"*, Proc. First Lake Louise Institute, J.M. Cameron et al., editors, World Scientific, Singapore (1986), p.353
44. A.M. Diamant-Berger et al., Phys. Letters B62, 485 (1976)
45. M. Zeller et al., AGS Experiment 777
46. L. Lyons et al., Z. Phys. C10, 215 (1981)

47. See reference 14
48. R.E. Shrock, Phys. Rev. D24, 1232 (1981)
49. T. Numao, Phys. Rev. D34, 2900 (1986)
50. T.P. Cheng and L.F. Li, Phys. Rev. Lett. 38, 381 (1977); Phys. Rev. D16, 1425 (1977)
51. F. Wilczek and A. Zee, Phys. Rev. Lett. 38, 531 (1977)
52. J.D. Bjorken and S. Weinberg, Phys Rev. Lett. 38, 622 (1977)
53. G.B. Gelmini and M. Roncadelli, Phys. Letters B99, 411 (1981)
54. Y. Chikashige et al., Phys. Letters B98, 265 (1981)
55. See reference 53
56. R.N. Mohapatra and P.B. Pal, Phys. Lett. B179, 105 (1986)
57. S.M. Bilenky and S.T. Petcov, Rev. Mod. Phys. 59, 671 (1987)
58. A. Santamaria et al., Phys. Rev. D36, 1408 (1987)
59. T.P. Cheng and M. Sher, Phys. Rev. D35, 3484 (1987)
60. O. Shanker, Nucl. Phys. B206, 253 (1982)
61. W. Buchmüller and D. Wyler, Phys. Letters B177, 377 (1986)
62. L. Lyons, *"An Introduction to the Possible Substructure of Quarks and Leptons"*, in Prog. Nucl. Part. Phys., D. Wilkinson, Ed., Pergamon Press, Vol. 10, 227
63. Y. Tomozawa, Phys. Rev. D25, 1448 (1982)
64. V. Visnjic-Triantafillou, Phys. Letters B95, 47 (1980); Phys. Rev. D25, 248 (1982)
65. F. Wilczek, Phys. Rev. Lett. 49, 1549 (1982)
D. Reiss, Phys. Letters B115, 217, (1982)
G. Gelmini et al., Nucl. Phys. B219, 31 (1983)
66 D. Chang and G. Senjanovic, Phys. Letters B188, 231 (1987)
67. A.A. Anselm et al., Leningrad Institute of Nuclear Physics, Preprint N° 1034 (February 1985)
68. A. Jodidio et al., Phys. Rev. D34, 1967 (1986)
69. T. Goldman et al., Phys. Rev. D36, 1543 (1987)
70. See reference 15
71. G. Bélanger and J. Ng, TRIUMF Preprint TRI-PP-86-39 (June 1986)
72. J.H. Lopes and J.A. Martins Simoes, Phys. Rev. D35, 3428 (1987)
73. O. Shanker, Nucl. Phys. B185, 382 (1981)
74. Wei-Shu Hou and A. Soni, Phys. Rev. D35, 2776 (1987)
75. T. Cowan et al., Phys. Rev. Lett. 56, 444 (1986)
76. See reference 10
77. R. Eichler et al., Phys. Lett. B175, 101 (1986)
78. W. Buchmüller and D. Wyler, Nucl. Phys. B268, 621 (1986)
79. I.I. Bigi et al., Phys. Letters B166, 238 (1986)
80. A. Santamaria and J.W.F. Valle, Lawrence Berkeley Laboratory Preprint LBL-23896 and University of Valencia Preprint FTUV-7/87; see also A. Santamaria and J.W.F. Valle, Phys. Letters B195, 423 (1987)
81. B.A. Campbell, Phys. Rev. D28, 209 (1983)
82. J. Ellis and D.V. Nanopoulos, Phys. Letters B110, 44 (1982)
83. B.A. Campbell and K.A. Peterson, Phys. Letters B192, 401 (1987)
84. R.R. Volkas and G.C. Joshi, Phys. Rev. D35, 1050 (1987)
85. see for instance J.W.F. Valle, Phys. Letters B196, 157 (1987) and references therein
86. B.A. Campbell et al., Int. Journal Mod. Phys. A2, 831 (1987)
87. A. Masiero et al., Phys. Rev. Lett. 57, 663 (1986)
88. F. Borzumati and A. Masiero, Phys. Rev. Lett. 57, 961 (1986)
89. E. Ma, Phys. Letters B191, 287 (1987)
90. S.M. Barr and A. Masiero, Phys. Rev. Lett. 58, 187 (1986)
91. See reference 79
92. Y. Kizukuri, Phys. Letters B185, 183 (1987)
93. G. Eilam and T.G. Rizzo, Phys. Letters B188, 91 (1987)
94. M.J.S. Levine, Phys. Rev. D36, 1329 (1987)
95. J. Bernabéu and A. Santamaria, Phys. Letters B197, 418 (1987)
96. J. Bernabéu et al., Phys. Letters B187, 303 (1987)

An Upper Limit for the Electron Antineutrino Mass from Tritium β-Decay

W. Kündig, M. Fritschi, E. Holzschuh, R.E. Pixley, and H. Stüssi

Physics Institute, University of Zurich, CH-8001 Zurich, Switzerland

The endpoint region of the tritium β-spectrum has been measured with 27 eV resolution, using a magnetic spectrometer. The tritium activity was implanted into a thin layer of carbon. The neutrino mass determined is consistent with zero with an upper limit of 18 eV/c^2, which includes instrumental and statistical uncertainties as well as uncertainties due to energy loss in the source and the final electronic states. This result is in contradiction to the finite mass of 30 eV/c^2 reported by the ITEP-group. In the discussion a review of the present status of the various tritium experiments is given.

1. Introduction

The question of a nonzero neutrino mass has received considerable attention since Lubimov *et al.* [1] reported in 1980 evidence for a finite electron antineutrino mass of 35 eV/c^2. The importance of this result has prompted many groups to start measurements aimed at checking the ITEP value (Institute of Theoretical and Experimental Physics in Moscow).

On Feb. 24 1987 the first naked-eye supernova since the invention of the telescope has been discovered in the Large Magellanic Cloud. The $\approx M_\odot/30$ of neutrinos emitted during the ≈ 1 s event caused neutrino 'bursts' in the Kamiokande-, the IMB- (Irvine-Michigan-Brookhaven), the Baksan and the Mont Blanc-detectors. As to now many questions about the neutrino observations, like the 4h 43' time difference between the Mont Blanc- and the Kamiokande Detector, are still open. From the observation that the low energetic neutrinos were somewhat delayed and the assumption that the ν production time is small compared to the observed time spread, it is possible to estimate a mass limit:

$$m_\nu c^2 \simeq \sqrt{\frac{2\,c\,(t_1 - t_2)\,E_2^2\,E_1^2}{d\,(E_2^2 - E_1^2)}}\,,$$

where E_i are the neutrino energies, t_i the corresponding traveling times, and $d = 163000$ light years the supernova distance. From the observed time- and energy spread an upper limit between 25 and 30 eV/c^2 is estimated [2].

Beside supernova experiments, a measurement of the endpoint region of the tritium β-spectrum still appears to be the most sensitive, direct and model independent method to determine the rest mass of the electron antineutrino.

The present ITEP value is: $m_\nu = 30^{+2}_{-8}\,\text{eV}/c^2$ [3]. The result from our 1985 measurements was [4]: The neutrino mass determined is consistent with zero with an upper limit of 18 eV/c^2, a result in clear contradiction with the ITEP claim. In the meantime results on neutrino mass measurements have been reported by the Tokyo and the Los Alamos group [5,6]. The quoted uncertainties ($\simeq 30\,\text{eV}/c^2$) do not yet allow to decide between the ITEP and the Zurich result.

In this talk several problems related to this difficult experiment will be presented, but no new results will be reported. Experimental improvements and the investigation of the systematic error will be mentioned. In the last section the various results are compared to each other.

2. The Tritium β-Decay

Tritium decays with a halflife of 12.3 y: $^3\mathrm{H} \rightarrow {}^3\mathrm{He}^+ + \mathrm{e}^- + \bar{\nu}$. The reason for this choice is the unusually low endpoint energy ($E_0 \simeq 18.6$ keV) and in part the relatively simple initial and final atomic states. The spectrum is given by Fermi's golden rule and is proportional to the electron and neutrino density of states:

$$N(E) = A\,F(E)\,p\,W\,\varepsilon^2\sqrt{1-\frac{(m_\nu c^2)^2}{\varepsilon^2}}\,,$$

where p, E, and W are momentum, kinetic and total energy of the electron, $\varepsilon = E_0 - E$ is the total neutrino energy, and E_0 the endpoint energy. The recoil energy of about 3 eV is of no importance. The Fermi function $F(E)$, which takes the Coulomb interaction between the shielded nucleus and the outgoing electron into account is [7]:

$$F(E) = AS\frac{\tilde{W}}{pW}e^{\pi\tilde{\eta}}\,\tilde{p}^{2\gamma_1+1}\,|\Gamma(\gamma_1+i\tilde{\eta})|^2\left(1-2\alpha WR+\frac{13}{15}\alpha^2-\frac{\alpha R}{W}\right).$$

The radiation correction is:

$$\begin{aligned} S \;=\; 1+\frac{\alpha}{2\pi}&\left[3\ln(\frac{m_p}{m_e})-\frac{3}{4}+4\left(\frac{\operatorname{arctanh}\beta}{\beta}-1\right)\left(\frac{W_0}{3W}-\frac{11}{6}+\ln(2\varepsilon)\right)\right.\\ &\left.+\frac{4}{\beta}L(\frac{2\beta}{1+\beta})+\frac{\operatorname{arctanh}\beta}{\beta}\left(2+2\beta^2+\frac{\varepsilon^2}{6W^2}-4\operatorname{arctanh}\beta\right)\right]\\ &+2\alpha^2\ln(\frac{m_p}{m_e})\left(1+\frac{2\alpha}{\pi}\left(3\ln 2-\frac{3}{2}+\frac{\pi^2}{3}\right)\right)+\ldots. \end{aligned}$$

W and W_0 are the total electron- and endpoint energies in $m_e c^2$ units, $p = (W^2-1)^{1/2}$ the momentum, $\beta = v/c = p/W$, and $\gamma_1 = (1-4\alpha^2)^{1/2}$. The energy and the momentum for the shielded nucleus are: $\tilde{W} = W - \alpha^2$, $\tilde{p} = (\tilde{W}^2-1)^{1/2}$. $\tilde{\eta} = 2\alpha\tilde{W}/\tilde{p} = 2\alpha/\tilde{\beta}$ is the Coulomb factor, $R = 0.0044821$ the nuclear radius in units $\hbar/m_e c$ and m_p is the proton mass. The Spence function is defined as:

$$L(x) = \int_0^x \frac{\ln(1-z)}{z}dz = -\sum_{i=1}^{\infty}\frac{x^i}{i^2}\,.$$

In a good approximation one may use for $E >\approx 1$ keV:

$$\begin{aligned} F(E) \;\simeq\; 2\pi\tilde{\eta}\left(1-e^{-2\pi\tilde{\eta}}\right)^{-1}&\left(1+2.3594993\times10^{-3}\beta^{-1}-4.0728615\times10^{-4}\beta^{-2}\right.\\ &\left.+2.3249667\times10^{-5}\beta^{-3}-5.5784863\times10^{-7}\beta^{-4}-0.00676678\,\mathrm{eV}\,\varepsilon^{-1}\right). \end{aligned}$$

Since $F(E)$ varies only by $\approx 1\%$ over the interesting upper 2 keV end of the spectrum, the errors introduced by the approximations in $F(E)$ are negligible.

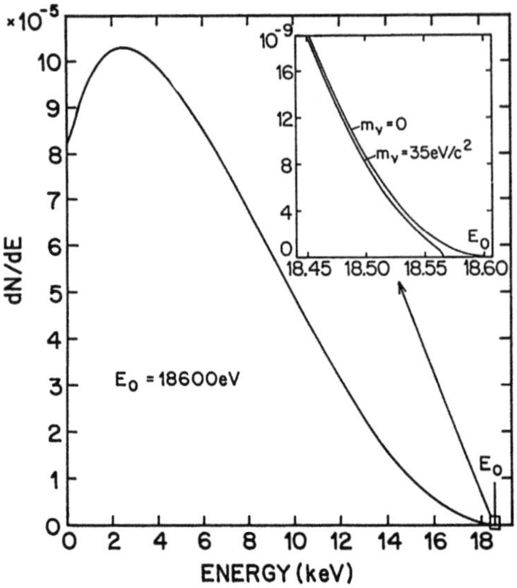

Figure 1: β–spectrum of ^{3}H. In the insert the spectrum near the endpoint is shown at a 2500 times expanded scale. The area is normalized to one.

The neutrino mass is determined from a measurement of the spectral shape near the endpoint (see Fig. 1).

One of the main difficulties of the experiment is the extremely low counting rate near the endpoint. Since the spectral area in this region is proportional to ε^3, the counting time of the experiment to determine a mass limit is approximately proportional to m_ν^{-6}! If a $35\,\mathrm{eV}/c^2$ limit can be reached with a given spectrometer in 10 days, it takes 50 years to reach $10\,\mathrm{eV}/c^2$ with the same spectrometer, of course assuming that there are no background and systematic errors.

There are three major sources for systematic errors common to all tritium experiments. These are shape and width of the spectrometer resolution function, energy loss when the β-particles leave the source, and the electronic final states. A fit to the measured β-spectrum gives little indication for an inadequate representation of any of these distributions and independent knowledge is therefore of prime importance for a reliable determination of m_ν.

3. Spectrometer

The instrument employed consisted of a toroidal field magnetic spectrometer of the Tretyakov type with 2648 mm average source detector distance, modified with a radial, electrostatic retarding field around the source. Figure 2 shows a cross section of the approximately cylindrically symmetric arrangement with some electron trajectories. The magnetic field is produced by 36 rectangular current loops (3,4). The electrons leaving the source (1) are decelerated by putting the source on positive high voltage with respect to a grid (2) near ground potential. The accepted solid angle and trajectories are limited by apertures (6,7). The detector (5) is a position sensitive proportional counter, also on high voltage (15 kV) with respect to a second grid of the same type as at the source. Position is determined by charge division using a resistive anode wire. The diameter

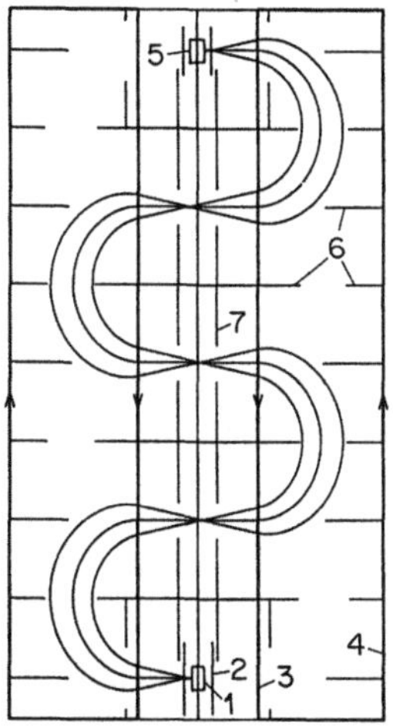

Figure 2: Scale drawing of the spectrometer.

of the detector is 6 cm with 10 cm active length. The source assembly consists of 10 rings, each with 5 cm diameter and 1 cm width, i.e. the total active source surface is 157 cm^2. To compensate for the source length a gradient voltage, proportional to the spectrometer dispersion, is applied along the rings. Spectra are recorded by stepping the source high voltage while keeping the analyzing energy, E_{mag}, of the magnetic spectrometer constant. For the present investigation $E_{mag} = 2.2$ keV was used, which gives 27 eV resolution (FWHM) and 1.1 eV/mm dispersion.

In summary, the spectrometer has the following advantages:

- The modest relative resolution (1%) and the small dispersion in the magnetic part require a mechanical precision which can easily be achieved. Also the magnetic and electric fields are known analytically, allowing a reliable Monte Carlo calculation of the resolution function. This is an important point since a resolution determination from measurements of conversion lines is complicated by shake-off effects etc. [8].
- The particles entering the detector have constant energy during a run, implying constant detection efficiency.
- Only electrons emitted approximately perpendicular to the source are detected.
- A position sensitive detector has sufficient resolution, i.e. no energy defining slits are necessary and the luminosity is high.
- The resolution is simply determined by the magnet setting and can be chosen for a best compromise between high count rate and high resolution.

4. Sources and Energy Loss Spectrum

The sources were prepared by implantation of T_2^+ ions into carbon, evaporated onto an Al backing. The thickness of the carbon layer was about 2000 Å. The implantation energies were in the range 185 to 400 eV per ion, giving typically an activity of 0.3 mCi cm^{-2}.

The dominant cause of energy loss of β-particles traversing the carbon substrate is due to excitation of plasmons which are well represented by a known parameterization procedure [9]. The required tritium profiles in the sources were measured with 50 Å resolution (FWHM) using a nuclear recoil technique similar to that described by Ross *et al.* [10].

Since the uncertainties in the energy loss function (profile and energy loss parameterization) contribute the largest error to our result, a considerable effort is presently underway to get an improved source profile determination and from energy loss measurements on carbon foils a better characterization of the loss function. A newly installed nitrogen cooling system for the spectrometer should allow the use of thinner sources, and thus reduce the energy loss uncertainties, in addition the use of monomolecular film sources is planned.

5. Measurements and Data Analysis

The data so far available were obtained in four runs using three sources. The measurements were performed with 27 eV resolution (FWHM) in an energy range of about 17.6 keV to 19.3 keV with 5 eV steps. To improve the mass sensitivity, this range was divided into several counting regions with more time spent near the endpoint. A single up-down pass required about one hour. The total measuring time was 27 days. Event data were recorded on magnetic tape and later binned into β-spectra. Figure 3 shows the total dataset used in the evaluation.

The β-spectrum was assumed to have the form:

$$N(E) = A\,F(E)\,p\,W\,(1+\alpha\varepsilon_0)\sum_i w_i\,\varepsilon_i^2\sqrt{1-\frac{(m_\nu c^2)^2}{\varepsilon_i^2}} + B\,,$$

where $\varepsilon_i = E_{0_i} - E$. The sum runs over all electronic final states (FS) with branching fractions w_i and endpoints E_{0_i}. Each term in the sum is set to zero for E above the point

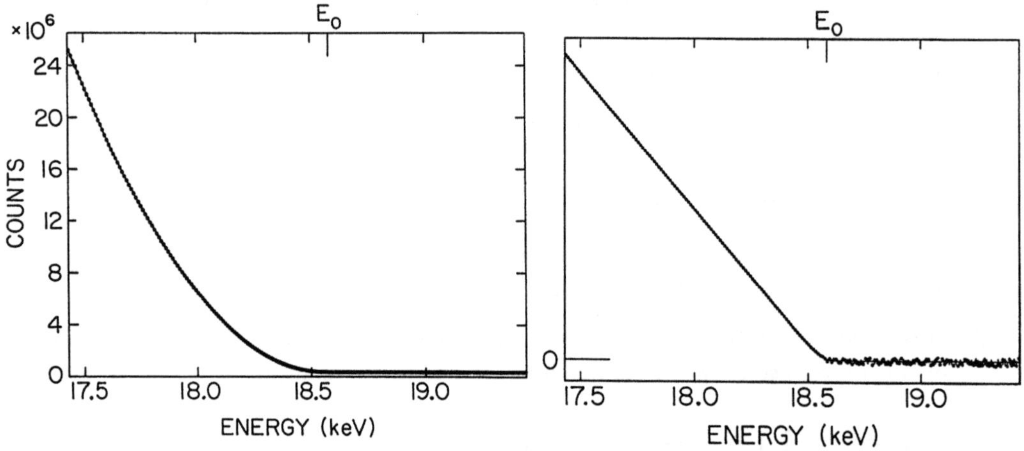

Figure 3: Measured tritium β-spectrum. The original data of the measured four runs were, after the fitting, binned and added. The vertical scale is correct for the central region of the spectrum. It is ten times less away from the endpoint region, and adjusted at the end for the different energy ranges measured. At right the Kurie-plot of the total dataset is shown. The fitted curve is for $m_\nu = 0$.

where the term first reaches zero. The E_{0_i} are defined by $E_{0_i} = E_{0_0} - E_{ex_i}$, where E_{0_0} corresponds to the endpoint of the electronic groundstate. E_{ex_i} are the helium excitation energies measured from the groundstate. Backscattering in the source substrate is taken into account by the parameter α, which enters in this approximate form by a convolution of a constant backscattering distribution with a β-spectrum proportional to ε_0^2. The fit function is obtained by convoluting $N(E)$ with the spectrometer resolution function and the energy loss spectrum and by adding a constant background term B. A small correction factor ($< 7\%$ variation) for the known changing spectrometer acceptance as a function of retarding voltage has been included. The free parameters are α, E_{0_0}, B, an overall normalization A, and the neutrino mass squared m_ν^2. The latter was allowed to take on nonphysical, negative values, in which case the square root in $N(E)$ was replaced by its first order expansion $\mid 1 - (m_\nu c^2)^2/(2\varepsilon_i \mid \varepsilon_i \mid) \mid$. This gave symmetric, parabola-like curves for χ^2 versus m_ν^2. Our final result is essentially independent of this assumption.

Concerning the FS for our sources we argue as follows. The extremely small diffusion rate of hydrogen isotopes implanted in carbon and the stability of our sources strongly suggest that C–T bonds are formed. Kaplan *et al.* [11] calculated the FS for a variety of molecules with C–T bonds and found very little variation in terms of the groundstate fraction w_0 and the mean excitation energy $\overline{E}_{ex}$. By fitting various FS distributions we found that just these two quantities are important, whereas details of the FS influence the result very little. We thus take CH_3T, the simplest case, as an adequate representation for the FS of our sources. The corresponding fit results for the four spectra and the combined data set are listed in table 1. As can be seen, good fits were obtained and m_ν^2 is compatible with zero within one sigma. The nonstatistical variation in the E_{0_0} values, caused by uncertainties in the average source-detector distance mainly due to the use of different detectors, is irrelevant for the mass determination.

Table 1: Fit results for four runs with three different sources and the combined dataset using CH_3T final states. Errors indicated are one standard deviation. UL is the statistical upper limit at the 95% confidence level.

Run/ Source	χ^2/DOF	$E_{0_0} - 18500$ [eV]	α [keV^{-1}]	m_ν^2 [(eV/c^2)^2]	UL [(eV/c^2)^2]
1/A	328.8 / 337	82.6 ± 0.3	0.018 ± 0.004	$+140 \pm 130$	356
2/B	345.6 / 356	84.5 ± 0.2	0.020 ± 0.002	-9 ± 140	218
3/B	349.9 / 356	84.4 ± 0.2	0.022 ± 0.002	-22 ± 140	202
4/C	343.7 / 317	77.6 ± 0.2	0.021 ± 0.002	-85 ± 93	70
all	1370.2 / 1369	–	–	-11 ± 63	95

The $m_\nu = 35\,\mathrm{eV}/c^2$ hypothesis was tested by fitting all data with fixed values $m_\nu = 0$ and $m_\nu = 35\,\mathrm{eV}/c^2$, giving $\chi^2 = 1370.2$ and $\chi^2 = 1773.2$, respectively, with 1370 degrees of freedom (DOF). As can be seen in Fig. 4, the latter case can be ruled out without question. In Fig. 5 χ^2 as a function of m_ν^2 of the total dataset is shown.

Systematic errors for m_ν^2 were investigated by fitting the data with changed input parameters of the fitted function. The effects of varying each parameter separately are listed in table 2. The direction of the changes was chosen such that m_ν^2 was increased. Making all changes in table 2 simultaneously shifts m_ν^2 up to $398\,(\mathrm{eV}/c^2)^2$, but increases

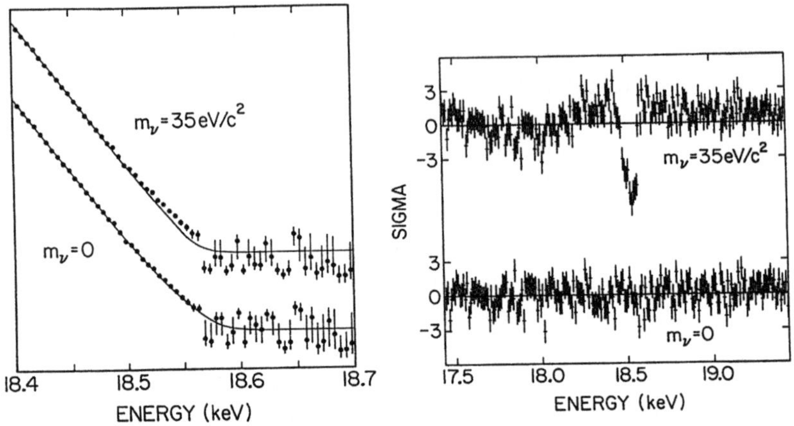

Figure 4: Kurie-plot of the data from run #4 fitted with $m_\nu = 0$ and $m_\nu = 35\,\mathrm{eV}/c^2$. All other parameters ($E_{0_0}$, B, A, α) are fitted. Notice the strong correlation of m_ν with the background B in the fitted curve. At right the differences between the fitted function and the data for $m_\nu = 0$ and $m_\nu = 35\,\mathrm{eV}/c^2$ for the total dataset are plotted. The difference figure was obtained by an appropriate binning after the fit.

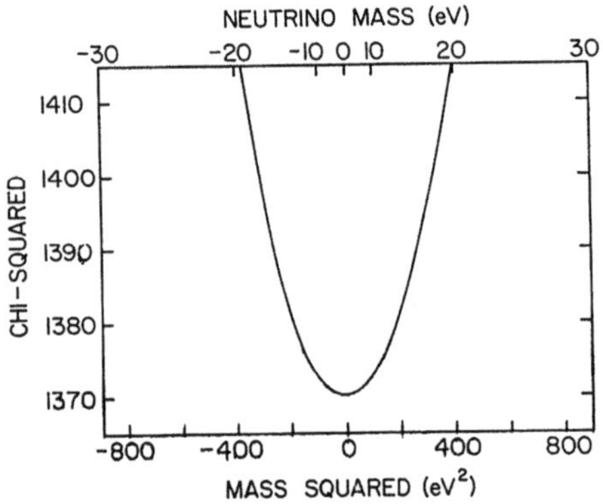

Figure 5: Plot of χ^2 versus $(m_\nu c^2)^2$ for the combined dataset, with all other free parameters being fitted. The number of degrees of freedom for these fits is 1370. Assumed are the final states of CH_3T.

Table 2: Fitted values of m_ν^2 and the corresponding shift Δm_ν^2 when the input parameters of the fitted function are changed. Shown are only changes which increase m_ν^2. The complete set of data was fitted, with DOF = 1369.

Input parameters	Relative change %	χ^2	m_ν^2 $[(\mathrm{eV}/c^2)^2]$	Δm_ν^2 $[(\mathrm{eV}/c^2)^2]$
best set	–	1370.2	−11	–
spectrometer resolution	+ 10	1370.2	52	63
mean source thickness	+ 15	1373.8	120	131
ground state fraction	−2.5	1370,5	30	41
mean excitation energy	+ 6	1373.2	111	122
width of excited states	+30	1386.3	52	63

χ^2 by 28 units and is thus not acceptable. Therefore we treat the shifts of m_ν^2 in table 2 as independent errors and add them quadratically. Assuming the best fitted value for m_ν^2 would be zero (and not $-11\,(\mathrm{eV}/c^2)^2$), a statistical UL of $106\,(\mathrm{eV}/c^2)^2$ for m_ν^2 at the 95% confidence level was calculated. Adding both numbers linearly (rather than quadratically) we arrived at our result, $m_\nu^2 < 310\,(\mathrm{eV}/c^2)^2$, or

$$m_\nu < 18\,\mathrm{eV}/c^2 .$$

In conclusion we find no indication of a nonzero mass for the electron antineutrino, which is in strong contradiction to the results of Ref. 1. We see no possible source of errors in our experiment large enough to account for this discrepancy. The measurements done since last year confirm this result. The error limits will probably be reduced in the near future.

6. Discussion

In the following the measurements of the four groups reporting results are compared.

ITEP: The ITEP source is tritiated valine (an amino acid) evaporated onto a backing. The spectrometer resolution and energy loss is determined by measurements of a conversion line source. The tritium sources are rather thick and there appears to be a problem with the right amount of the energy loss fraction (about 70 %). They therefore measured the tritium spectrum over a wide energy range (3.4 keV) and required that the fitted mass value does not systematically vary when only part of the spectrum is analysed. Thus effectively the energy loss is determined from the tritium spectrum, a potentially dangerous method. In this way they obtained $m_\nu = 30 \pm 2$ eV/c^2. A 'model-independent' mass range $17 < m_\nu < 40$ eV/c^2 was derived by varying the final states with the condition that the fitted endpoint is compatible with the ^{3}H–^{3}He mass difference.

Zurich: The Zurich group uses a magnetic spectrometer of basically the same type as first built at ITEP. However, it is operated in a mode requiring only moderate mechanical precision which can easily be achieved, and which thus allows a reliable calculation of the resolution function by a Monte Carlo method. Uniform sources were prepared by implanting of T_2^+ into an evaporated carbon layer. The thickness of the sources was measured by a nuclear recoil method and the energy loss distribution was determined by using the measured single interaction distribution and the mean free path. The final states where taken as calculated for the methane molecule. This is justified by the well known fact that the carbon-hydrogen bond is insensitive to the chemical environment and by the calculations in Ref. 10. Measurements with three sources gave consistently values for m_ν^2 compatible with zero within one sigma and a statistical upper limit of 106 $(\mathrm{eV}/c^2)^2$ (95 % C.L.). A systematic error of 204 $(\mathrm{eV}/c^2)^2$ for m_ν^2 was determined by varying all distributions used in the analysis within estimated uncertainties. Adding both numbers linearly the result is $m_\nu < 18$ eV/c^2.

Tokyo: This group uses the Cd salt of the tritiated arachidic acid ($CH_3(CH_2)_{18}COOH$). The source consists of two monomolecular layers and is about 50 Å thick. Such a thin source is necessary since their spectrometer accepts only electrons emitted at a small angle (22.5^0) to the surface making the source effectively thicker (131 Å). They have found a clever way to measure the combination of resolution and energy loss. This is done by preparing a test source identical to the tritium source except that the natural Cd is replaced by ^{109}Cd and that it does not contain tritium. In the analysis the same final states calculations were used as by the ITEP group. The result for m_ν^2 is close to zero with a statistical upper limit of 832 $(eV/c^2)^2$ (95 % C.L.). The systematic error from resolution is given to be 188 $(eV/c^2)^2$. Their present result [5] is $m_\nu < 32$ eV/c^2. This does not yet include errors from the final states. To improve statistics the group plans to use a source with a larger area and a new type of detector.

Los Alamos: This experiment is distinguished by a gaseous T_2 source [6]. The T_2 is enclosed in a 3.7 m long tube, differentially pumped at the ends. The decay electrons are transported by a magnetic guiding field into a magnetic spectrometer. The obvious advantage is that the final states and the cross section for inelastic scattering within the source are know accurately. The spectrometer resolution was derived from a measurement of the K-conversion line of ^{83}Kr gas (line shape not known experimentally). Uncertainties for the energy loss result from electrons trapped in local minima of the guiding field. The given systematic errors for m_ν^2 from resolution and energy loss are 87 $(eV/c^2)^2$ and 80 $(eV/c^2)^2$, respectively. A quadratic shape correction factor ($N(E) \propto (1 + a\varepsilon_0^2) \times \mathit{Spectrum}$) in the fitting function is still not explained. The fitted value of m_ν^2 is close to zero with a one sigma error of 453 $(eV/c^2)^2$. The final result is $m_\nu < 27$ eV/c^2 (95 % C.L.). To improve statistics the group plans to do long time measurements using a new Si-detector instead of a proportional counter. The problem with the trapped electrons should be overcome by a gradient superimposed on the guiding field.

References

1. V.A. Lubimov *et al.*, Phys. Lett. B 94, 266 (1980).
2. K. Sato and H. Suzuki, Phys. Rev. Lett. 58, 2722 (1987).
3. S. Boris *et al.*, Phys. Rev. Lett. 58, 2019 (1987).
4. M. Fritschi *et al.*, Phys. Lett. B 173, 485 (1986).
5. H. Kawakami *et al.*, Phys. Lett. B 187, 198 (1987).
6. J.F. Wilkerson *et al.*, Phys. Rev. Lett. 58, 2023 (1987).
7. H. Behrens and W. Bühring, *Electron Radial Wave Functions and Nuclear Beta-decay*, (Clarendon Press, Oxford, 1982).
8. W. Kündig *et al.* in Fourth Morion Workshop (La Plagne, France, 1984) edited by J. Tran Thanh Van (Frontieres, Paris, 1984).
9. R.E. Burge and D.L. Misell, Phil. Mag. 18, 251 (1968).
10. G.G. Ross *et al.*, J. Nucl. Mat. 128, 730 (1984).
11. I.G. Kaplan *et al.*, DAN USSR 279, 1110 (1984); Phys. Lett. B 112, 417 (1982); Phys. Lett. B 161, 389 (1985).

Measurements of the Tau-Neutrino Mass

K.R. Schubert

Institut für Hochenergiephysik der Universität Heidelberg,
Schröderstraße 90, D-6900 Heidelberg, Fed. Rep. of Germany

The tau lepton has been disovered 1975 at the e^+e^- storage ring SPEAR at Stanford [1] and has been confirmed soon afterwards at the DORIS ring at DESY. All experimental studies of the τ and its neutrino ν_τ have been performed at e^+e^- storage rings, in addition to the two above also at CESR (Cornell), PETRA (DESY), and PEP (Stanford). The τ is found to be a pointlike spin-1/2 particle with mass (1784 ± 3) *MeV* [2]. Its production cross section is shown in fig. 1, and with the known integrated luminosities of the five storage rings over the last 12 years we estimate that the world production of τ leptons has been about $3 \cdot 10^6$ so far.

The τ is unstable and decays with a mean life time of $(3.07 \pm 0.09) \cdot 10^{-13}$ *s* [3] into various final states [2]. The dominant decays are $\tau^- \to \pi^- \nu_\tau$, $\tau^- \to \rho^- \nu_\tau$, $\tau^- \to e^- \overline{\nu_e} \nu_\tau$, and $\tau^- \to \mu^- \overline{\nu_\mu} \nu_\tau$; their measured rates agree well with the expectations of the standard weak interaction theory.

From the absence of τ^+ production in neutrino beams with $\overline{\nu_e}$ and $\overline{\nu_\mu}$, from the absence of decays as $\tau \to e + hadrons$, $\tau \to \mu + hadrons$, $\tau \to 3\mu$, and from the equality of the two partial rates $\Gamma(\tau \to \mu\nu\nu)$ and $\Gamma(\tau \to e\nu\nu)$ we know that

$$\nu_\tau \neq \nu_e,\ \nu_\mu,\ \overline{\nu_e},\ \overline{\nu_\mu}\ .$$

The tau-neutrino is therefore one of the 12 fundamental fermions of the standard model:

$$\begin{pmatrix}\nu_e\\ e^-\end{pmatrix} \begin{pmatrix}\nu_\mu\\ \mu^-\end{pmatrix} \begin{pmatrix}\nu_\tau\\ \tau^-\end{pmatrix} \quad \begin{pmatrix}u\\ d\end{pmatrix} \begin{pmatrix}c\\ s\end{pmatrix} \begin{pmatrix}b\\ t\end{pmatrix} .$$

As for the electron- and for the mu-neutrino, the mass of the tau-neutrino is experimentally compatible with zero. In absence of any strong theoretical arguments for zero mass, the experimental sensitivity has to be steadily increased, and table 1 summarizes the efforts which have been undertaken so far for measuring $m(\nu_\tau)$.

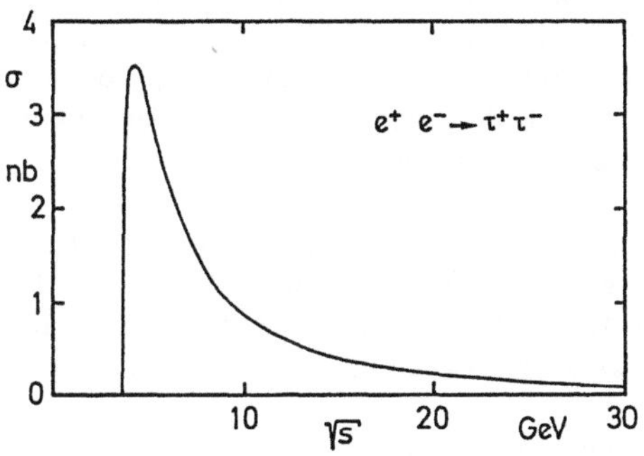

Fig. 1:

Production cross section of τ pairs in e^+e^- annihilation as function of the centre-of-mass energy.

Table 1: Summary of $m(\nu_\tau)$ Measurements.

Experiment	Year	Ref.	Method	Result
DELCO	1979	4	$E(e)$	$< 250\ MeV$ (95%CL)
MARK-II	1982	5	$E(\pi)$	$< 250\ MeV$ (95%CL)
MARK-II	1984	6	$m(4\pi)$	$< 164\ MeV$ (95%CL)
DELCO	1985	7	$m(KK\pi)$	$< 157\ MeV$ (95%CL)
MARK-II	1985	8	$m(4\pi)$	$< 143\ MeV$ (95%CL)
MARK-II	1985	9	$m(5\pi)$	$< 125\ MeV$ (95%CL)
ARGUS	1985	10	$E(3\pi)$	$< 70\ MeV$ (95%CL)
HRS	1986	11	$m(5\pi, 6\pi)$	$< 84\ MeV$ (95%CL)
CLEO	1987	12	$E(3\pi)$	$< 85\ MeV$ (95%CL)
HRS	1987	13	$m(5\pi, 6\pi)$	$< 76\ MeV$ (95%CL)
ARGUS	1987	14,15	$m(4\pi)$	$< 35\ MeV$ (95%CL)

There are essentially two different methods to obtain the tau neutrino mass, either from the energy spectrum or from the invariant mass spectrum of the tau decay products. Both will be described in the following using the two experiments which have obtained the smallest limit for each method. These two experiments have been performed with the ARGUS detector at the e^+e^- storage ring DORIS at DESY. The first [10] determines $m(\nu_\tau)$ from the energy spectrum of the 3π system in $\tau^\pm \to \pi^\pm\pi^+\pi^-\nu_\tau$ decays, and the second [14, 15] from the invariant mass spectrum of the 5π system in $\tau^\pm \to \pi^\pm\pi^+\pi^+\pi^-\pi^-\nu_\tau$ decays.

The energy spectrum method uses the fact that tau leptons are produced monochromatically in e^+e^- storage rings of equal beam energy. If the energy $\sqrt{s}$ of the storage ring is well known, each τ in the reaction $e^+e^- \to \tau^+\tau^-$ gets the energy $\sqrt{s}/2$ up to radiative corrections which are well calculable. In the ARGUS experiment [10], DORIS was operated at the energy of the $\Upsilon(2S)$ resonance, $\sqrt{s} = (10023.4\pm0.3)\ MeV$ [16]. The mean energy of all produced τ is therefore known to $\pm0.15\ MeV$. Their energy spread is however much larger owing to the beam energy resolution of the storage ring ($\approx \pm5\ MeV$) and radiative corrections. In the decays $\tau^\pm \to \pi^\pm\pi^+\pi^-\nu_\tau$, the momenta $\vec{p_i}$ of the three pions are measured with the magnetic spectrometer of the ARGUS detector, and from

$$E_i = \sqrt{p_i^2 + m_\pi^2}\ ,\ i = 1, 2, 3\ ,$$

$$E_\nu = \sqrt{s}/2 - E_1 - E_2 - E_3\ ,$$

$$m_{3\pi}^2 = (E_1 + E_2 + E_3)^2 - (\vec{p_1} + \vec{p_2} + \vec{p_3})^2\ ,$$

one obtains the neutrino energy E_ν and the invariant mass $m_{3\pi}$ of the three pion system in every observed decay. The τ leptons are produced without mean polarization. The angular distribution of the neutrinos in the τ rest system is therefore isotropic, and the E_ν spectrum in the the laboratory system is a box spectrum as sketched in fig. 2 if the tau leptons are monochromatic and if the three pion system has a fixed invariant mass.

Fig. 3 shows how the lower edge of the box spectrum depends on m_ν and on $m_{3\pi}$. For $m_{3\pi} = 1100\ MeV$, which corresponds to the most frequent case, a shift of the lower edge by $10\ MeV$ from its $m_\nu = 0$ value corresponds to $m(\nu_\tau) = 65\ MeV$. This value follows from kinematics alone and is indepent from any assumptions on the weak and strong interaction dynamics.

Using all their data on the $\Upsilon(2S)$ resonance, corresponding to an integrated luminosity of $\mathcal{L} \approx 40/pb$, the ARGUS group finds 1536 events of the type

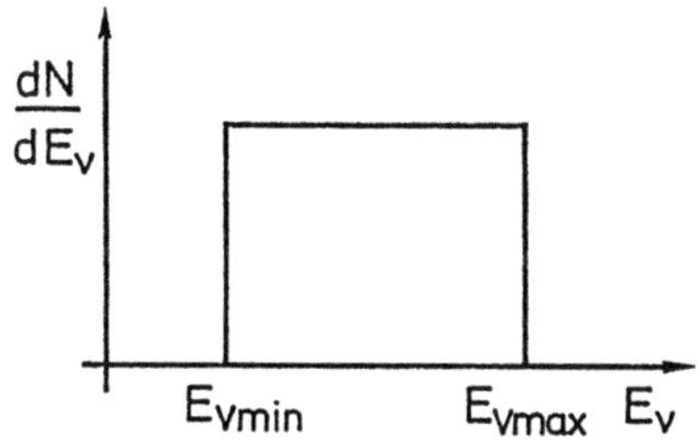

Fig. 2: Neutrino energy spectrum of monochromatic tau leptons in the quasi-two-body decay $\tau \rightarrow (\pi\pi\pi)\nu$ for fixed $m_{3\pi}$.

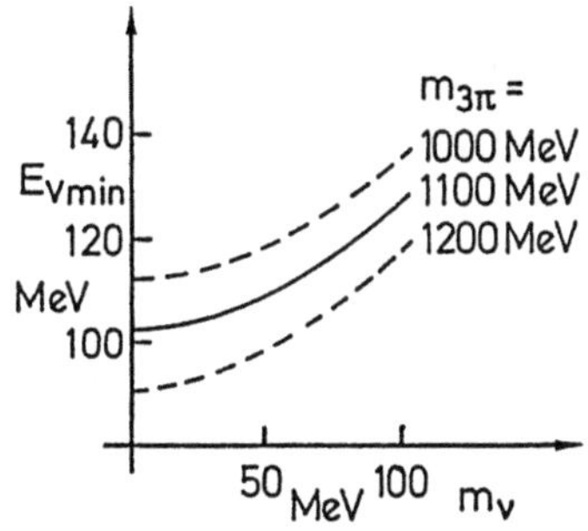

Fig. 3: Dependence of $E_{\nu min}$ on the tau neutrino mass for various $m_{3\pi}$ values

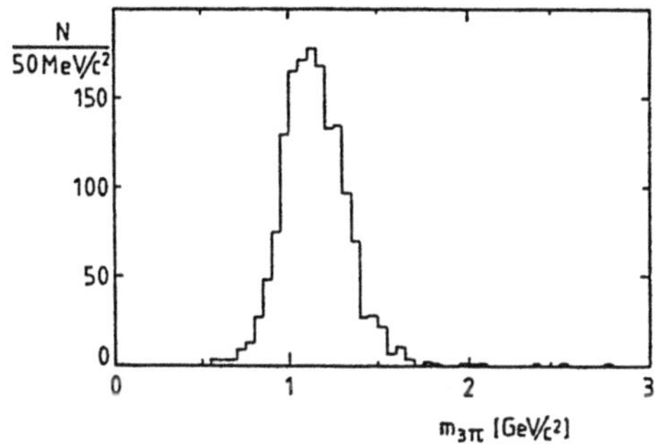

Fig. 4: Invariant mass spectrum of the 3π system in the 1536 $\tau \rightarrow \pi\pi\pi\nu$ decays of ref. 10.

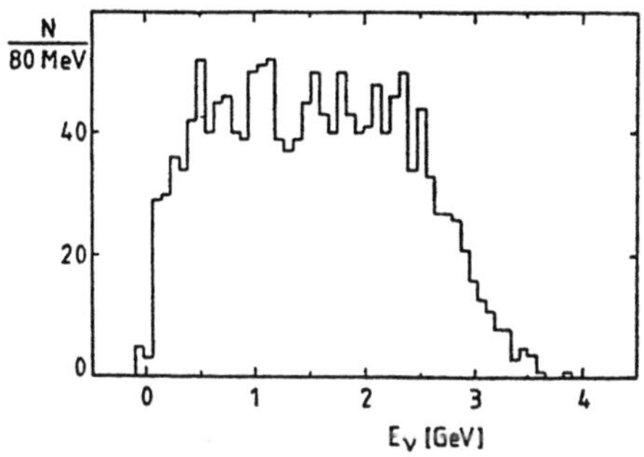

Fig. 5: Neutrino energy spectrum of the 1536 $\tau \rightarrow \pi\pi\pi\nu$ events in ref. 10.

$$e^+e^- \rightarrow \tau_1\tau_2 \ ; \quad \tau_1 \rightarrow \pi\nu,\ K\nu,\ e\nu\nu,\ \mu\nu\nu \ ; \quad \tau_2 \rightarrow 3\pi\nu \ .$$

Their $m_{3\pi}$ spectrum is shown in fig. 4, their E_ν spectrum in fig. 5. The background of non- 3π decays is less than 5%. The E_ν spectrum is only approximately box-like, since the $m_{3\pi}$ spectrum has a finite width. The E_ν spectrum can be fitted with the observed $m_{3\pi}$ distribution, the beam and spectrometer resolution, the spectrometer acceptance, and the known radiative corrections. In the region $-100\ MeV \leq E_\nu \leq 300\ MeV$, this fit depends critically on the tau neutrino mass as shown in fig. 6.

The statistical fit result is $m(\nu_\tau) < 56\ MeV$ with 95% confidence. Estimates of the systematic errors from the absolute spectrometer calibration, the storage ring energy calibration, the non-τ background, and from variations of the storage ring energy and spectrometer resolutions lead to the final result $m(\nu_\tau) < 70\ MeV$ with 95% confidence.

The 1987 ARGUS experiment [14, 15] uses the invariant mass method with $\tau \rightarrow (5\pi)\nu$ decays. This method relies only on the background-free selection of these τ decays and the spectrometer properties. It does not require precise knowledge of the τ energy as the previous method, but it has the disadvantage that it is in principle dependent on the decay dynamics. As will be shown at the end of this chapter, the result of the experiment does not depend on model assumptions for these dynamics.

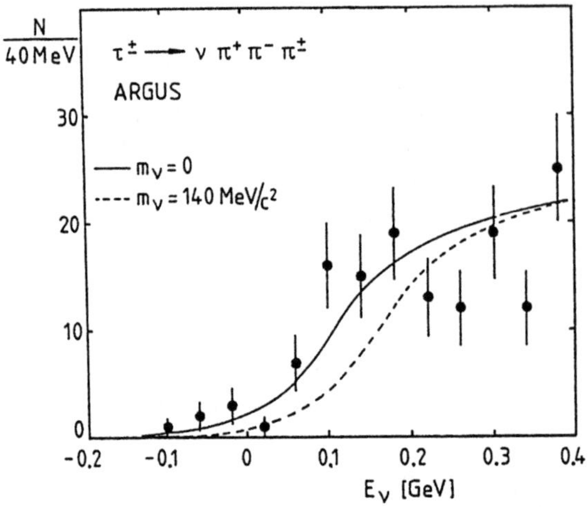

Fig. 6:

Lower edge region of the E_ν spectrum in fig. 5.

For using the 5π invariant mass method, the ARGUS group selects events of the type

$$e^+e^- \rightarrow \tau_1\tau_2 \ ; \quad \tau_1 \rightarrow e\nu\nu,\ \mu\nu\nu,\ \pi\nu,\ K\nu \ ; \quad \tau_2 \rightarrow \pi^\pm\pi^+\pi^+\pi^-\pi^-\nu$$

from all their data taken between 1983 and 1986. These data are taken at centre-of-mass energies between 9.4 and 10.6 GeV and correspond to an integrated luminosity of $\mathcal{L} \approx 200/pb$ which is a factor of 2.5 higher than for the data sample used in method 1. The sum BR_1 of branching ratios for the decays of τ_1 is about 0.5, the five-prong decay of τ_2 has a branching ratio $BR_2 \approx 6 \cdot 10^{-4}$ if decays with additional neutral pions are excluded. One therefore expects that

$$N_{12} \ = \ 2 \cdot \mathcal{L} \cdot \sigma(ee \rightarrow \tau\tau) \cdot BR_1 \cdot BR_2 \ = \ 94$$

events of the searched type have been produced in the ARGUS detector. A first selection – five charged particles in one hemisphere, one charged particle in the opposite hemisphere, no photons with $E_\gamma > 80\ MeV$ – leads to 224 candidates. The main background, $e^+e^- \rightarrow e^+e^-\gamma\gamma$ with both photons converting in the spectrometer into an e^+e^- pair, is removed by opening angle cuts between pairs of tracks. One of the remaining 50 candidates is shown in fig. 7, and fig. 8 shows the biplot of p(one-prong) vs. m(five-prong) for these candidates.

Background from $e^+e^- \rightarrow q\bar{q}$ events with six charged hadrons in the final state is quantitativily removed by rejecting all candidates above the straight line in fig. 8 as shown by various Monte Carlo simulations for these low-multiplicity multi-hadron events. The remaining background – photon-photon interactions with six charged particles in the detector – is efficiently removed by requiring that the missing momentum $\vec{p}_{miss}$ in the event is large,

$$|\vec{p}_{miss}| = |\sum_{i=1}^{6} \vec{p}_i| > 1.7\ GeV/c \ ,$$

and points into the spectrometer. There are 12 events which survive these cuts, their invariant mass distribution $dN/dm_{5\pi}$ is shown in fig. 9.

The mass-resolution functions of all 12 events are well described by Gaussian distributions with a typical width of $\pm 20\ MeV$. Both a simple phase-space model and a phase-space distribution weighted by the weak interaction matrix element [17] are used to describe the mass distribution in fig.9. The endpoint of the spectrum is directly $m(\tau) - m(\nu_\tau)$ if the $m_{5\pi}$ resolution is neglected.

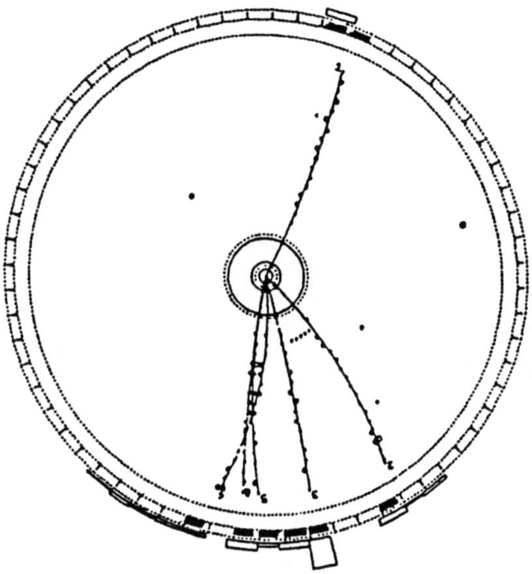

Fig. 7:

One event of the type $ee \rightarrow \tau\tau$ with one τ decaying into one charged particle and the other τ into five charged particles from the 1987 ARGUS experiment [14, 15].

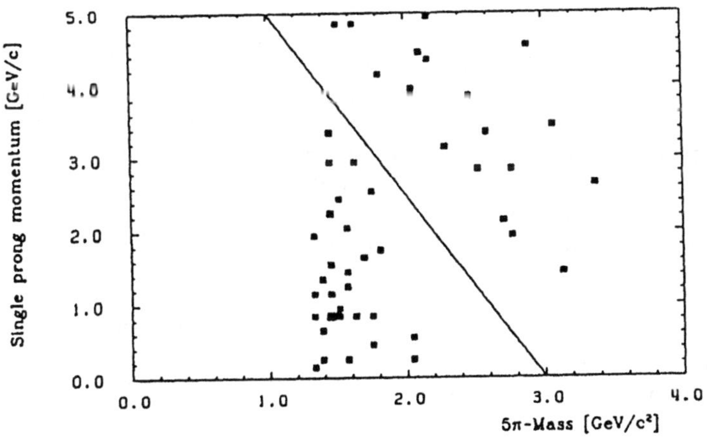

Fig. 8:

Momentum of the one-prong charged particle vs. invariant mass of the five charged particles in the five-prong τ decays from refs. 14,15.

Including the individual resolution function for each event in the maximum-likelihood-fit procedure leads to a fit result of $m(\nu_\tau) < 25\ MeV$ with 95% confidence. The result does not depend on the model for describing the invariant mass distribution but only on the mass of the three highest-m events and their resolution functions. The background in the 12 events is estimated to be much less than 1 event. In order to give a conservative $m(\nu_\tau)$ limit, the highest-m event has been removed from the fit. Including systematic errors from the τ mass and from variations in the resolution functions, this final fits gives $m(\nu_\tau) < 35\ MeV$ with 95% confidence.

In the frame of supersymmetric models [18], one expects that the mass of the three neutrino species are proportional to the square of the corresponding charged lepton masses,

$$m(\nu_\tau) \ : \ m(\nu_\mu) \ : \ m(\nu_e) \ = \ m_\tau^2 \ : \ m_\mu^2 \ : \ m_e^2$$

$$= \ 35\ MeV \ : \ 122\ keV \ : \ 2.9\ eV \ .$$

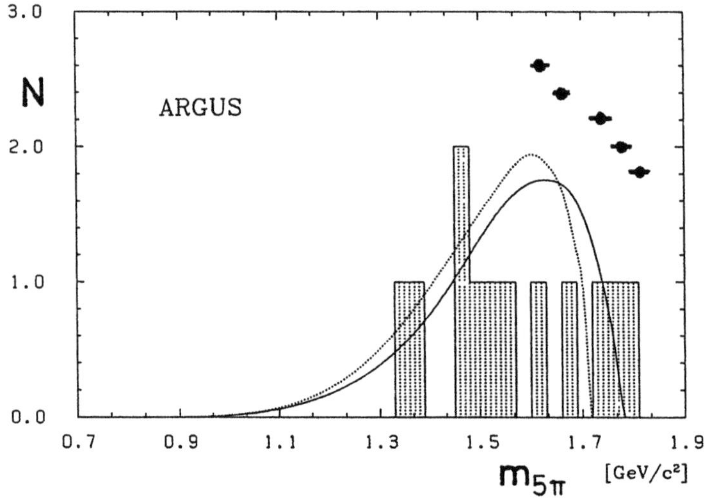

Fig. 9:

Distribution of the 5π invariant mass in $\tau^{\pm} \rightarrow \pi^{\pm}\pi^{+}\pi^{+}\pi^{-}\pi^{-}\nu$ decays from the 1987 ARGUS experiment [14, 15]. The solid curve corresponds to the expectation with $m_{\nu} = 0$ and the dashed curve to $m_{\nu} = 70\ MeV$. The experimental mass resolution of the five highest-m events is indicated by error bars.

Comparing this relation to the presently achieved experimental limits $m(\nu_{\mu}) < 250\ keV$ [2] and $m(\nu_e) < 18\ eV$ [19] shows that the tau neutrino mass determinations have already reached the best sensitivity of all three – at least in the case that the mass relation above is valid.

What improvements on the experimental $m(\nu_{\tau})$ sensitivity may be expected in the future? Since the energy spectrum method has a quadratic dependence of the measured quantity ($E_{\nu min}$) on the neutrino mass and the mass spectrum method a linear dependence ($m_{\nu} = m_{\tau} - m_{5\pi,6\pi}$), the mass spectrum method is more sensitive. With more statistics, it will be possible to reach the systematic limits of 5 to 10 MeV. Only e^+e^- storage rings are intensive τ sources, and $\sqrt{s} = (9-15)\ GeV$ is the best energy range for τ decay studies. Determination of the τ neutrino mass is therefore one of the motivating program points for a "B-Meson Factory", i.e. a new e^+e^- collider [20] in the energy range mentioned above and with a luminosity 10 to 100 times higher than available at DORIS now.

I would like to thank H.V. Klapdor for his invitation to present the ARGUS tau neutrino mass results at this Workshop. I also thank all my colleagues in the ARGUS collaboration for numerous discussions, especially with Yu. Zaitsev on method 1, B. Spaan on method 2, and with S. Weseler on the future.

REFERENCES

1 M. Perl et al, Phys.Rev.Lett. 35(1975)1489
2 Particle Data Group, Phys.Lett. 170B(1986)1
3 H.J. Gerber, Proc. Int. Europhysics Conf. on High Energy Physics, Uppsala 1987
4 W. Bacino et al (DELCO), Phys.Rev.Lett 42(1979)749
5 C.A. Blocker et al (MARK II), Phys.Lett. 109B(1982)119
6 C. Matteuzzi et al (MARK II), Phys.Rev.Lett. 52(1984)1869
7 G.B. Mills et al (DELCO), Phys.Rev.Lett. 54(1985)624
8 C. Matteuzzi et al(MARK II), Phys.Rev. D32(1985)800
9 P.R. Burchat et al (MARK II), Phys.Rev.Lett. 54(1985)2489
10 H. Albrecht et al (ARGUS), Phys.Lett. 163B(1985)404
11 S. Abachi et al (HRS), Phys.Rev.Lett. 56(1986)1039

12 S.E. Csorna et al(CLEO), Phys.Rev. D35(1987)2747
13 S. Abachi et al (HRS), Phys.Rev. D35(1987)2880
14 B. Spaan (ARGUS), Proc. 2nd Int. Symp. on Production and Decay of Heavy Flavous, Stanford 1987
15 H. Albrecht et al (ARGUS),"An improved upper limit on the ν_τ mass from the decay $\tau^- \rightarrow \pi^-\pi^-\pi^-\pi^+\pi^+\nu_\tau$", submitted to Phys.Lett., November 1987
16 D.P. Barber et al (DORIS, ARGUS, Crystal Ball), Phys.Lett. 135B(1984)498
17 Y.S. Tsai, Phys.Rev. D4(1971)2821
18 e.g. M.Gell-Mann et al, in "Supergravity", ed. by P. van Nieuwenkinsen, North-Holland 1979, p. 315
19 M. Fritschi et al (SIN), Phys.Lett. 173B(1986)485
20 R. Eichler et al, "Motivation and Design Study for a B-Meson Factory", SIN-PR-86-13(1986), unpublished

Experimental Limits on Radiative Neutrino Decay

L. Oberauer and F. von Feilitzsch

Physik-Department, Technische Universität München,
D-8046 Garching, Fed. Rep. of Germany

1. Introduction

Neutrino decay is an inevitable consequence of the hypothetical neutrino mass mixing, and the search for the decay is therefore an important part of the study of neutrino physics.

We present experimental limits on the neutrino lifetime for the radiative decay modes $\nu_H \rightarrow \nu_L + \gamma$ and $\nu_H \rightarrow \nu_L + e^+ + e^-$, where ν_H and ν_L are assumed to be massive neutrinos with $m_{\nu H} > m_{\nu L}$.

2. Theoretical Predictions for Neutrino Lifetimes

We consider the weak interaction in terms of the Weinberg-Salam model and assume that all left (right)-handed components are transformed as doublets (singlets) under the weak SU(2) transformation. The weak interaction eigenstate ν_e can be written as a superposition $\nu_e = \sum_i U_{ei}\nu_i$ of the mass eigenstates ν_i. This matrix U induces the decay of a massive neutrino ν_H into a lighter neutrino ν_L and a photon

$$\nu_H \rightarrow \nu_L + \gamma \tag{1}$$

as well as the decay into ν_L and an electron positron pair, if $m_{\nu H} - m_{\nu L} > 2m_e$

$$\nu_H \rightarrow \nu_L + e^+ + e^- \tag{2}$$

with the center of mass decay rates [1]

$$\Gamma\left(\nu_H \rightarrow \nu_L\gamma\right) = \frac{G_F^2 m_{\nu_H}^5}{128\pi^3}\frac{\alpha}{\pi}\frac{9}{16}\left(\sum_j U_{jH}U_{jL}^* \frac{m_{\ell j}^2}{M_w^2}\right)^2 \tag{3}$$

$$\Gamma\left(\nu_H \rightarrow \nu_L e^+ e^-\right) = \frac{G_F^2 m_{\nu_H}^5}{192\pi^3} \mid U_{eH} \mid^2 \tag{4}$$

where α is the fine structure constant, M_w represents the mass of the weak W-boson and $G_F^2 = 0.206 MeV^{-5} sec^{-1}$. All masses quoted in this contribution have to be read in units of MeV. Under the assumption that all charged leptons are much lighter than the W-boson, we expect (from Eqs.3 and 4) the electron positron decay width to be considerably larger than the gamma decay width.

3. Limits on Neutrino Lifetimes, Obtained in Laboratory Experiments

In searching for these decay modes an experiment has been performed [2], with a detector placed near the 2800 MW thermal power reactor at Gösgen (Switzerland), which can be used as an intense source of electron antineutrinos.

Reactor correlated events which might be attributed to either photons or electron positron pairs created by neutrino decay, have been looked for by comparing the counting rates during the reactor off and reactor on periods. The detector consisted of 30 lucite cells with outer dimensions $88cm \cdot 20cm \cdot 9cm$ filled with a liquid scintillator (NE235C). A detailed description of the experimental set up can be found in [2]. The experiment was performed between July 1, 1986 and August 11, 1986. 157.5 hours 'reactor on' and 248.5 hours 'reactor off' could be used for the off line analysis. The difference of the counting rates for reactor on minus reactor off was found to be consistent with zero, and limited to 6.1 counts/min (68 % CL) for an energy threshold of 0.384 MeV (0.63 counts/min for 1.499 MeV). These obtained experimental bounds were separately used for the analysis of both decay modes, taking into account the different detection efficiencies for γ-rays and for e^+e^--pairs.

3a. Neutrino Decay Mode $\nu_H \rightarrow \nu_L + \gamma$

If the decaying neutrino is dominantly coupled to electrons ($| U_{eH} |^2 \approx 1$), then only decay mode (Eq.1) is possible, because the value of m_{ν_e} is experimentally determined to be less than 45 eV [3]. The created particle ν_L might be some neutral fermion, possibly any lighter neutrino species.

In order to derive neutrino lifetime limits one has to calculate the expected laboratory photon spectrum and to compare it with the experimental bounds quoted above. From now on we will assume that the mass of the created neutrino ν_L is negligible (i.e. $m_{\nu_L} \ll m_{\nu_H}$). We can write the photon angular distribution dN in the center of mass system of the decaying neutrino

$$dN = \frac{1}{2}(1 + acos\theta)\, dcos\theta. \tag{5}$$

The factor a arises because of the unknown polarization of ν_L ($-1 < a < 1$). Consequently for a reactor which emits $n_\nu(E_\nu)$ neutrinos per second and per MeV, and a detector of volume V at a distance d (in our case d=64.7m) from the reactor core one expects

$$\left(\frac{m_\nu}{\tau_{c.m.}}\right) \frac{1}{cE_\nu} \frac{V n_\nu(E_\nu)}{4\pi d^2} \tag{6}$$

decaying neutrinos (per second and per MeV) in the detector. Here m_ν / E_ν arises because of the relativistic time dilation and $\tau_{c.m.}$ represents the c.m. neutrino lifetime. After the Lorentz transformation of the photon momentum from the c.m. into the laboratory system and combining Eqs.(5,6) one obtains the laboratory photon spectrum

$$\frac{dN}{dE_\gamma} = \left(\frac{m_\nu}{\tau_{c.m.}}\right) \frac{V}{c4\pi d^2} \int_{E_\gamma}^{\infty} \left(1 - a + \frac{2aE_\gamma}{E_\nu}\right) \frac{n_\nu(E_\nu) dE_\nu}{E_\nu^2}. \tag{7}$$

In order to calculate the expected counting rate in the detector one has to fold the photon spectrum with the efficiency for the detection of gamma rays. This has been estimated by a computer simulation for the applied discriminator threshold of 0.38 MeV.

From these calculations and the observed experimental limit upper neutrino lifetime limits as listed in Table 1 (68 % CL see ref.[2]) were obtained. These limits are more than one order of magnitude more restrictive than those quoted in ref.[4].

Table 1:

a	$m_\nu / \tau_{c.m.}$
-1	$\leq 4.5 \cdot 10^{-2} eV/sec$
0	$\leq 2.6 \cdot 10^{-2} eV/sec$
+1	$\leq 1.7 \cdot 10^{-2} eV/sec$

3b. Neutrino Decay Mode $\nu_H \rightarrow \nu_L + e^+ + e^-$

This decay mode (Eq.2) is possible, if heavy ($m_\nu > 1MeV$) neutrino mass eigenstates, subdominantly ($| U_{eH} |^2 \ll 1$) coupled to electrons exist.

The expected e^+e^- energy spectrum for the case of Dirac neutrinos can be calculated applying formulas which describe the c.m. double differential distribution of the created neutrino ν_L [5]. The detailed calculation for the energy spectrum can be found in ref.[2,4].

The number of heavy neutrinos in the detector follows by multiplying the spectrum of reactor neutrinos with the square of the relevant mixing parameter. The total number of decaying neutrinos is then given by

$$\left(\frac{m_\nu}{\tau_{c.m.}}\right) \frac{1}{cE} \frac{V n_\nu(E_\nu)}{4\pi d^2} \mid U_{eH} \mid^2 . \tag{8}$$

The integral over the e^+e^- laboratory total energy spectrum is therefore proportional to the quantity $m_\nu \Gamma \mid U_{eH} \mid^2$, using $\Gamma = 1/\tau_{c.m.}$. In order to obtain upper limits on $m_\nu \Gamma \mid U_{eH} \mid^2$ this integral has to be compared with the measured differential counting rates. With the use of the calculated expression for $\Gamma(m_\nu)$ (Eq.4), we show upper limits on $\mid U_{eH} \mid^2$ as a function of m_ν in Fig.1, together with results from other experiments. Our limits range from $2.0 \cdot 10^{-2}$ to $7.8 \cdot 10^{-4}$ for masses m_ν from 1.5 MeV to 4.0 MeV [2].

For $m_\nu > 10MeV$ limits on $\mid U_{eH} \mid^2$ are obtained from a study of the $\pi^- \rightarrow e^- + \bar{\nu}_e$ decay [6,7,8], while for $m_\nu > 1GeV$ limits on the same parameter can be found in ref.[9].

Cosmological observations like the abundance of primordial deuterium and the so far confirmed agreement of the microwave background with a Planck spectrum are the basis of astrophysical calculations with the aim to find constraints on the mixing parameter $\mid U_{eH} \mid^2$ [10-12]. High energy electrons and positrons, emitted due to neutrino decay, scatter with background photons and create high energy photons. Most of them are rapidly thermalized by pair creation and Compton scattering with thermal electrons. However some photons with an energy exceeding the threshold for photofission may destroy the light elements synthesized in the early universe and may

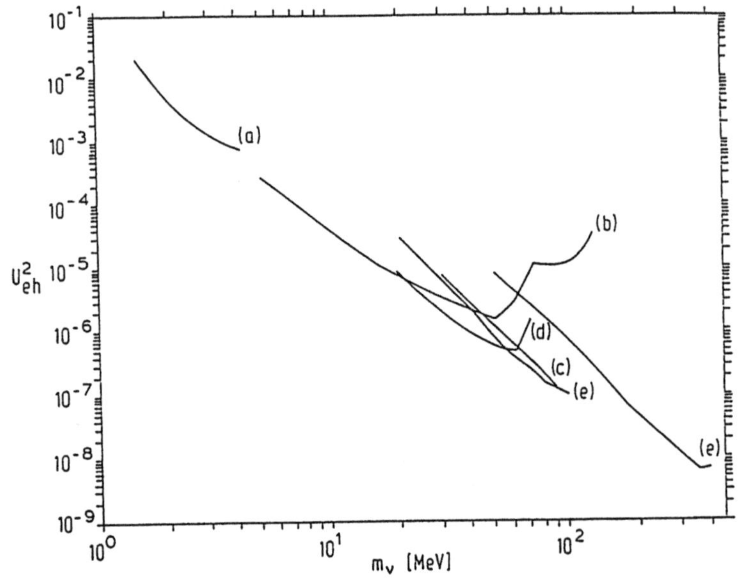

Fig.1: Laboratory upper limits on the mixing parameter $\mid U_{eH} \mid^2$ as a function of $m_{\nu h}$. Curve (a) corresponds to ref.[2], (b) to ref.[6a], (c) to ref.[6b], (d) to ref.[8], (e) to ref.[7].

distort the spectrum of the background radiation. Thus heavy neutrinos have to decay away before the termination of nucleosynthesis and the decoupling of radiation in order to meet the observed abundance of primordial deuterium and the background spectrum. Calculations, based on this arguments lead to *lower limits* on $|U_{eH}|^2$. The exact size of these cosmological constraints is still under discussion [13,14], they are, however, in the range of the limits we report here. If further laboratory experiments will provide sufficiently improved *upper limits* on $|U_{eH}|^2$, the existence of massive neutrinos in the mass range 1-100 MeV could be excluded, although one must not forget that the cosmological lower limits are highly depending on the understanding of the history of the universe.

4. Limits on Neutrino Decay $\nu_H \rightarrow \nu_L + \gamma$, Derived from SN 1987a

Lifetime limits for the gamma decay mode have been deduced from several astrophysical observations [15,16]. However, these constraints are based on assumptions about neutrino sources like the sun, former supernova explosions and white dwarfs, which have not yet been verified by experimental means.

The detected neutrino burst from SN 1987a at Feb.23,1987 together with the simultaneously measured γ -flux offer the extraordinary chance to derive very stringent and essentially model independent limits on the decay $\nu_H \rightarrow \nu_L + \gamma$ [17] for all generations of neutrinos.

A limit for the γ-flux in the energy range $4.1MeV \rightarrow 6.4MeV$ of $0.3\gamma/cm^2sec$ during a 10 sec interval encompassing the neutrino burst was obtained with the "Solar Maximum Mission" (SMM) satellite [18]. The observed duration of ten seconds for the detected neutrino pulse indicates the electron neutrino mass to be lower than $20eV$ which is confirmed by laboratory experiments [19]. This implies that a possible γ burst originating from the decay (Eq.1) should not extend these ten seconds.

The measured energy spectrum of the neutrino burst attributed to the SN 1987a by K.HIRATA et al.[20] and R.M.BIONTA et al.[21] is consistent with a Fermi Dirac distribution at a temperature $2MeV < T_{\bar{\nu}_e} < 4.5MeV$ [22]. This is in very good agreement with most of the present models for gravitational stellar collapses.

The analysis for the lifetime limit for $\bar{\nu}_e$ is related to Eqs.6,7 and is explained in detail in ref.[17]. It should be emphasized, that the resulting bound ($T_{\bar{\nu}_e} = 3MeV$)

$$\tau/m_{\bar{\nu}_e} > 8.3 \cdot 10^{14} sec/eV$$

is merely based on the two experimental facts quoted above and fairly insensitive to variations of the neutrino temperature [17].

Clearly, no μ- or τ-neutrinos emitted from SN 1987a could be identified with terrestial detectors. However their spectrum can be calculated from standard weak interaction and Fermi-statistics as long as $m_{\nu_i} << T_{\bar{\nu}_e}$, were m_{ν_i} corresponds to the mass eigenstates ν_i to which a neutrino with a given flavor is dominantly coupled [23]. If the neutrino masses hold $m_{\nu_{\mu,\tau}} < 20eV$ and provided that these neutrino flavors did not decay inside the supernova interior we deduce from the ν-spectra mentioned above and the γ flux reported a lifetime limit of

$$\tau/m_{\nu_i} > 3.3 \cdot 10^{14} sec/eV$$

at a temperature of $T_{\bar{\nu}_e} = 3MeV$. For higher $m_{\nu_{\mu,\tau}}$ this limit changes due to the increasing duration of the expected γ pulse. The corresponding limits are depicted in Fig.2 for masses up to $1MeV$.

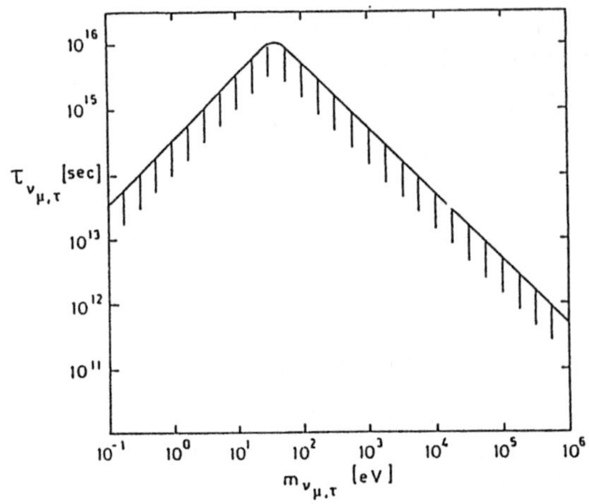

Fig.2: Lower limits on μ, τ-neutrino lifetime as a function of $m_{\nu\mu,\tau}$ up to 1MeV, as obtained from SN 1987a.

References:

[1] E. Sato and M. Kobayashi, Progr. Theor. Phys. 58, 1775 (1977)
[2] L. Oberauer et al., accepted for publ. in Phys. Lett. B
[3] V.A. Lubimov et al., Phys. Lett. 94B, 266 (1980) ; M. Fritschi et al. Phys. Lett. 173B, 485 (1986)
[4] P. Vogel, Phys. Rev. D.30, 1505 (1984)
[5] R.E. Shrock, Phys. Rev. D24, 1232 (1981); 24, 1275 (1981)
[6a] D.A. Bryman et al., Phys. Rev. Lett 50, 1546 (1983)
[6b] D.A. Bryman et al., Phys. Rev. Lett 50, 7 (1983)
[7] G. Bernadi et al., Phys. Let 166B, 479 (1986)
[8] N. de Leener-Rosier et al., Phys. Lett. 177B, 228 (1986)
[9] Charm Coll., Phys. Lett. 166B, 473 (1986)
[10] S. Sarkar,A.M. Cooper, Phys. Lett. 148B, 347(1984)
[11] M. Kawasaki, K. Sato, Phys. Lett. 169B, 280(1986)
[12] M. Kawasaki et al., Phys. Lett. 178B, 71(1986)
[13] D. Lindley, Astrophys.J. 294(1985)
[14] S. Sarkar, Mini-review Bari Conf. 1985, unpublished.
[15] R. Cowsik, Phys. Rev.Lett. 39, 784(1977)
[16] G. Raffelt, Phys. Rev. D31, 3002(1985)
[17] F. von Feilitzsch, L. Oberauer, to be published
[18] W.T. Vestraud, A. Gosh, E. Chupp, IAU-Circular No. 4338,4340,4365
[19] M. Fritschi et al., Phys. Lett. 173B, 485(1986)
[20] K. Hirata et al., Phys. Rev. Lett. 58, 1490(1987)
[21] R.M. Bionta et al., Phys. Rev. Lett. 58, 1494(1987)
[22] J. Arafune and M. Fukugita, to be publ. in Phys. Rev. Lett. 59, 367(1987) and D.N. Schramm, XXIInd Recontre de Moriond, March 87
[23] J.R. Wilson et al., Ann. N.Y. Acad. Sci. 470, 267(1986)

KARMEN: Neutrino Physics at ISIS*

G. Drexlin[1], *H. Gemmeke*[1], *G. Giorginis*[1], *W. Grandegger*[1], *J. Kleinfeller*[1], *R. Maschuw*[1], *P. Plischke*[1], *F. Raupp*[1], *F.K. Schmidt*[1], *J. Wochele*[1], *B. Zeitnitz*[1], *E. Finckh*[2], *W. Kretschmer*[2], *D. Vötisch*[2], *J.A. Edgington*[3], *T. Gorringe*[3], *N.E. Booth*[4], and *A. Dodd*[5]

[1]Kernforschungszentrum und Universität Karlsruhe, Institut für Kernphysik I, P.O.Box 3640, D-7500 Karlsruhe, Fed. Rep. of Germany

[2]Physikalisches Institut, Universität Erlangen-Nürnberg, Erwin-Rommel-Str. 1, D-8520 Erlangen, Fed. Rep. of Germany

[3]Physics Department, Queen Mary College, Mile End Road, London E14 NS, United Kingdom

[4]Department of Nuclear Physics, Oxford University, Keble Road, Oxford, OX1 3RH, United Kingdom

[5]Particle Physics Department, Rutherford Appleton Laboratory, Chilton, Didcot, OX11 0QX, United Kingdom

1. Introduction

KARMEN denotes an experimental program of neutrino physics to be performed at the spallation neutron source ISIS of the Rutherford Appleton Laboratory |1|. It is based on the 'beam dump' neutrino source of the 800 MeV, 200 µA proton beam from a rapid cycling synchrotron (50 Hz) bombarding a UD_2O spallation target. Monoenergetic ν_μ with $E_{\nu_\mu} = 29.8$ MeV emerge from π^+ decay at rest whereas the subsequent μ^+ decay provides ν_e and $\bar{\nu}_\mu$ with energies up to 52.8 MeV (see fig.1a).

Due to the different lifetimes of π^+ (26 nsec) and μ^+ (2.2 µsec) there are prompt ν_μ within the first 500 nsec correlated to the proton double pulse (duty cycle 10^{-5}) whereas after 0.5 µsec one is left with $\bar{\nu}_\mu$ and $\bar{\nu}_e$ only from the slower μ^+ decay (duty cycle 2.5×10^{-4}) (see fig.1b).

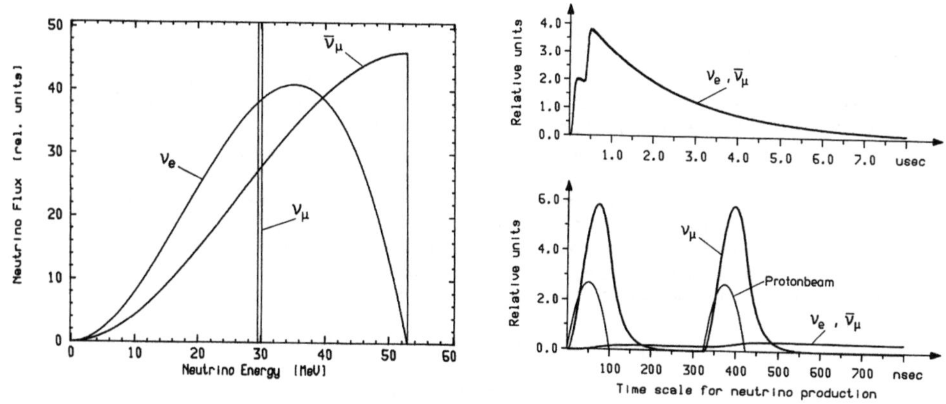

Fig.1: Energy spectrum (a) and time structure (b) of the ISIS beam dump neutrino source.

*Work supported in part by the Bundesministerium für Forschung und Technologie, Bonn, FRG

This distinct time structure with its low ν-duty cycle allows very effective suppression of background as well as stringent signatures on events induced by the different neutrinos.

The notoriously low cross section of neutrino reactions particularly at low energies require large target masses as well as heavy shielding against any background radiation. At ISIS, adjacent to the target station with its biological shield of effectively 4.2 m of steel, a 6000 tonnes neutrino blockhouse with 2 m thick steel walls and a 3 m thick steel roof has been set up. A 56 t scintillation calorimeter with good energy and time resolution and reasonable spatial resolution is currently being installed. It is dedicated to look for two different appearance oscillations $\nu_\mu \rightarrow \nu_e$ and $\bar{\nu}_\mu \rightarrow \bar{\nu}_e$ simultaneously, to study the $^{12}C(\nu_e\ e^-)^{12}N$ reaction and to observe the neutral current excitation $^{12}C(\nu_\mu, \nu_\mu')^{12}C$ (15.1 MeV). We will also try to deduce another value for the ν_e-e scattering total cross section with respect to the NC - CC interference problem. At a later stage of the experiment it is intended to install another 50 t detector but with much higher granularity to enable differential cross section measurements of the neutrino electron scattering.

2. ν-oscillations

Neutrino oscillations require massive neutrinos as well as mixing of their mass eigenstates |2|. For a two component composition of mass eigenstates m_1 and m_2 the vacuum ocsillation probability is given by

$$P(\nu_x \rightarrow \nu_y) = \sin^2 2\theta \sin^2 \left[\frac{\Delta m^2[eV^2]\ L[m]}{E[MeV]}\right]$$

with $\Delta m^2 = |m_1^2 - m_2^2|$, and $\sin^2 2\theta$ denoting the mass mixing.

Evidence of appearance oscillations $\nu_\mu \rightarrow \nu_e$ or $\bar{\nu}_\mu \rightarrow \bar{\nu}_e$ would require statistically significant detection of ν_e or $\bar{\nu}_e$ in excess of any inherent occurence or faking background. In the 56 t scintillation calorimeter ν_e will be detected via the reaction

$$\nu_e + {}^{12}C \rightarrow {}^{12}N + e^- - 17.3\ \text{MeV}$$

followed by the decay

$$^{12}N \rightarrow {}^{12}C + e^+ \ (16.3\ \text{MeV e.p.}) + \nu_e; \quad (\tau = 16\ \text{msec}).$$

For oscillation of the monoenergetic ν_μ to ν_e this would mean a spatially correlated delayed coincidence between a 12.7 MeV monoenergetic electron during the prompt ν_μ-time window and a positron with up to 16.3 MeV during the beam pause where no neutrinos are present. This should provide a very stringent almost background free event signature. The same signature applied to the actually present ν_e from μ^+ decay and extrapolated to the appropriate ν_μ time- and energy window serves for normalization and provides a flux independant measurement of the oscillation probability $P(\nu_\mu \rightarrow \nu_e)$.

A similar delayed coincidence signature will also exist for a possible $\bar{\nu}_\mu \rightarrow \bar{\nu}_e$ appearance oscillation: $\bar{\nu}_e$ which are not present in the original neutrino beam will be detected by the process

$$\bar{\nu}_e + {}^1H \rightarrow n + e^+ - 1.8\ \text{MeV}$$

followed by

$$\text{Gd}\ (n,\langle 3\gamma \rangle)\ 8\ \text{MeV}.$$

A high energy positron up to 51 MeV will be detected within the $\bar{\nu}_\mu$ time window of 0.5 to 5 μsec. With some delay the neutron after thermalization would give rise to, on average, three gamma quanta with sum energy of 8 MeV from capture by a small amount of Gd brought into the detector.

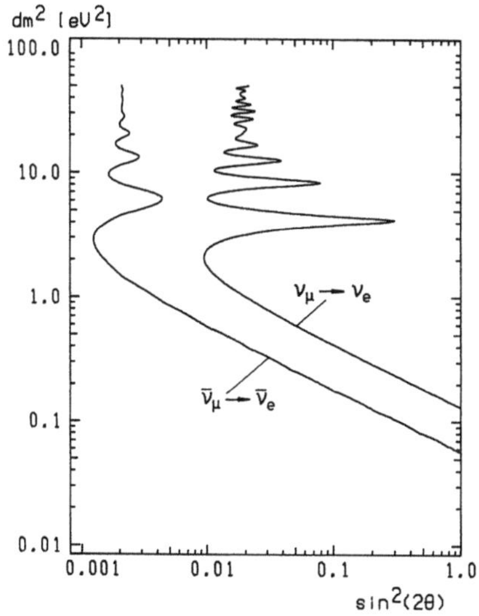

Fig.2: Sensitivities for neutrino oscillations at at KARMEN.

Fig.2 shows the sensitivities of these oscillation experiments to the mass mixing and the mass difference in terms of a two component neutrino composition. The figures are for two years of measuring time at full ISIS intensity (approximately 4 years actual time). Without overinterpreting these curves we believe that it is a realistic estimate to be sensitive to the mixing $\sin^2 2\theta$ slightly below 10^{-2} and to Δm^2 slightly below 10^{-1} eV^2 even with a background rate of 0.1/day.

3. ν-nucleus interaction

Apart from the charged current reaction $^{12}C(\nu_e,e^-)^{12}N$ |3| the scintillation calorimeter will also be used to look for the neutral current excitation $^{12}C(0^+,0)(\nu_\mu, \nu_\mu')^{12}C(1^+,1)$, 15.1 MeV above ground state. At 'beam dump' energies this transition can be shown to be exclusively isovector axialvector |4| thus a measurement would probe selectively one specific electroweak coupling. Comparison to the same excitation by ν_e would prove the μ-e universality assumption for the ν-coupling to the neutral current. The signature of such processes would be the detection of a monoenergetic 15.1 MeV gamma quantum in the calorimeter during the short ν_μ- or $(\bar{\nu}_\mu, \nu_e)$ time window, as the $(1^+,1)$ state decays to more than 90% back to the ground state of ^{12}C. Other transitions are either particle unstable not being detectable or cascade down via those states yielding much lower γ-energies. Thus although only a single prong event, this specific transition should clearly be detectable with a high resolution calorimeter.

4. ν-e scattering

The main target of a second fine grain calorimeter currently under investigation (a 20 ℓ LArTPC is operated at KfK |5|) will be the measurement of the differential cross section of the neutrino electron scattering. This then should not only resolve the problem of the NC-CC interference of the weak interaction |6,7| but also contribute to the question of a possible magnetic moment of the neutrino |8|. As the e.m. cross section due to a magnetic moment μ_ν,

$$d\sigma/dy = \pi r_e^2 \mu_\nu^2 (1-y)/y; \; y = Te/E_\nu,$$

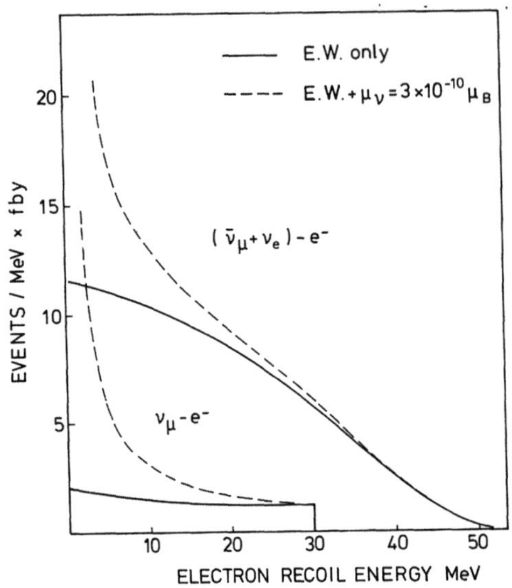

Fig.3: Neutrino electron scattering at KARMEN II, sensitivity to neutrino magnetic moment.

only depends on the neutrino energy transfer y, the small 'electroweak' cross section at low energies advantageously enhances the magnetic moment sensitivity compared to high energy experiments. Fig.3 shows the corresponding values for ν_μ-e and ($\bar{\nu}_\mu$ + ν_e)-e scattering respectively. It indicates that an adequate detector should have the lowest possible threshold at least at about 5 MeV minimum electron recoil energy.

5. The KARMEN detector

Finally the 56 t scintillation calorimeter currently being installend at RAL shall be described (see fig.4). A large steel vessel 6 m high, 3.2 m wide and 3.5 m long is filled with 65 000 ℓ mineral oil based liquid scintillator PPP (KfK/MERCK). The 150 mm thick head walls contain 1120 bores each, to keep the overall 2240 phototubes looking into the scintillator. The central detector consists of 512 optically separated modules of 17.7x18.1 cm cross section and 350 cm length. Each module is viewed by two 3" phototubes XP 3462 from either end. The optical segmentation consists of very thin (2x1.5 mm) double lucite sheets glued together at its rim containing Gd-coated paper in between (7 mg/cm^2 Gd ≙ 0.1 weight %). The thin air gap between the lucite sheets provides total internal reflection to guide the light towards the phototubes whereas the Gd_2O_3 paper serves for neutron detection. The central detector with an active mass of more than 98% is surrounded by a layer of half width modules serving as active inner anticounter to define a 'contained event'. 20 cm thick steel slabs providing the mechanical strength of the detector are used as passive inner shield followed by a 3 cm Ne110 scintillator active outer anticounter. This sandwich anticounter system predominantly is designed to suppress neutrals entering the detector from the last 20 cm of passive shielding by a factor of 10^{-2}.

The calorimeter is self triggering from each of its 512 modules with a 70 nsec event time followed by a fast (< 400 nsec) 'valid event' decision from a timing logic and a trigger processor. A 30 µsec pretrigger activates the detector for stopped muons which when decaying could fake a ν-induced event during the relevant time windows.

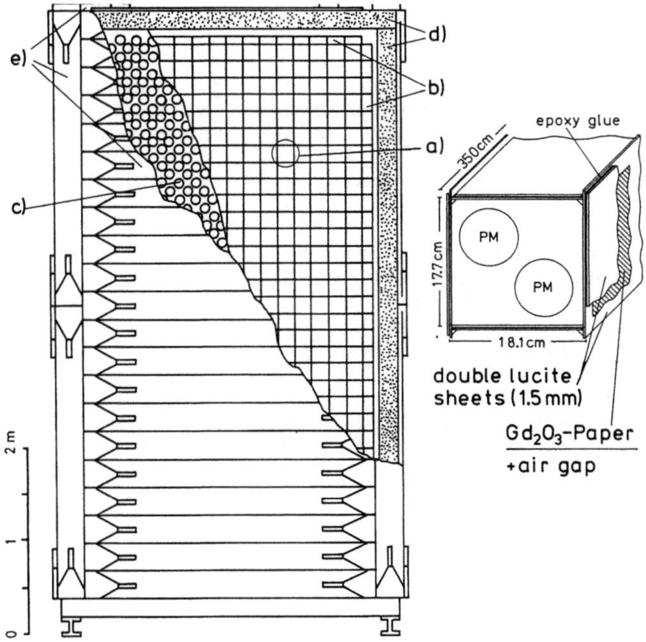

Fig.4: The KARMEN neutrino detector
a) 512 central modules 17.7x18.1x350 cm
b) inner active shield, 96 half width modules
c) 2240 PM 3", VALVO XP3462
d) passive inner shield 20 cm steel
e) outer active shield 3cm NE110 plastic scintillator.

The properties of the detector system have been tested in various prototypes of 200 ℓ (1 module), 1000 ℓ (9 modules) and 8000 ℓ (35 central + 24 outer modules) size. The position dependant light output has been measured with triggered cosmics and can completely be reproduced by Monte Carlo calculations. As expected any light output is linearly correlated to the energy deposit in the detector. The same linear correlation holds for the time difference measured between the ends of a module and the actual position of light emission providing a spatial resolution of about 10 cm. From cosmic ray measurements as well as with the UV-laser calibration system the energy resolution was determined to be $\sigma(E)/E = .12/\sqrt{E(MeV)}$. Finally a neutron detection efficiency of 30% could be deduced from a mock up experiment with tagged neutrons in the 1000 ℓ detector |9|.

The detector is currently in its final stage of assembly. All phototubes have been attached, the electronics and data handling system has been installed and is running through debugging procedure. Insertion of the delicate optical segmentation has just started. The calorimeter will be commissioned for data taking after the ISIS shut down period in spring 1988.

7. References

|1| B.Zeitnitz, Prog. in Part. and Nucl. Phys. Vol. 13, (445) 1985.

|2| For a review of recent accelerator neutrino oscillation experiments, see W.Y.Lee in 12th Int.Conf. on Neutrino Physics and Astrophysics, Sendai 1986, p.157

|3| Recent refined calculations of the transition $^{12}C(\nu_e\ e^-)^{12}N$ g.s. by T.W.Donnelly, to be published; see also |4|

|4| T.W.Donnelly, R.D.Peccei, Phys.Rep. 50 (1-85) 1979 and T.W.Donnelly, Prog.in Part. and Nucl.Phys. Vol.13, (183) 1986

|5| G.Giorginis, D.Mann, K.H.Ottmann, S.Wölfle, H.Gemmeke, R.Maschuw, B.Zeitnitz, to be published in NIM for Phys.Res.A 1987

|6| B.Kayer, F.Fishbach, S.P.Rosen, H.Spivok, Phys.Rev.D20 (20) 1979

|7| H.H.Chen et al., in 12th Int.Conf. on Neutrino Physics and Astrophysics, Sendai 1986, p.392

|8| D.Grosser, Phys.Lett. 166B (340) 1986

|9| H.Gemmeke, W.Grandegger, R.Maschuw, P.Plischke, B.Zeitnitz, to be published in IEEE Trans. of Nucl.Sci. NS35 Vol. 1, 1987

Search for Right-Handed Currents in Nuclear β-Decay

V.A. Wichers, T.R. Hageman, J. van Klinken, H.W. Wilschut, and D. Atkinson

K.V.I. and State University of Groningen, Zernikelaan 25,
NL-9747 AA Groningen, The Netherlands

Constraints are reported on right-handed currents based on a precise comparison of the longitudinal β-polarizations P_F and P_{GT} in Fermi and Gamow-Teller transitions. Measurements with the matching isotopes ^{26m}Al and ^{30}P gave P_F/P_{GT} = 1.003 ± 0.004.

1 Introduction

The phenomenological V-A structure of charged-current weak interactions is a built-in feature of the standard model. Maximum violation of parity remains therefore an open question, and its verification is left as a fundamental issue to experimental physics. The experimental status of absolute β-polarization measurements is [1]: $P_F = 0.97 \pm 0.19$ and $P_{GT} = 1.00 \pm 0.01$. The higher accuracy in P_{GT} is due to the fact that P_{GT} can be studied off-line with electrons using Mott-polarimeters. In contrast the fast superallowed Fermi transitions must be studied with less sensitive positron-polarimeters. The relatively large uncertainty in P_F allows large deviations from the V-A value $P_\beta = -1$.

The interest in a possible existence of V+A currents revived [2] with the advent of left-right symmetric models of electro-weak interactions [3]. In these models both W_R and W_L exist. The gauge-boson mass eigenstates $W_{1,2}$ are mixtures of W_L and W_R: $W_1 = W_L \cos\zeta - W_R \sin\zeta$, $W_2 = W_L \sin\zeta$ +

$W_R \cos\zeta$. The structure of the leptonic current in β-decay is thus determined by the mixing-angle ζ and by the squared mass-ratio $\delta \equiv (M_1/M_2)^2$. In case of a light right-handed neutrino, P_F and P_{GT} deviate from the V-A value -1: $P_F \cong -1 + 2(\delta-\zeta)^2$, $P_{GT} \cong -1 + 2(\delta+\zeta)^2$. Note that P_F and P_{GT} become different. As a consequence the ratio $R \equiv P_F/P_{GT} \cong 1+8\delta\zeta$ is also sensitive to V+A currents. This observation led us to consider relative P_β-measurements, which offer better prospects for high accuracies than the anyhow more difficult absolute measurements. The reason is that systematic errors cancel in the ratio. Only one pair of isotopes is suitable for our purposes: ^{26m}Al (0^+-0^+; E_0=3.211 MeV, $t_{1/2}$=6.35 s), and ^{30}P (1^+-0^+; E_0=3.205 MeV, $t_{1/2}$=149.8s). These isotopes both decay by positron emission, and have nearly equal endpoints.

2 Apparatus

The present polarimeter [4] is based on Bhabha-scattering of positrons with spin-alligned electrons. The complete device yields 1.5% statistical accuracy in the polarization-ratio per day. This high statistical power is achieved through the combination of a large polarization sensitivity (≈0.1 for symmetric collisions), and the acceptance of a wide range in scattering kinematics. The transmission of the Mini-Orange spectrometer is centered at ≈1.8 MeV and has a width of ≈1.8 MeV. The five detectors cover a large solid angle of the foil with both symmetric collisions (between outer detectors), and asymmetric ones (between the central and an outer detector).

Instrumental and spurious effects are suppressed in the present polarimeter design by introducing the rotational symmetry of Mini-Orange spectrometer, and of the scatterer and detector assembly. It is also important that the magnetic scatterer is operated at a small holding field.

The complete setup consists of four polarimeters and a dual system for target transport. The striving for a symmetric arrangement was extended to the concept of a multi-polarimeter arrangement, symmetric with respect to the beam-axis and with respect to the plane of the targets (Fig. 2). Effects from possible beam-fluctuations, or from misallignment of the targets are of opposite sign for opposed polarimeters, and cancel to first order in the combined result.

The ^{26m}Al and ^{30}P sources are produced on-line in (p,n)-reactions on ^{26}Mg and ^{30}P, respectively, using 9 MeV protons from the cyclotron of the Vrije Universiteit of Amsterdam. A ^{26}Mg and a ^{30}Si target are mounted on opposite wings of a target-rotor, which is driven by a stepping-motor.

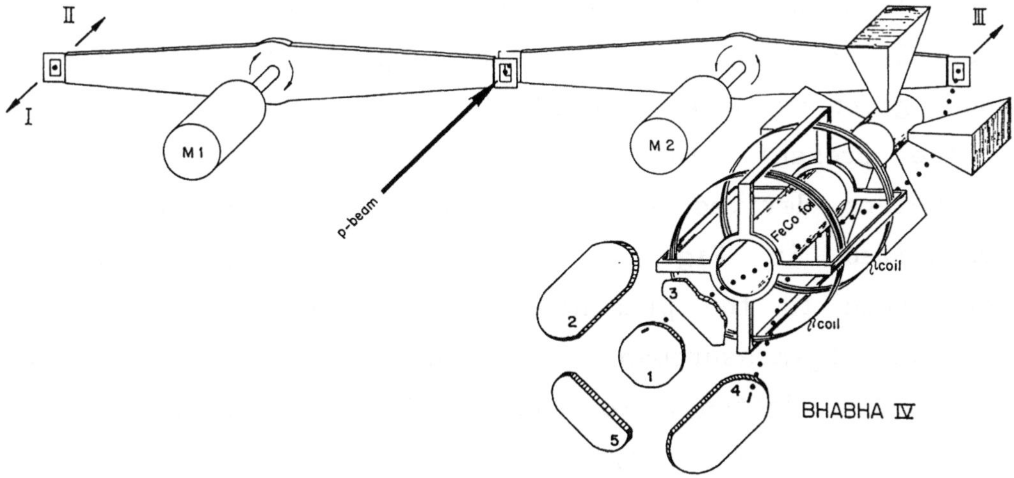

Fig. 1. Sketch of the fourfold Bhabha polarimeter with details on polarimeter IV.

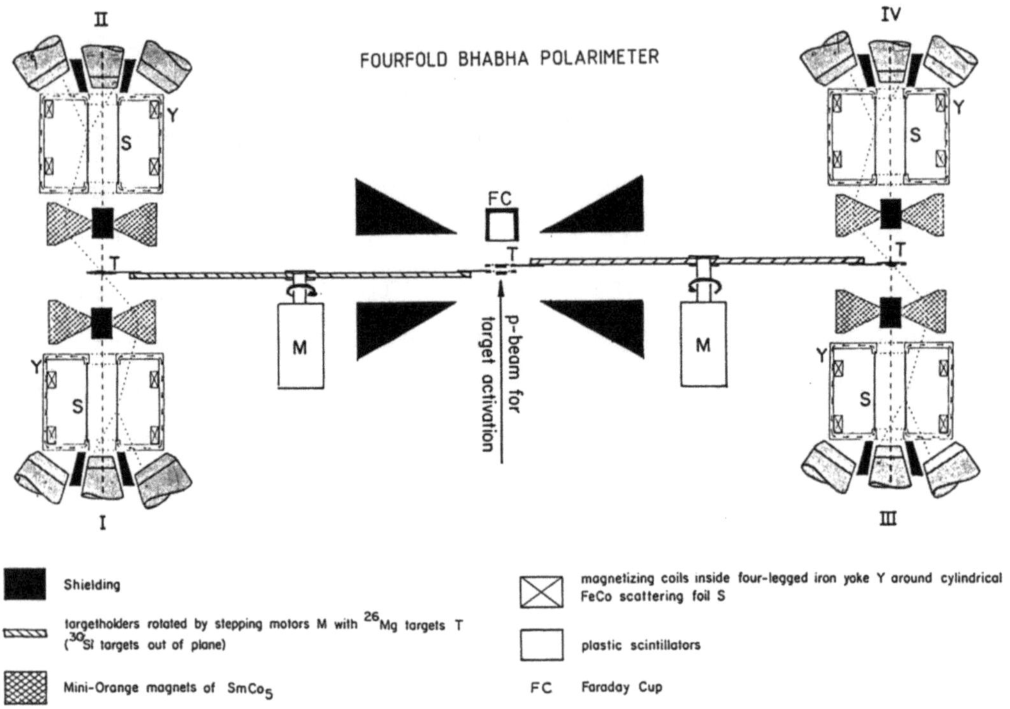

Fig. 2. Horizontal view of the arrangement.

3 Result

The final result is R = 1.003 ± 0.004, consistent with the V-A value R = 1. The result for R translates directly into the 90% confidence bounds on the right-handed parameters δ and ζ given by $-0.49 < \delta\zeta\times10^3 < 1.2$

The present status of constraints on right-handed parameters derived from β- and μ-decay experiments is summarized in Fig. 3. The 90% confidence contours correspond to the present result (bold curves [4]), to previous P_β-measurements (dashed), to an analysis of ^{19}Ne-decay (dotted), to the precise measurement of the e^+-endpoint in μ^+-decay (1), and to the Michel-parameter (2) (see [5] and references quoted therein).

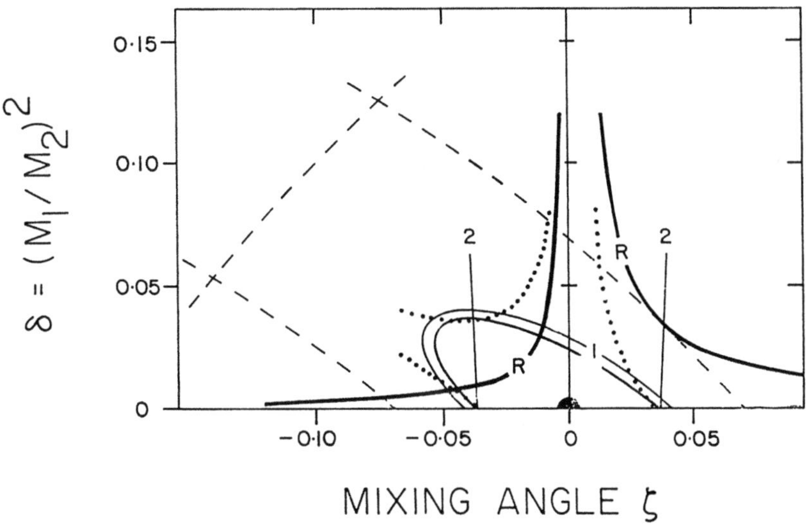

Fig. 3. Summary of bounds on right-handed currents from β- and μ-decay.

The bounds in Fig. 3 are valid in case of Dirac neutrinos or light right-handed Majorana neutrinos. Searches for $(\beta\beta)_{0\nu}$-decay provide accurate bounds on V+A currents if neutrinos are Majorana particles [6]. These experiments constrain the parameter combinations $\langle\delta\rangle \equiv \delta v_e$, and $\langle\zeta\rangle \equiv \zeta v_e$, when the predominantly right-handed neutrino has a mass $\geq$ 100 MeV/c^2, and where v_e may be viewed as the mixing-angle of left- and right-handed neutrinos [7]. Therefore, experiments in β- and μ-decay are complementary to $(\beta\beta)_{0\nu}$-decay searches in constraining right-handed currents.

References

1 J. van Klinken, F.W.J. Koks, and H. Behrens, Phys. Lett. 79B, 199 (1978)

2 M.A.B. Bég *et al.*, Phys. Rev. Lett. 38, 1252 (1977)

3 See R.N. Mohapatra, this conference for a review

4 V.A. Wichers *et al.*, Phys. Rev. Lett. 58, 1821 (1987)

5 A. Jodidio *et al.*, Phys. Rev. D34, 1967 (1986)
6 T. Tomoda, A. Faessler, K.W. Schmid, F. Grümmer, Nucl. Phys. B153, 1 (1985)
7 M. Doi, T. Kotani and E. Takasugi, Prog. Theor. Phys. Suppl. 83, 1 (1985)

The Neutrino as a Tachyonic Non-charged Light Magnetic Monopole?

J.J. Steyaert

Institut de Physique Nucléaire, Université Catholique de Louvain, B-1348 Louvain-la-Neuve, Belgium

It has been shown by CHODOS et al. [1] that ν_μ could be a tachyon and results on ν_e by ETH [2] could have a tachyon interpretation, i.e. a negative squared mass as a result of kinematics.

These interpretations are hampered by the common believe that tachyon do not exist. Sporadic searches were negative.

First of all we want to let you feel that perhaps the tachyon could exist.

We performed in LLN experiments about attenuation of gamma rays in copper. The obtained spectra were interpreted as an indirect evidence of tachyon existence, thanks to a new effect, the tachyo-electric effect we think we discovered. It is a very simple effect where a tachyon pair transfers its 4-momentum to an electron in matter. This is a special case of reactions already considered by TERLETSKII [3] where the absorbing particle does not change its rest mass (M = M').

The clue that this effect is present, and that maybe tachyons exist, is the shape of the spectrum.

Supposing a momentum distribution

$$N(p) = \frac{C}{p^2 + \gamma^2} \tag{1}$$

by tachyo-electric effect we obtain a distribution of the kinetic energy T of recoiling electrons hit by a pair of tachyons of "mass" m_t :

$$N(T) = \frac{C}{2T(T + m_e) + 4m_t^2 + \gamma^2 - 2(T^2 + 2Tm_e)^{1/2}(T^2 + 4m_t^2)^{1/2}} \tag{2}$$

This almost lorentzian shape fits the experimental spectra very well on the high energy side and "explains" the background on low energy side. The maximum is given by

$$T = \frac{2m_t^2}{m_e} \tag{3}$$

A direct evidence for tachyon existence was obtained at LLN cyclotron, CYCLONE, with a beam of 4.43 MeV γ's from the reaction $^{12}C(p,p')^{12}C^*$. The γ were attenuated in a 20 cm long copper cylinder by a factor of about 330 and detected by a plastic scintillator in time relation with a beam-pickoff via a Time to Amplitude converter.

This time of flight device showed clearly the v = c behaviour of the attenuated γ rays. A copper plate of 0.5 mm was put in front of the plastic scintillator. The discriminator was set up in such a way that electrons of energy higher that 1 MeV were recorded.

By difference the role of the copper plate could be obtained. A clear supra-luminal signal was found.

This encouraging signal needed to be confirmed by changing the experimental conditions.

With the help of T. Knöpfle and colleagues of M.P.I. in Heidelberg the experiment was performed with a Tandem Emperor fitted with a buncher and rebuncher giving 12 MeV protons pulses of width 150 ps.

This time the detector was a BaF_2, borrowed from G.S.I. Darmstadt, of resolution 600 ps in time and about 300 keV in energy.

Again, via a time of flight method, a supra-luminal signal was found. More information could be obtained as two parameters (E, t) recording was performed.

It should be noted that no copper was used and the intensity of tachyons was high, about the same as the γ ray signal, mainly when a 4 cm thick aluminium plate was placed close to the carbon target.

A difference of energy of about 300 keV between γ and tachyon signal could be attributed to a coupling of tachyon which earth magnetic field. The tachyon we detect could thus be a light magnetic monopole. This point of view is suggested by the strong coupling to the photon. More experiments are planned.

Tachyon pair creation could well be an electro-magnetic phenomena.

This is suggested by the possible creation process.

Why to so speak about the tachyon in a neutrino workshop ?

Simply because these two particles are probably related.

On the theoretical side, it was shown by LOCHAK [4] that the Dirac equation has a solution with a massless magnetic monopole, and that a non-linear version of that same equation has plane wave solutions in term of bradyons ($v < c$), luxons ($v = c$) and tachyons ($v > c$).

Moreover, the Dirac quantization conditions $g = \frac{137}{2} ne$ could be obtained with a minimal hypothesis and Lochak proposes that the case n=0 could well be the neutrino.

A there are many neutrinos, at least 3, one could wander if there are many tachyons !

Most of the experiments we performed shown a tachyon of "mass" $m_t = 209$ keV, the parameter in $p_t^2 - E_t^2 = m_t^2$.

More difficult experiments could show $m_t \simeq$ 10-20 keV and probably $m_t \sim 1$ GeV.

The consequences of having such a huge amount of tachyons, as soon as energetic photons are present, are immense, mostly in astrophysics.

It has been shown by SUDARSHAN [5], then by YAMAMOTO [6], that tachyons play a role in elastic scattering.

It so happen that the characteristic momentum of 209 keV of the tachyon we know best corresponds to a recoil energy of 93 eV for the proton, the prominent constituent of the Universe, and to a speed of 133 km/s for the hydrogen atom.

Remarking that most of the astrophysical speeds of jets, clusters, galaxies ranges in the domain 10 to 1000 km/s, we believe that tachyon should seriously be considered as a fundamental particle at work in the cosmos.

The Dark Matter Problem of the galaxies is maybe just due to hydrogen clouds elastically scattered at a critical speed, preventing them to orbit only under gravitationnal force. The flat "rotational" curve of most galaxies point maybe in that direction. We have no time to go very deep in the proposal [7] that the "nuclear-atomic" reaction

$$H + H \rightarrow D + \nu \tag{4}$$

could be important among hydrogen clouds. The D recoils with a speed of 230 km/s which is about the "flat" speed of Sb galaxies like NGC 4565 and M31 [8]. When the gas content decreases the speed decreases and becomes less flat, probably due to pure gravitational force among dominant stars.

The inconsistencies in the $H^+ + H^-$ problem [9] are probably related to what happen deep in the sun.

The Solar Neutrino Problem is maybe just caused by an extra heat transfer due to elastic scattering of 100 eV protons that are certainly present in a kT = 1500 eV hot plasma. The WIMPS [10] hypothesis was conceived on that purpose. The suggested cross section of about 10^{-36} cm^2 for elastic scattering could mimic a cross section of 10^{-14} cm^2 times a probability factor of $(10^{-13}/10^{-8})^3 = 10^{-15}$. A factor of $10^{-36}/10^{-29} = 10^{-7}$ could be shared by $m_{proton}/m_{WIMPS} \sim 10^{-1}$ and (resonance width) /1500 eV $\sim 10^{-6}$. At top of resonance the tachyon acts as a catalyser to improve heath transfer and, maybe, produce D by reaction (4).

Last but not least, the solar wind origin and the corona heating to about 100 eV is maybe also related to the universal tachyon.

References

1 A. Chodos , A.I. Hauser, V.A. Kostelecky, Phys. Lett. 150B, 431 (1985).

2 These Proceedings.

3 Ya.P. Terlestkii, Kinetics of Reactions of Tachyon Production and Absorption, JINR-P2 11149 (Dubna) and DESY-L-Ueb-218 (1977).

4 G. Lochak, Int. Journ. Theo. Phys. 24, 1019 (1985).

5 J. Dhar, E.C.G. Sudarshan, Phys. Rev. 174, 1808-1815 (1968).

6 Hiroshi Yamamoto, Progr. Theor. Phys. 55, 2004 (1976).

7 J. Steyaert, Quasar Energy from Frozen Fusion via Massive Neutrinos ? In The Big Bang and Georges Lemaître, ed. by A. Berger, p. 135, (Reidel, 1984).

8 S.M. Faber, J.S. Gallagher, Ann. Rev. Astron. Astrophys. 17, 145 (1979).

9 V. Sidis et al., Phys. Rev. A, 27, 2442 (1983).
S. Szucs et al., J. Phys. B 17, 1621 (1984).
F. Brouillard, In Electronic and atomic collisions, ed. by J. Eichler, I.V. Hertel, N. Stalterfaht, p. 347 (Elsevier, 1984).

10 S. Raby, G.B. West, Nucl. Phys. B292, 793 (1987).

Part III

Double Beta Decay and Neutrino Mass

Possible Test of Grand Unification in the Double Beta-Decay*

A. Faessler

Institut für Theoretische Physik, Universität Tübingen,
D-7400 Tübingen, Fed. Rep. of Germany

The more successful grand unified theories predict that the neutrino is identical with its antiparticle and therefore is a Majorana neutrino which violates lepton number conservation. Such a neutrino should have a finite mass and also a small right handed weak interaction. If the double neutrinoless beta decay is observed with the full decay energy in the two electrons, it would establish that the electron neutrino is a Majorana particle. It is shown that the relativistic corrections of the nucleonic wave functions are essential for determining an upper limit of the right handedness from the measured lower limit of the life-time against the neutrinoless double beta decay. The upper limit for the right handedness of the weak interaction derived from the lower limit of the life-times against the neutrinoless beta decay is $|\langle\eta\rangle| < 1.8 \times 10^{-8}$ and the upper limit for the neutrino mass is $|\langle m_\nu \rangle| < 1.9$[eV]. To test the nuclear structure part of the theory the double beta decay with two neutrinos is calculated and compared with the known experimental values. It is shown that it is important to take into account in the intermediate nucleus not only the particle-hole, but also the particle-particle interaction which acts destructively and reduces the double beta decay with two neutrinos by about a factor 1/100 and one obtains a reasonable agreement with the experimental measured data. Applying the same improved description to the β^+-decay in proton rich nuclei, one can explain the long standing puzzle of the quenching of the Gamow-Teller strength in agreement with the data.

1. INTRODUCTION

The neutrino is the only fermion from which we do not know if it is different from its antiparticle ($\nu \neq \bar{\nu}$) and therefore a Dirac particle or if it is identical with its antiparticle ($\nu = \bar{\nu}$) and therefore a Majorana particle.

In 1955 the physicists did believe that this question was solved in favour of a Dirac neutrino by the experiment of Davis[1]. He was using antineutrinos from the beta decay at a reactor to induce the inverse beta decay in ^{37}Cl for which one needs in the standard model a neutrino:

$$n \rightarrow p + e^- + \bar{\nu}$$

$$\nu + {}^{37}Cl \rightarrow {}^{37}Ar + e^- \qquad (1)$$

Since he did not observe the formation of ^{37}Ar by the inverse beta decay he assumed that the antineutrino $\bar{\nu}$ is different from the neutrino ν and thus the neutrino is a Dirac particle. But this conclusion was outdated already two years later. In 1957 Lee and Yang proposed[2] to test if in the weak interaction parity is conserved, and in 1957 Wu et al.[3] did find that indeed parity is violated in the beta decay and soon it turned out that it is violated maximaly and the weak interaction is purely left handed. Therefore, the neutrino created in the beta decay of the neutron in eq. (1) has a positive helicity (is right handed), while the neutrino needed for the inverse beta decay must have a negative helicity and therefore needs to be left handed.

---

* Supported by the Bundesministerium für Forschung und Technologie under contract number 06 TÜ 778/9

Thus, even if the neutrino is a Majorana particle, reaction (1) would be forbidden due to helicity mismatch of the two reactions. Thus, we know already since about 30 years that the problem, if the neutrino is a Dirac or if it is a Majorana particle is not solved. But why are we just today discussing this question so intensively? The reason is that the grand unified theories from which we think that they are most successful predict that the neutrino is a Majorana particle[4]. Measurements of the lower limit of the proton life-time seem to exclude SU5 and thus one concentrated on SO10, which is also a subgroup of dynamical groups discussed in superstring theories. SO10 which has first been proposed by Fritsch and Minkowski[5] predicts not only in improved versions[4] that the neutrino is a Majorana particle, but automatically it predicts also that it has a mass and a weak right handed interaction. As we will see below these facts allow the double beta without neutrinos. Or inversely: the existence of the double neutrinoless beta decay would establish that the neutrino is a Majorana particle.

2. THE THEORETICAL DESCRIPTION OF THE 0ν DOUBLE BETA DECAY

Figure 1 shows the double beta decay with two neutrinos in the standard model.

Figure 2 shows the diagram for the double neutrinoless beta decay. This is naturally possible if the neutrino is identical with its antiparticle since at the first vertex, one would emit an antineutrino and at the second vertex one needs to absorb a neutrino in the standard model. Since we have only two particles in the continuum in the process in fig. 2, the phase space is bigger by a factor 10^6 compared to fig. 1. Thus, even if the matrix element squared is reduced by a factor 10^{-6}, one has the same transition probability for the neutrinoless double beta decay as for the double beta decay with two neutrinos shown in fig. 1.

But even if the neutrino is a Majorana neutrino, the process in fig. 2 cannot happen since for a pure left handed weak interaction theory, the emitted neutrino must be right handed (positive helicity), while the absorbed neutrino must be left handed (negative helicity). But grand unified theories predict also that the neutrino has a

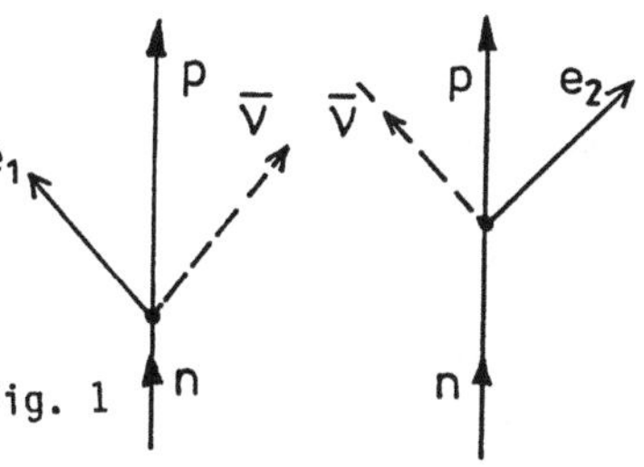

Fig. 1

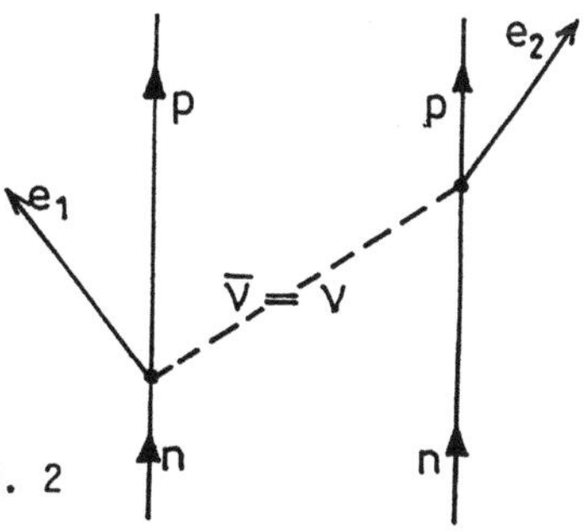

Fig. 2

Fig. 1 Double beta decay with two neutrinos in the standard model. Two neutrons inside the nucleus are transformed by two beta decays into protons. Two electrons and two neutrinos are emitted into the continuum. The kinetic energy available from the Q value of the reaction has to be distributed over the four particles emitted into the continuum. The phase space is proportional to the product of the momentum squared of all the particles in the continuum. Thus, the fact that the available kinetic energy has to be distributed over four particles, reduces the phase space drastically.

Fig. 2 Diagrams for the double neutrinoless beta decay with a Majorana neutrino. By having only two particles in the final states in the continuum, the phase space is increased by a factor of about 10^6 compared to fig. 1. Even with a Majorana neutrino this process is only possible if the neutrino has a finite mass and/or also a right handed weak interaction.

mass and a slight right handed weak interaction. With a finite mass the neutrino has not any more a good helicity and the interference term between the leading helicity and the small admixtures allows a double neutrinoless beta decay. The same happens if we have also right handed vector bosons which allow a right handed weak interaction. If the first vertex is a left handed V-A interaction and the second vertex is a right handed V+A interaction, the process in fig. 2 is also allowed. Thus, the double neutrinoless beta decay can either happen due to the finite mass of the neutrino or due to a small admixture of right handed leptonic currents.

$$H_W(x) = \frac{G\cos\vartheta_C}{\sqrt{2}} [j_{L\mu}(x)J_L^\mu(x) + \kappa j_{L\mu}(x)J_R^\mu(x) + \eta j_{R\mu}(x)J_L^\mu(x) + \lambda j_{R\mu}(x)J_R^\mu(x)] + \text{h.c.} \tag{2}$$

$$j_{L\mu}(x) = \bar{e}\gamma_\mu(1-\gamma^5)\nu$$

$$j_{R\mu}(x) = \bar{e}\gamma_\mu(1+\gamma^5)\nu$$

$$J_L^\mu(x) \cong \sum_n \tau_+(n)\delta(x-x_n)(g_V, g_A\vec{\sigma}(n))$$

$$J_R^\mu(x) \cong \sum_n \tau_+(n)\delta(x-x_n)(g_V, -gA\vec{\sigma}(n))$$

$$g_V = 1.0 ; \qquad g_A = 1.254$$

G is the Fermi coupling constant and ϑ_C the Cabbibo angle. $j_{L\mu}(x)$ and $j_{R\mu}(x)$ are the left and the right handed leptonic currents. The capital J's are the hadronic current operators given here in the non-relativistic reduction summed over all neutrons in the nucleus. We shall later on see that this non-relativistic representation is not good enough and that we have to include the small relativistic amplitudes of the nucleons. The first term of the weak interaction Hamiltonian with a left handed current on the leptonic and on the hadronic side can only contribute to the double neutrinoless beta decay due to the finite mass of the neutrino. The same is true for the second term proportional to κ. Since κ must be extremely small, it cannot be observed relative to the strength 1 of the first term which has the same selection rules. The third term proportional to η with a right handed leptonic current and a left handed hadronic current at one of the vertices in fig. 2 would allow the double neutrinoless beta decay even if the mass of the neutrino would be zero. The same is true for the term proportional to λ with right handed leptonic and hadronic currents.

Table 1 shows the various contributions to the double neutrinoless beta decay transition amplitude going from a 0^+ initial state to a 0^+ final state in an even mass nucleus.

If one considers only the first term in the weak interaction Hamiltonian (2), the double neutrinoless beta decay can only occur due to the finite mass of the neutrino. If one considers in fig. 2 a contribution with a left handed and a right handed vertex, the transition operator on the neutrino side can be the energy or the momentum of the neutrino.

$$\begin{aligned} (1-\gamma^5)(E_\nu\gamma^0 - \vec{p}_\nu\vec{\gamma} + m_\nu)(1-\gamma^5) &= 2(1-\gamma^5)m_\nu \\ (1-\gamma^5)(E_\nu\gamma^0 - \vec{p}_\nu\vec{\gamma} + m_\nu)(1+\gamma^5) &= 2(1-\gamma^5)(E_\nu\gamma^0 - \vec{p}_\nu\vec{\gamma}) \end{aligned} \tag{3}$$

These equations are proportional to the neutrino propagator. The first shows on the left and the right hand side terms which project on left handed neutrinos. The commutation operators of the γ-matrices let only a term survive which is proportional to the neutrino mass m_ν and thus, the neutrino transition operator for two left handed vertices is proportional to the neutrino mass m_ν. If we project on one side on a left handed and on the other side on a right handed neutrino terms proportional to the neutrino mass and the neutrino momentum survive. As discussed above, the neutrino momentum is changing the parity and since we have a transition from a 0^+ to a 0^+ state, we have to have a second operator which compensates this parity change. Haxton[6] prefers to compensate this on the electron side, while we showed[7] that the matrix element is by about a factor 200 larger if we compensate the negative parity of the neutrino transition operator by the nuclear part of this transition operator. This can be done by including the small relativistic amplitudes of the nucleon wave functions. In the normal beta decay the small relativistic amplitude of the nucleon wave function leads to the second forbidden beta decay since the transition matrix element is proportional to $p_\nu/M_N c$. Because the Q-value of a beta decay is typically of one or two MeV, this ratio of the neutrino momentum over the momentum corresponding to the rest mass of the nucleon is of the order of 10^{-3}. If one squares the matrix elements, this leads to a transition probability which is about 6 orders of magnitudes smaller than the allowed Fermi or Gamow-Teller transition. In the neutrinoless double beta decay the neutrino is an intermediate virtual particle and thus one has to integrate over all momenta from zero to infinity. This integration is only limited by the momentum which can be transferred to a nucleon inside a nucleus, so that it still stays in the nucleus. This is twice the Fermi momentum $2p_F$ = 560 MeV/c. Therefore, one finds that the above ratio of the momenta is not any more small, but of the order of unity. In addition, one gets from weak magnetism a factor involving the spin g-factor of the proton minus the spin g-factor of the neutron which multiplies this ratio by a factor of about 9.4 as indicated in the last line of table 1. Taking into account a partial cancellation of the diagrams with different time orderings of the two vertices using the second last line of table 1, one finds that the contribution from the relativistic small amplitudes (recoil term) yields a matrix element which is about a factor 200 larger and therefore a transition probability leading by a factor 40 000 over the term calculated by Haxton.

To calculate the nuclear matrix elements we use an extended way of shell model calculations which are based on an angular momentum and particle number projected Hartree-Fock-Bogoliubov approach. This approach is called MONSTER which stands for "Model Of Number and Spin Projected Two Quasiparticle Wave Functions in Realistic Model Spaces". In addition, one has naturally to take into account short range Brückner correlations[7] and the finite size of the nucleons. This is done by a momentum dependence of the vector and axial vector coupling constant[7].

Experimentally the double neutrinoless beta decay has not yet been found. For several transitions exist lower limits of the half lives. The best measurements for the double neutrinoless beta decay from ^{76}Ge to ^{76}Se have been performed by Avignone[8] and Caldwell[9]. The most restrictive lower limits for the half life of ^{76}Ge against the double neutrinoless beta decay have been given by Caldwell[10] at the Osaka and at the Berkeley conferences. Fig. 3 shows the limitations for the neutrino mass and the right handedness of the weak interaction imposed by the lower limit given by Caldwell at the Osaka Conference[10] of $\tau^{0\nu}_{1/2} > 3.9 \times 10^{23}$ years.

The results of fig. 3 give not directly an upper limit of the mass of the neutrino or the right handedness parameter η or λ defined in eq. (2), but for a special linear combination of the electron, μ and τ neutrino masses weighted by the eigenvector of the mass matrix.

$$\langle m_\nu \rangle = \sum_i m_i U^2_{ei} , \quad \langle \lambda \rangle = \lambda \sum_i U_{ei} V_{ei} , \quad \langle \eta \rangle = \eta \sum_i U_{ei} V_{ei} \qquad (4)$$

Table 1 Nature of the different transition operators for the neutrino (ν), the electron (e) and the nuclear (nucleus) part of the double beta decay transition. The first column shows the nature of the two vertices. L•L indicates a left handed vertex at each transition of a neutron into a proton shown in fig. 2. The neutrino transition operator is proportional to the mass of the neutrino m_ν and has therefore positive parity (π) indicated in brakets. The two electrons (e) are emitted with the largest probability in orbital angular momentum L=0 states. The emission of a P wave electron would reduce the transition probability by almost a factor 1/100. On the nuclear side, we have at each vertex either a Fermi transition ($1_1 \cdot 1_2$) or at both vertices Gamow-Teller transitions ($\vec{\sigma}_1 \cdot \vec{\sigma}_2$). If one of the leptonic vertices is right handed and proportional to the parameter η in eq. (2), the neutrino transition operator can be either the neutrino energy E_ν or the neutrino momentum p_ν. The neutrino energy has a positive parity and thus the electron and the nuclear part can be the same as for two left handed leptonic vertices. But if the neutrino transition operator is proportional to the neutrino momentum, one has to compensate the negative parity of the neutrino momentum p_ν by a negative parity either in the electron transition operator or in the nuclear transition operator. Haxton[6], who first was considering the momentum for the neutrino transition operator, chose to compensate the negative parity by emitting one electron in a P wave. This naturally reduces the transition probability drastically. We choose to compensate the negative parity in the nuclear part of the operator by including small relativistic amplitudes. Since the neutrino is virtual its momentum has to be integrated over all values p_ν and thus, this term is not small but it turns out to be the leading term. It is already for the matrix element by about a factor 200 larger than all the over terms, for the L•ηR combination.

	$\gamma(\pi)$	$e(\pi)$	nucleus (π)
L•L	$m_\nu(+)$	SS(+)	$1_1 \cdot 1_2$; $\vec{\sigma}_1 \cdot \vec{\sigma}_2$ (+)
L•ηR	$E_\nu(+)$	SS(+)	$1_1 \cdot 1_2$; $\vec{\sigma}_1 \cdot \vec{\sigma}_2$ (+)
$\eta<<1$	$P_\nu(-)$	SP(-)	$1_1 \cdot 1_2$; $\vec{\sigma}_1 \cdot \vec{\sigma}_2$ (+)
		SS(+)	$\sigma_1 \cdot (g_P^S - g_\eta^S) p_\nu / Mc$ (-)

Table 2 Upper limits obtained from the nuclear matrix elements calculated in Tübingen[7] and by Haxton[6] in connection with the lower limit of the half life for the neutrinoless double beta decay of ^{76}Ge given by Caldwell in Osaka[10] $\tau_{1/2}^{0\nu} > 3.9 \times 10^{23}$ years.

	Tübingen[7]	Haxton[6]
$\lvert\langle m_\nu \rangle\rvert$	<1.7eV	<2.5eV
$\lvert\langle \lambda \rangle\rvert$	$<2.9\times10^{-6}$	$<5.4\times10^{-6}$
$\lvert\langle \eta \rangle\rvert$	$<3.9\times10^{-8}$	$<8.1\times10^{-6}$

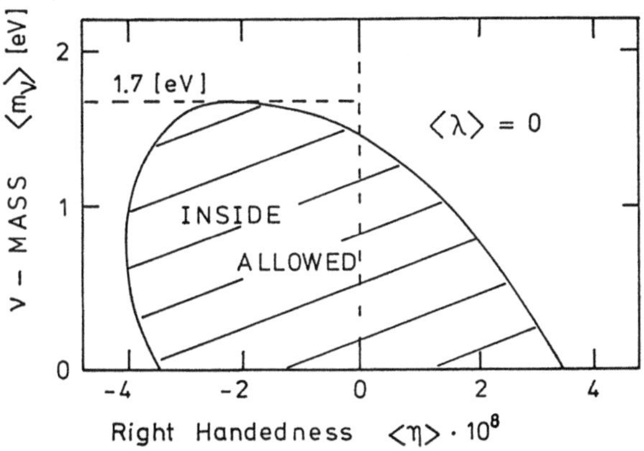

Fig. 3 Average values of the neutrino mass and the parameter for the right handedness of the weak interaction allowed by the lower limit of the half life of the double neutrinoless beta decay of ^{76}Ge given by Caldwell[10] $\tau^{0\nu}_{1/2}$ > 3.9 x 10^{23} years. The bras and the kets around the neutrino mass and the right handedness parameter η of the weak interaction are defined in eq. (3). The right handedness parameter η and λ are defined in eq. (2).

Fig. 3 and table 2 show that the inclusion of the small relativistic amplitudes of the nucleon wave function gives for the right handedness <η> a value which is by more than 2 orders of magnitudes more stringent than the best value obtained before by Haxton. The result means that the right handedness <η> is by at least a factor of around 4×10^{-8} smaller than the left handed part of the weak interaction. This result yields more stringent lower limits on the mass of right handed vector bosons combined with the mixing angle between the right handed and the left handed vector bosons.

3. THE TWO NEUTRINO DOUBLE BETA DECAY

The weak point of the upper limit for the neutrino mass and the right handedness of the weak interaction <η> and <λ> is the fact that the nuclear part of the calculation cannot be tested since there exists no measurements of the double neutrinoless beta decay probability. We have only lower limits of the half life. Since there are measurements available for the two neutrino double beta decay with the geochemical method[11] and since recently also by a laboratory measurement[12], one could try to calculate with the same methods the double beta decay with two neutrinos and compare them with the data. This would be an interesting test of the nuclear part of the calculation. If one performs this calculation, one finds that the calculated transition probability for the double beta decay with two neutrinos (2ν ββ) is by a factor 10 to 100 too large. If this discrepancy cannot be explained to be typical for the 2ν ββ decay, it naturally casts doubt on the reliability of the calculations of the 0ν ββ transition probabilities calculated in chapter 2.

The nuclear many-body methods for calculating the double beta decay are either shell model calculations[6] or extended shell model calculations based on the Hartree-Fock-Bogoliubov approach called MONSTER[7] or they are based on the random phase approach (RPA). Let's try to analyse the 2ν ββ decay in the RPA approximation.

Fig. 4 explains why the 2ν ββ decay amplitude is so drastically reduced. Therefore the small effects which normally do not play a major role can affect the 2ν ββ transition probability. If one looks to the β^+(p→n) amplitude, which is by the Pauli blocking so drastically reduced, one finds the neutron-particle proton-hole excitations indicated in fig. 5 on the left hand side. The matrix elements involved in these diagrams are Pauli suppressed by a factor $(u_n v_p)^2$ = $(\text{small})^4$. The second diagram shown in fig. 5 shows the particle-particle excitation relative to an odd-

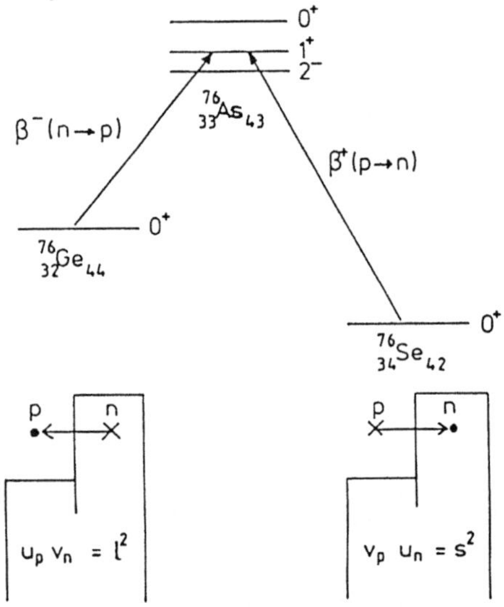

Fig. 4 The upper part shows the way how in the random phase approximation (RPA) the 2ν ββ decay is calculated. One starts with an RPA proton-particle neutron-hole calculation to the intermediate ^{76}As nucleus and also in the same way from the ^{76}Se ground state by neutron-particle proton-hole excitations to the same intermediate nucleus. For the Fermi transitions the β^-(n→p) amplitude moves just a neutron into the same proton level and the β^+(p→n) amplitude moves a proton into the same neutron level. For the Gamow-Teller transitions it can also involve a spin flip, but the orbital part remains the same. One immediately realizes that the occupation and nonoccupation amplitudes favour the β^- amplitude, but disfavour the β^+ amplitude. There one has a transition from an unoccupied to an occupied single particle state, which is two fold small (s^2) first by the fact that the occupation amplitude for the proton v_p and secondly that the unoccupation amplitude for the neutron state u_n are both small. Therefore the 2ν ββ is drastically reduced.

odd-mass nucleus leading also to the intermediate nucleus ^{76}As. These particle-particle matrix elements are not so drastically reduced. They are proportional to $(u_n u_p)^2$ = (small)^2 (large)^2. These particle-particle excitations increase drastically the ground state correlations corresponding to ^{76}Se and therefore one can admix, to excite the intermediate nucleus ^{76}As, also backgoing diagrams corresponding to an excitation of a neutron-hole and a proton-particle proportional to $v_n u_p$= (large)^2. Therefore the β^+(p→n) amplitude is very drastically affected by allowing also particle-particle (or hole-hole) excitations and therefore also particle-particle (or hole-hole) ground state correlations for ^{76}Se. The neutron particle proton hole force in the isovector channel is repulsive while the particle-particle force is attractive. Therefore both excitations tend to cancel each other and therefore the amplitude β^+ is drastically reduced and by that one obtains agreement with the experimental data.

We calculated the 2ν ββ decay half lives for the following $0_i^+ \rightarrow 0_f^+$ transitions: ^{76}Ge → ^{76}Se, ^{82}Se → ^{82}Kr, ^{128}Te → ^{128}Xe and ^{130}Te → ^{130}Xe. We take the full 3ħω and 4ħω major oscillator shells for the description of the nuclei with A = 76, 82 and a model space consisting of $1p_{3/2,1/2}$, the full 4ħω shell, $0h_{11/2,9/2}$, and $1f_{7/2,5/2}$ sub-shells for A = 128, 130. The single particle energies are calculated with the Coulomb-corrected Woods-Saxon potential. As a realistic two-body interaction we use the nuclear matter G-matrix calculated from the Bonn one boson-exchange potential. In order to take into account the renormalization for the finite basis, we multiply

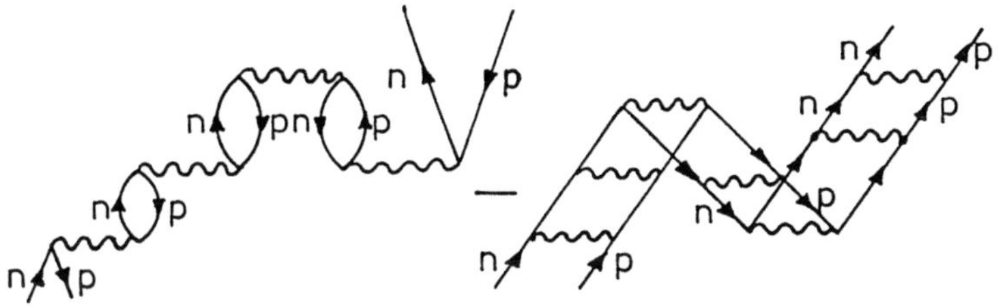

Fig. 5 The left hand side shows the isovector neutron-particle proton-hole excitations leading from ^{76}Se to ^{76}As. The particle-hole matrix elements are proportional to the unoccupation amplitude for the neutrons and the occupation amplitude for the protons squared $(u_n v_p)^2 = (\text{small})^4$. The ground state correlations and the collectivity due to these excitations are therefore extremely weak. The transition probability is further reduced by the Pauli principle since the transition amplitude is proportional to $v_p u_n = (\text{small})^2$. The right hand side shows particle-particle excitations where the matrix elements have the Pauli factors $(u_n u_p)^2 = (\text{small})^2 (\text{large})^2$. Normally this particle-particle ground state correlations and excitations are neglected, but due to the weakness of the neutron particle proton hole excitations they play a major role in this case. They increase the ground state correlations so much that the backgoing diagrams for the excitations which have an amplitude proportional to $u_p v_n = (\text{large})^2$ play a decisive role. Since the isovector particle-hole force is repulsive and the particle-particle (and the hole-hole) force is attractive the inclusion of the particle-particle (and hole-hole) correlations tends to quench the 2ν $\beta\beta$ transition probability.

the pairing matrix elements of this force by factors g^p_{pair} and g^n_{pair}. These factors lie close to unity and deviate at most less than 20 % from the value obtained from the reaction matrix of the Bonn potential. These factors are adjusted to the odd-even mass differences. We also multiply the particle-hole matrix elements by a factor g_{ph} which we fix to the energy of the Gamow-Teller giant resonance (GTGR) in the intermediate odd-nucleus. This factor is also normally close to unity. The biggest deviation found is 30 %. To show the influence of the particle-particle correlations, we multiply also the particle-particle matrix elements by a factor g_{pp}. $g_{pp} = 0$ switches off the particle-particle correlations while $g_{pp} = 1$ gives the values predicted by the theory including these correlations. Fig. 6 shows the Gamow-Teller matrix element for the different transitions as a function of the factor g_{pp} with which we multiplied the particle-particle matrix elements. One sees that for $g_{pp} = 1$ the matrix element is practically reduced to zero and thus one obtains the quenching of the 2ν $\beta\beta$ decay probability needed.

One sees from fig. 7 that the reduction of the 2ν $\beta\beta$ transition probability or the increase of the half-lives yields a reasonably good agreement with the data for the particle-particle strength predicted by the reaction matrix elements calculated from the Bonn potential ($g_{pp} = 1$). (Since there is a cancellation point, the half-lives tend always to infinity for a value around $g_{pp} = 1$. But the theory always agrees with the data for a value $g_{pp} \cong 1$.)

Vogel and Zirnbauer[21] have recently also studied the reduction of the 2ν $\beta\beta$ decay using a phenomenological zero range (δ-)force. They also found a strong quenching due to the particle-particle correlations, but at a strength of only 40 % suggested by the Pandya transformation from the particle-hole matrix elements. Recently Muto and Klapdor, in a preprint from October 1987, confirmed our results that for a realistic interaction the experimental quenching can be obtained for the correct particle-particle strength[22].

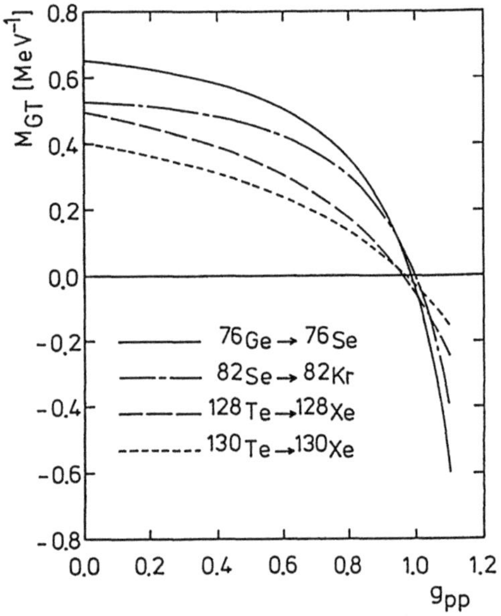

Fig. 6 The calculated[13] 2ν $\beta\beta$ decay matrix elements M_{GT} as function of the strength g_{pp} of the Brückner reaction particle-particle matrix elements of the Bonn potential. One sees that including the full particle-particle correlations, one gets cancellation between the particle-hole and the particle-particle amplitudes which reduces the transition amplitude drastically and yields a value close to zero if one uses the full reaction matrix elements corresponding to the Bonn potential (g_{pp} = 1).

4. THE PARTICLE-PARTICLE CORRELATION FOR THE 0ν $\beta\beta$ DECAY

After we found in chapter 3, due to the particle-particle correlations, a strong reduction of the 2ν $\beta\beta$ decay probability, one naturally is asking if this effect is not also reducing the 0ν $\beta\beta$ decay probability in the same way. Fig. 8 shows the diagrams for the 2ν $\beta\beta$ and the 0ν $\beta\beta$ decay. The difference between the two diagrams is that for the two neutrino double beta decay the vertices 1 and 2 are not related to each other, and therefore there is no r_{12} dependence of the transition operator. This is different for the 0ν $\beta\beta$ decay due to the exchange of the neutrino. The Fermi, the Gamow-Teller and the recoil term are multiplied all with an r_{12} dependence.

$$\hat{O}_F = H_F(r_{12})\ \tau_+(1)\ \tau_+(2)$$

$$\hat{O}_{GT} = H_{GT}(r_{12})\ \tau_+(1)\ \tau_+(2)\ \vec{\sigma}(1)\cdot\vec{\sigma}(2) \qquad (5)$$

$$\hat{O}_{Recoil} = H_R(r_{12})\ \tau_+(1)\ \tau_+(2)\ [\vec{\sigma}(1)\cdot(g_p^S - g_n^S)\ \frac{\vec{p}_\nu}{M_N c}\ ]$$

In eq. (5) the transition operators for the 0ν $\beta\beta$ decay with their r_{12} dependence on the distance between the two vertices is indicated. If one expands the dependence on r_{12} in multipoles depending on r_1 and r_2, one sees that one is not only exciting Fermi and Gamow-Teller states, but also other natural and unnatural multipolarities. But the particle-particle ground state correlations are not so important for the higher multipole states and thus, they are not so drastically reducing the transi-

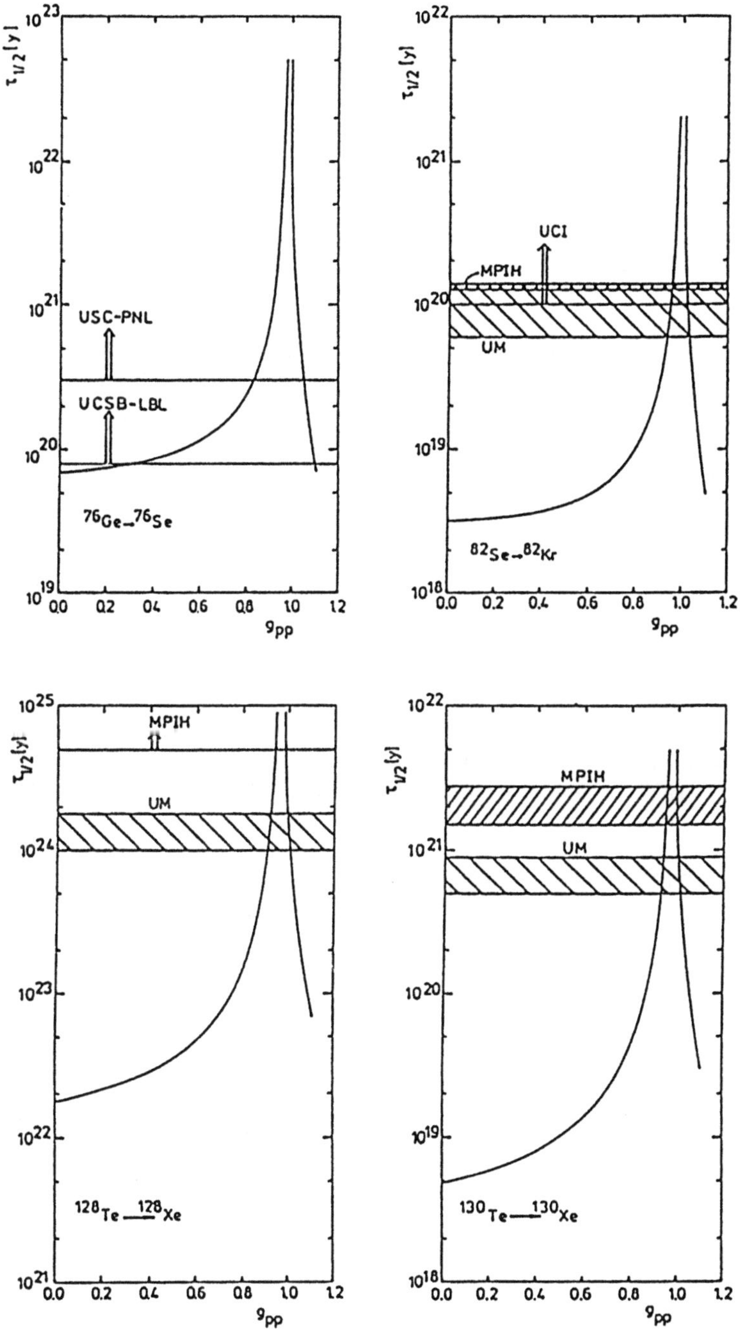

Fig. 7 The calculated 2ν $\beta\beta$ decay half-lifes $\tau_{1/2}$ as a function of g_{pp}. Experimental half-lifes are indicated either by hatched regions or by lines with arrows (for lower limits), and are taken from ref. 8, 9, 11, 12 and 14. For $^{82}Se \rightarrow {}^{82}Kr$ Moe and coworkers[11] measured recently for the first time a half-life of $\tau_{1/2} = (1.2 + 0.8 - 0.3)\cdot 10^{20}$ years in the laboratory.

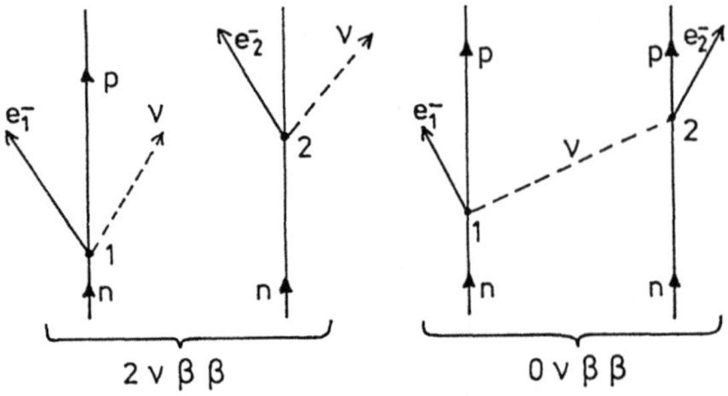

Fig. 8 The left hand side shows the diagram for the 2ν $\beta\beta$ decay probability while the right hand side gives the diagram for the 0ν $\beta\beta$ decay probability. The transition for 2ν $\beta\beta$ does not depend on the distance between vertex 1 and 2, r_{12}, but the 0ν $\beta\beta$ transition operator is dependent on the distance r_{12}. A multipole expansion of this dependene on the distance yields an intermediate excitation of other states than just only the Fermi and the Gamow-Teller states.

tion probabilities for the 0ν $\beta\beta$ decay as for the 2ν $\beta\beta$ decay. Fig. 9 shows the transition matrix elements without and with the higher multipolarities for the Fermi (F), the Gamow-Teller (GT) and the recoil (R) term as a function of the particle-particle strength g_{pp}. If one compares the matrix elements without the particle-particle correlations (g_{pp} = 0) and the particle-particle correlations predicted by the Bonn potential (g_{pp} = 1), one finds for the recoil term, which is the leading one, only a reduction of the order of 30 % if one includes the higher multipole states. Without the higher multipole states one finds for the GT and the F transition amplitudes the same reduction as discussed for the 2ν $\beta\beta$ decay probability. But with the higher multipolarities also for the GT and the F transitions the quenching is drastically reduced. The upper limits for the neutrino mass (average in the way discussed above) and the right handedness $\langle\eta\rangle$ are shown in fig. 10 for four 0ν $\beta\beta$ decays.

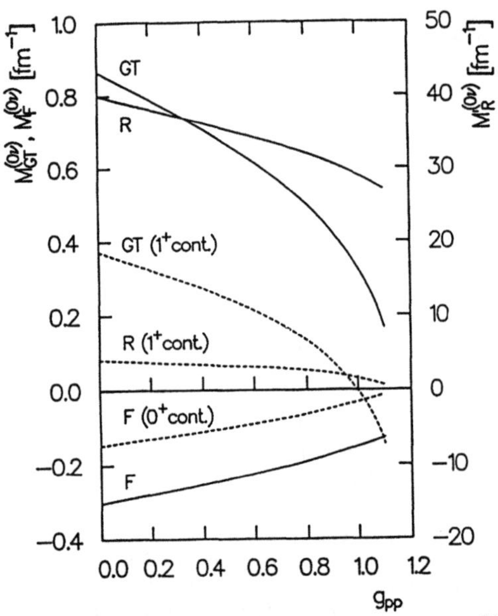

Fig. 9 Transition amplitudes for the $^{76}Ge \rightarrow {}^{76}Se$ $0\nu\beta\beta$ decay as a function of g_{pp} with and without (only 0^+ and 1^+) higher multipole states for the intermediate nucleus ^{76}As.

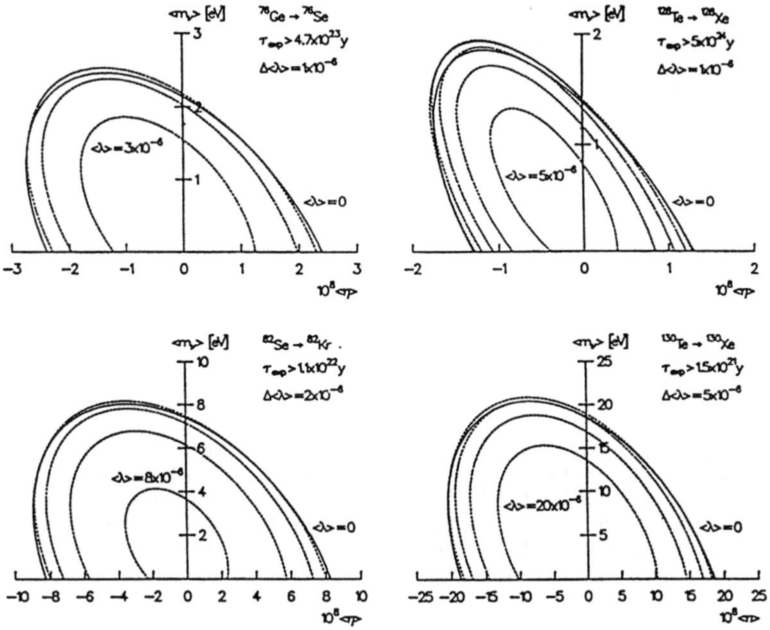

Fig. 10 The allowed regions (inside the ellipses) deduced from the experimental bounds (Caldwell, Berkeley Conference[10] and refs. 11 and 15).

The results of fig. 10 which include now the particle-particle correlations show that these correlations do not affect very much the results for the 0ν $\beta\beta$ decay.

Recently we received a preprint from Vogel and collaborators[23] in which they are also investigating if the particle-particle correlations have an influence on the 0ν $\beta\beta$ decay. Their answer is "yes". The difference to our answer which is "no" (or more exactly only "minor") comes from the fact that they did use a zero-range δ-force where the higher mulitpoles are drastically changed compared to a realistic force used in our calculations.

5.PARTICLE-PARTICLE CORRELATIONS AND THE SINGLE β^+ DECAY

The same reduction which one finds for the β^+-branch of the 2ν $\beta\beta$ decay one should also find for the β^+ decay from neutron deficient nuclei. Indeed it is well known[16,17] that the β^+ decay in neutron deficient nuclei are drastically quenched compared to the theoretical results.

Fig. 11 shows the reduced β^+ decay probability as a function of the particle-particle strength g_{pp} for the β^+ decay of ^{148}Dy → ^{148}Tb. The solid line is calculated using a Saxon-Woods potential with parameters from Bohr and Mottelson[18] and the dashed line using a Saxon-Woods potential with parameters from Bertsch[19]. (The single particle energies given by Bertsch in ref. 19 are wrong, but the parameters of the Saxon-Woods potential seem to be correct.).

We studied further the β^+ decay of ^{150}Er and ^{152}Yb. Fig. 12 shows the particle-particle strength g_{pp} fitted to the experimental transition probability for these β^+ transitions including the one of ^{148}Dy. Independently, if one takes the axial coupling constant g_A = 1.25 of the free nucleon or the quenched value g_A = 1.00 often the used in nuclei, one finds a particle-particle strength which is close to unity. This is a big triumph of the theory since one should obtain agreement for g_{pp} = 1 corresponding to the reaction matrix elements of the Bonn potential.

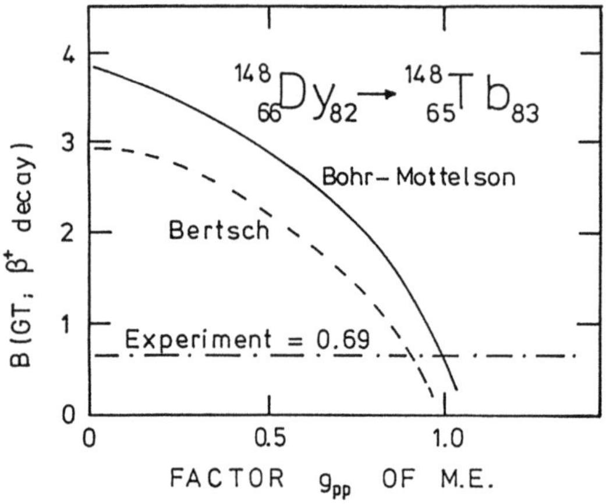

Fig. 11 Reduced β^+ decay probability B(GT) as a function of the particle-particle strength g_{pp} with which the particle-particle reaction matrix elements of the Bonn potential are multiplied. The solid line is calculated with single particle wave functions of a Saxon-Woods potential with parameters of Bohr and Mottelson[18] and the dashed line with Saxon-Woods parameters of Bertsch[19]. One finds a drastic reduction for the physical value of the particle-particle strength g_{pp} = 1. Around this value the theory agrees with the experimental data B(GT) = 0.69. The single-particle basis used in the RPA calculation is the full N=4 and 5 and the $i_{13/2}$ and $i_{11/2}$ shells.

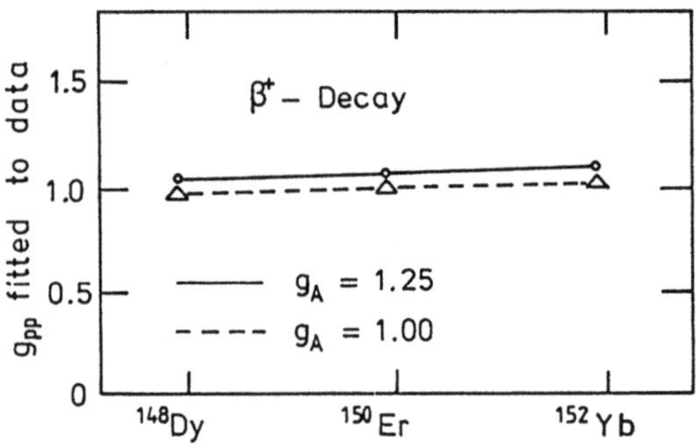

Fig. 12 Particle-particle strength g_{pp} with which the particle-particle reaction matrix elements of the Bonn potential are multiplied fitted to the experimental β^+ transition probability[16,17] for the β^+ decay of ^{148}Dy, ^{150}Er and ^{152}Yb. The solid line is the g_{pp} fitted with the axial vector coupling constant of the free nucleon g_A = 1.25, while the dashed line gives the results for the quenched axial vector coupling constant g_A = 1.00 which is often used in nuclei.

The same effect which yields agreement for the 2ν ββ decay can also explain the quenching in the β^+ decay of neutron deficient nuclei.

6. CONCLUSIONS

The double neutrinoless β decay (0ν ββ) can distinguish if the neutrino is a Dirac particle, that means if the neutrino is different from the antineutrino, or if it is a Majorana particle and therefore identical with its antiparticle. Only in the case of a Majorana particle, the double neutrinoless β decay (0ν ββ) is possible. Grand unified theories predict that the neutrino is a Majorana particle, especially the ones which are built on the SO(10) group. But being a Majorana particle the neutrino has to have a finite mass and for SO(10) also slight right handed weak interaction. Since the 0ν $\beta\beta$ decay needs either a finite mass of the neutrino or a right handedness of the weak interaction, on can derive from the lower limit of the half-lives of 0ν $\beta\beta$ decays upper limits for an averaged neutrino mass $\langle m_\nu \rangle \lesssim 1.9$eV and for the right handedness of the weak interaction $\langle \eta \rangle \lesssim 1.8\times10^{-8}$.

If one tries to test the calculation by the known 2ν $\beta\beta$ decays one finds that the theory is by a factor 10 to 100 larger than the experimental data. But if one includes the particle-particle correlations of protons with neutrons, which are attractive, they cancel by a large part the neutron-particle and proton-hole correlations which are repulsive. Including these additional ground state correlations one finds a strong quenching of the 2ν $\beta\beta$ transitions in agreement with the experimental data.

For the 0ν $\beta\beta$ transition amplitude this quenching is for the leading recoil term only about a factor 0.7 or 30 %. This difference between the 2ν $\beta\beta$ and 0ν $\beta\beta$ decay stands from the fact that in the 0ν $\beta\beta$ one has for the transition operator a dependence on the distance between the two vertices and thus, by a multipole expansion one gets a transition operator which excites also higher multipoles in the intermediate nucleus. These higher multipoles are not strongly quenched and thus the 0ν $\beta\beta$ transitions are not so drastically affected by the particle-particle correlations.

The quenching of the β^+ branch in the 2ν $\beta\beta$ decay can also be tested in the single β^+ decay of neutron deficient nuclei. Including the particle-particle correlations one finds also agreement for these transitions which could not be understood before.

At the end I would like to thank Prof. O. Civitarese, Dr. J. Suhonen, Dipl.-Phys. Th. Taigel and Dr. T. Tomoda for the collaboration on the above problems.

REFERENCES

1. R. Davis, PHys. Rev. 97 (1955) 766
2. T.D. Lee, C.N. Yang, Phys. Rev. 104 (1956) 254
3. C.S. Wu, E. Ambler, R.W. Hayward, D.D. Hopper, R.P. Hudson, Phys. Rev. 105 (1957) 1413
4. P. Langacker, Phys. Rep. 72 (1981) 185
5. H. Fritsch, P. Minkowski, 73 (1981) 67
6. W.C. Haxton, G.J. Stephenson, Progr. Part. Nucl. Phys. 12 (1984) 409
7. T. Tomoda, A. Faessler, K.W. Schmid, F. Grümmer, Nucl. Phys. A452 (1986) 591
8. F.T. Avignone III, R.L. Brodzinski, J.C. Evans, Ir., K. Hensley, H.S. Miley and J.H. Reeves, Phys. Rev. C34 (1986) 666
9. D.O. Caldwell, R.M. Eisberg, D.M. Grumm, D.L. Hale, M.S. Witherell, F.S. Goulding, D.A. Laudis, N.W. Madden, D.F. Malone, R.H. Pehl, A.R. Smith, Phys. Rev. D33 (1986) 2737
10. D.O. Caldwell, Osaka Conference June 1986 ($\tau^{0\nu}_{1/2}(^{76}_{23}Ge)>3.9\times10^{23}$ years); Berkeley Conference July 1986 ($\tau^{0\nu}_{1/2}(^{76}Ge)>4.7\times10^{23}$ years)
11. T.Kirsten, in Proceedings of the International Symposium on Nuclear Beta Decay and Neutrinos, Osaka, Japan 1986; Edited by T. Kotani, H. Ejiri, E. Takasugi (World Scientific, Singapore, 1986, p. 81)
12. S.R. Elliott, A.A. Hahn, M.K. Moe, preprint October 1987 and Frankfurter Allgemeine Zeitung, 7. Oktober 1987
13. O. Civitarese, A. Faessler, T. Tomoda, Phys. Lett. B194 (1987) 11
14. O.K. Manuel, Proc. Int. Symp. on Nuclear Beta Decay and Neutrinos, Osaka, Japan, June 1986; eds. T. Kotani, H. Ejiri, E. Takasugi (World Scientific, Singapore, 1986, p. 71)
15. S.R. Elliott, A.A. Hahn, M.K. Moe, Proc. Int. Symp. on Weak and Electromagnetic Interactions in Nuclei, July 1986, ed. H.V. Klapdor (Springer, Berlin 1986) p. 692
16. E. Nolte et al., Z. Phys. A306 (1982) 223
17. P. Kleinheinz et al., Phys. Rev. Lett. 55 (1985) 2664

18. A. Bohr, B.R. Mottelson, Nucl. Structure, Vol. 1 (Benjamin, New York, 1969)
19. G. Bertsch, The Practitioners Shell Model (American Elsevier, New York, 1972)
20. T. Tomoda, A. Faessler, Phys. Lett. B
21. P. Vogel, M.R. Zirnbauer, Phys. Rev. Lett. 57 (1986) 3148
22. K. Muto, H.V. Klapdor, preprint October 1987
23. J. Engel, P. Vogel, M.R. Zirnbauer, preprint MAP-95, June 1987

Nuclear Structure Effects on the Suppression of Two-Neutrino Double Beta Decay

K. Muto and H.V. Klapdor*

Max-Planck-Institut für Kernphysik, P.O. Box 103980, D-6900 Heidelberg, Fed. Rep. of Germany

Neutrinoless double beta decay is expected to clarify the fundamental nature of the neutrino. In deducing quantities such as the neutrino mass, the nuclear part of the transition amplitude has to be calculated theoretically. A test of nuclear structure calculations is provided by the two-neutrino double beta (2ν $\beta\beta$) decay which is free from unknown physical quantities on the particle physics side. It is shown, by applying the proton-neutron quasiparticle RPA model, that the 2ν $\beta\beta$ decay rate is strongly suppressed by particle-hole correlations in the ground states of parent and daughter nuclei. Mainly responsible for the ground-state correlations is the particle-particle interaction. We have adjusted its strength to experimental Gamow-Teller strengths of β^+ decays of nuclei with mass number A = 60-160 and obtain with this strength calculated 2ν $\beta\beta$ half-lives consistent with experiment. The long-standing discrepancy between theoretical and experimental half-lives by a large factor up to two orders of magnitude seems to be resolved.

1. Introduction

The nuclear double beta ($\beta\beta$) decay [1] is one of the rarest processes in nature with a half-life of the order of $\sim 10^{20}$ years or longer. It can be observed, when a nucleus is stable against ordinary $\beta^{\pm}$ decay because of energy conservation or very strong suppression of energetically allowed decays (such as ^{48}Ca and ^{96}Zr) and also against alpha-particle emission in the case of heavy nuclei. The $\beta\beta$ decay takes place through two decay modes (see fig. 1): the two-neutrino (2ν) mode in which two neutrinos are emitted together with two electrons, and the neutrinoless (0ν) mode in which no neutrino emerges out of the nucleus,

$$2\nu: \ (A,Z) \rightarrow (A,Z+2) + 2e^- + 2\bar{\nu} \qquad (1a)$$

$$0\nu: \ (A,Z) \rightarrow (A,Z+2) + 2e^- \qquad (1b)$$

The 2ν $\beta\beta$ decay occurs in the framework of the standard electroweak theory [2], irrespective to the nature of neutrinos, and can be described by successive β transitions via virtual intermediate states. On the other hand, the 0ν $\beta\beta$ decay is forbidden in the standard theory. It can occur when the following conditions are satisfied:

---

* On leave from Department of Physics, Tokyo Institute of Technology, Tokyo, Japan.

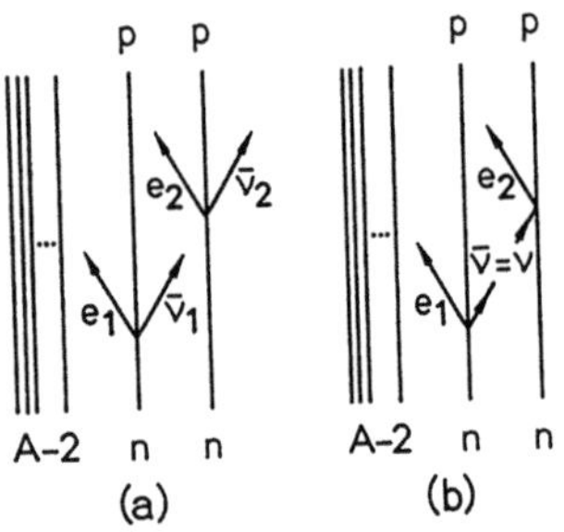

Fig. 1 Two decay modes of the nuclear $\beta\beta$ decay possible in the two-nucleon mechanism; (a) two-neutrino mode and (b) neutrinoless mode.

(i) The neutrino is a Majorana particle: In the first transition of fig. 1(b) an antineutrino appears, $n \rightarrow p + e^- + \bar{\nu}$, and it is a neutrino that is absorbed in the second transition, $\nu + n \rightarrow p + e^-$. Thus, the neutrino must be identical with its antiparticle.

(ii) Helicity matching: If the (anti)neutrino is massless, the neutrino has negative helicity (left-handed) and the antineutrino positive helicity (right-handed), with the consequence that the 0ν mode does not occur with only the left-handed weak interactions. This helicity mismatch can be avoided either by a finite mass of the neutrino which implies admixture of the opposite helicity, or by the presence of right-handed interactions. Observation of the 0ν $\beta\beta$ decay implies in <u>any</u> case that at least the electron neutrino is a massive Majorana neutrino [3].

The 0ν mode $\beta\beta$ decay is thus of particular interest. It is expected to give an answer to the questions concerning the nature of neutrinos, Dirac or Majorana, massless or massive. Grand unified theories based on the SO(10) gauge group [4] imply the existence of Majorana neutrinos and also of right-handed weak interactions. The 0ν decay is at present the most sensitive test of the existence of Majorana neutrinos, and yields at present the sharpest limits on the mass of the electron neutrino.

Determination of the neutrino mass from observed $\beta\beta$ half-lives requires the knowledge of nuclear matrix elements, which have to be estimated theoretically. The 2ν $\beta\beta$ decay provides a test of such a nuclear structure calculation. Until recently, however, theoretical half-lives of the 2ν mode systematically underestimated the experimental values by large factors up to two orders of magnitude. It was argued that the calculated nuclear matrix elements of the 0ν mode could therefore have large uncertainties. Fig. 2 compares experimental half-lives [5-9] for four typical nuclei with those from systematic calculations by Haxton, Stephenson and Strottman [10], Klapdor and Grotz [11] and Vogel and Fisher [12]. The discrepancies are quite large for ^{130}Te particularly for the results of [10,12]. It is also clear that the calculations overestimate the decay rate of ^{82}Se. (A recent counter measurement by Elliott, Hahn and Moe [7] gave a half-life of ^{82}Se consistent with the geochemical measurements [8,9]). The long half-lives indicate that the nuclear matrix elements are strongly suppressed, and that the calculations could not reproduce the full degree of suppression.

The calculation of Klapdor and Grotz [11] showed, however, already a remarkable improvement particularly for ^{128}Te and ^{130}Te and showed, more generally, the way to the solution of the problem. They included in their calculation besides pairing and spin-isospin forces, also another important part of the proton-neutron interaction,

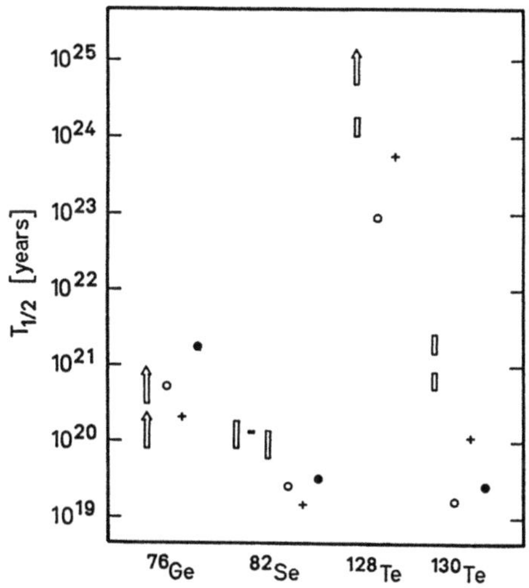

Fig. 2 Experimental and theoretical (before the end of 1986) half-lives of 2ν $\beta\beta$ decays. The experimental values [5-9] are denoted by a square, or an arrow when only a lower limit was obtained. The theoretical values are taken from three systematic calculations: open circles from Haxton, Stephenson and Strottman (shell model) [10], crosses from Klapdor and Grotz (projected BCS) [11], and filled circles from Vogel and Fisher (quasiparticle RPA, but without the particle-particle force) [12].

the quadrupole-quadrupole force. This force especially in the isotopes $^{128,130}Te$ and $^{128,130}Xe$ leads to low-energetic 2^+ phonons, which are admixed to the 0^+ ground states and lead to destructive contributions to the 2ν $\beta\beta$ matrix elements. While the spin-isospin force used in the usual RPA calculations is 'tailored' to describe Gamow-Teller (GT) particle-hole (ph) forces, the quadrupole-quadrupole force contains on the other hand a component of particle-particle (pp) forces. It was therefore natural to investigate ph and pp forces separately.

This has been done by Vogel and Zirnbauer [13] by a calculation based on the proton-neutron quasiparticle random phase approximation (QRPA), with a zero-range force. It was demonstrated by subsequent QRPA calculations by Civitarese, Faessler and Tomoda [14] and Muto and Klapdor [15] using different realistic nucleon-nucleon interactions, that inclusion of the pp force leads to a suppression mechanism which can solve the earlier discrepancies between calculation and experiment. In particular, in the calculation of [15] the pp interaction strength, which is critical for the prediction of the 2ν $\beta\beta$ half-lives, was fixed to reproduce the GT strength observed in β^+ decay of a number of nuclei with A = 60-160 and with this pp strength the QRPA calculation yielded half-lives in remarkable agreement with the experimental values. These calculations will be described in some detail in sections 3,4.

2. The decay rate formula for the 2ν mode

The 2ν mode $\beta\beta$ decay is described in second-order perturbation of the weak interaction. Nuclear transitions are dominated by GT transitions via $J^\pi = 1^+$ intermediate states. Higher-order terms, such as the nucleon recoil terms and contributions from the right-handed currents, are neglected. In addition, since the kinetic energy released in the decay, $T_0 \leq 4$ MeV for all possible $\beta\beta$ emitters, is shared by four leptons in the continuum, both electrons and neutrinos are assumed to be emitted in S wave, and finite wave-length effects can be neglected.

The decay rate of a $0_i^+ \rightarrow 0_f^+$ transition is given by

$$\omega_{2\nu} = \frac{G^4 g_A^4}{32\pi^7} \int_{m_e}^{T_o+m_e} F(Z,e_1)k_1 e_1 de_1 \int_{m_e}^{T_o+2m_e-e_1} F(Z,e_2)k_2 e_2 de_2$$

$$\times \int_o^{T_o+2m_e-e_1-e_2} \nu_1^2 \nu_2^2 d\nu_1 \sum_{aa'} A_{aa'} \tag{2}$$

where G is the Fermi coupling constant and $g_A/g_V = -1.25$. The Fermi function F(Z,e) takes into account distortion of the electron wave function due to the charge of the nucleus, and $k = \sqrt{e^2-m_e^2}$ is the electron momentum. Nuclear GT transition matrix elements and energy denominators of the perturbation are confined in $A_{aa'}$,

$$A_{aa'} = \langle 0_f^+ \| t_-\sigma \| 1_a^+ \rangle \langle 1_a^+ \| t_-\sigma \| 0_i^+ \rangle \langle 0_f^+ \| t_-\sigma \| 1_{a'}^+ \rangle \langle 1_{a'}^+ \| t_-\sigma \| 0_i^+ \rangle$$

$$\times \frac{1}{3} (K_a K_{a'} + L_a L_{a'} + \frac{1}{2} K_a L_{a'} + \frac{1}{2} L_a K_{a'}) \tag{3}$$

where

$$K_a = \frac{1}{E_a + e_1 + \nu_1 - E_i} + \frac{1}{E_a + e_2 + \nu_2 - E_i}$$
$$L_a = \frac{1}{E_a + e_1 + \nu_2 - E_i} + \frac{1}{E_a + e_2 + \nu_1 - E_i} \tag{4}$$

The quantity $A_{aa'}$ contains energies of both nuclear intermediate state and leptons.

The nuclear matrix element is defined as

$$M_{GT} = \sum_a \frac{\langle 0_f^+ \| t_-\sigma \| 1_a^+ \rangle \langle 1_a^+ \| t_-\sigma \| 0_i^+ \rangle}{E_a + T_o/2 + m_e - E_i} \tag{5}$$

The sum of lepton energies $e + \nu$ in the denominator is replaced by $T_o/2 + m_e$. This is in general a good approximation. The integrand of eq. (2) becomes large when a pair of electron and neutrino carries a half of the available kinetic energy. With eq. (5) the decay rate is factorized into two parts,

$$\omega_{2\nu} = F^{2\nu} |M_{GT}|^2 \tag{6}$$

The nuclear part is separated from the phase-space integral $F^{2\nu}$, and considerations on nuclear structure effects are focused on the nuclear matrix element.

3. Calculation of the 2ν $\beta\beta$ matrix elements M_{GT} by the quasiparticle RPA

The proton-neutron quasiparticle RPA [16,17] describes excitations of a charge-changing mode, such as the GT β transitions, from the ground state of an even-even nucleus (A,Z) to states with J^π in the odd-odd daughter nucleus (A,Z±1). A quasiparticle basis is defined by the Bogoliubov transformation from a nucleon basis $\{c_{jm}^+, c_{jm}\}$,

$$a^{+}_{jm} = u_j c^{+}_{jm} - v_j (-1)^{j+m} c_{j-m}$$
$$a_{jm} = u_j c_{jm} + v_j (-1)^{j+m} c^{+}_{j-m} \tag{7}$$

with $u_j^2 + v_j^2 = 1$, where v_j^2 is the nucleon occupation probability in the orbital j. The QRPA equation can be derived from the equation of motion under assumptions analogous to those of the usual RPA. It is a non-hermitian matrix equation of the form

$$\begin{bmatrix} A & B \\ -B & -A \end{bmatrix} \begin{bmatrix} X \\ Y \end{bmatrix} = \omega \begin{bmatrix} X \\ Y \end{bmatrix} \tag{8}$$

where ω are energy eigenvalues, and X and Y are forward-going and backward-going amplitudes of the eigenstates, respectively. Elements of the submatrices for states of a proton-neutron pair are given by

$$\begin{aligned} A_{pn,p'n'} = {} & \delta(pn,p'n')(\epsilon_p + \epsilon_n) \\ & + (u_p u_n u_{p'} u_{n'} + v_p v_n v_{p'} v_{n'}) \langle pn|V|p'n' \rangle \\ & + (u_p v_n u_{p'} v_{n'} + v_p u_n v_{p'} u_{n'}) \langle pn^{-1}|V|p'n'^{-1} \rangle \end{aligned} \tag{9a}$$

$$\begin{aligned} B_{pn,p'n'} = {} & (u_p u_n v_{p'} v_{n'} + v_p v_n u_{p'} u_{n'}) \langle pn|V|p'n' \rangle \\ & - (u_p v_n v_{p'} u_{n'} + v_p u_n u_{p'} v_{n'}) \langle pn^{-1}|V|p'n'^{-1} \rangle \end{aligned} \tag{9b}$$

where ϵ are quasiparticle energies, and p and n specify proton and neutron states, respectively. It is noted that the pp interaction is not an extra interaction being added to this model, but naturally appears together with the ph interaction by the Bogoliubov transformation. The ph and pp interactions reduce in the limit of the nucleon picture, $u_j v_j \to 0$, to those acting between particle-hole states and between two-particle (two-hole) states, respectively.

In the QRPA, the nuclear matrix element of a $0^+_i \to 0^+_f$ transition of the 2ν mode $\beta\beta$ decay is expressed as

$$M_{GT} = \sum_{a,b} \frac{\langle 0^+_f \| t_- \sigma \| 1^+_b \rangle \langle 1^+_b | 1^+_a \rangle \langle 1^+_a \| t_- \sigma \| 0^+_i \rangle}{E_a + T_o/2 + m_e - E_i} \tag{10}$$

Of the two reduced matrix elements of the Pauli spin operator, the second one describes β^- GT transition $0^+_i \to 1^+_a$, and the other is the inverse of the β^+ transition $0^+_f \to 1^+_b$,

$$\langle 1^+_a \| t_- \sigma \| 0^+_i \rangle = \sum_{p,n} \langle p \| \sigma \| n \rangle (u_p v_n X^{pn}_a - v_p u_n Y^{pn}_a) \tag{11a}$$

$$\langle 0^+_f \| t_- \sigma \| 1^+_b \rangle = \sum_{p,n} \langle p \| \sigma \| n \rangle (\bar{v}_p \bar{u}_n \bar{X}^{pn}_b - \bar{u}_p \bar{v}_n \bar{Y}^{pn}_b) \tag{11b}$$

The term X comes from excitation of a particle-hole pair, and the Y term from destruction of one of the pairs existing in the RPA ground state. The overlap of intermediate 1^+ states is given by

$$\langle 1_b^+ | 1_a^+ \rangle = \sum_{pn} (\bar{X}_b^{pn} X_a^{pn} - \bar{Y}_b^{pn} Y_a^{pn}) \tag{12}$$

The quantities without (with) bar are defined with respect to the initial (final) 0^+ state.

We calculate half-lives of $0_i^+ \rightarrow 0_f^+$ transitions for the $\beta\beta$ emitters ^{76}Ge, ^{82}Se and $^{128,130}Te$. A model space of full $3\hbar\omega$ and $4\hbar\omega$ oscillator major shells is assumed for A = 76 and 82 systems, and 1p, full $4\hbar\omega$ and 1f-0h shells for A = 128 and 130. Single-particle energies are obtained by solving the eigenvalue equation of a nucleon in the Coulomb-corrected Woods-Saxon potential [18]. The occupation amplitudes, u_j and v_j, are computed in the BCS approximation, and then the QRPA equation for $J^\pi = 1^+$ is solved for both 0_i^+ and 0_f^+ ground states. We adopt as a nucleon-nucleon interaction the G-matrix of the Paris potential [19] fitted to a sum of Yukawa terms [20]. The same interaction is used consistently for the BCS and RPA calculations.

The realistic interaction, which was constructed from a phase-shift analysis of nucleon-nucleon scattering, could be renormalized in a nucleus, and additionally in the restricted model space. The renormalization could differ in the BCS and RPA approximations. In order to adjust the interaction strengths, we introduce four renormalization factors to be multiplied on corresponding interactions. These factors, however, are not expected to deviate much from unity. In the BCS calculation, the renormalization factors of the pairing interaction are fixed for proton and neutron systems separately so as to reproduce experimental even-odd mass differences [21]. In the RPA, we introduce two factors, g_{ph} multiplied on the ph interaction, which is mainly responsible to mixing of particle-hole states in the odd-odd intermediate nucleus, and g_{pp} for the pp interaction, which is important for correlations in the ground state of the even-even nuclei (g.s. correlations). The ph interaction strength is adjusted so that the RPA calculation reproduces the excitation energy of the GT giant resonance observed in charge-exchange (p,n) reactions on the $\beta\beta$ emitters [22]. The values of g_{pair}^p, g_{pair}^n and g_{ph} thus fixed are close to unity, as expected, with deviations of at most 30%.

The nuclear matrix elements of $\beta\beta$ decay are calculated with the adjusted interaction strengths, but as function of the renormalization factor of the pp interaction, g_{pp}, which is not so far fixed. Fig. 3 shows the behaviour of the nuclear matrix elements. As g_{pp} is increased, the matrix element decreases and crosses zero at around $g_{pp} \sim 1$. This is nice for the calculation with a realistic interaction, and implies a long half-life consistent with experiment. It is, however, seen that the matrix element decreases so rapidly near the point of crossing that the value of g_{pp} is very critical in predicting the half-life with satisfactory reliability. The rapid change comes from the g.s. correlations enhanced by the pp interaction. Fig. 4 shows that the probability (amplitude squared) of the g.s. correlations, which is quite small when the pp interaction is switched off ($g_{pp} = 0$), increases exponentially as a function of g_{pp}.

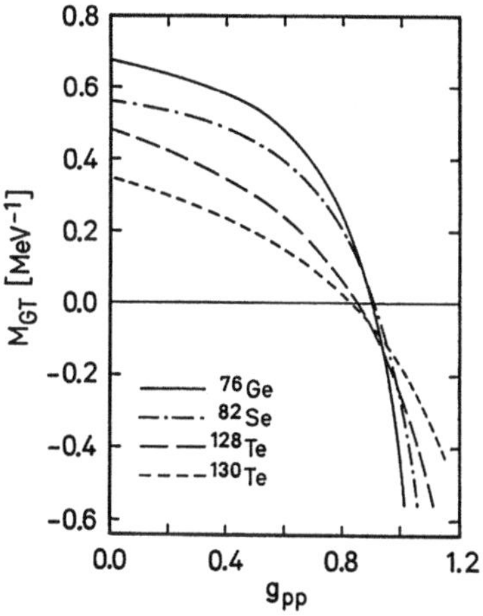

Fig. 3 The 2ν $\beta\beta$ decay nuclear matrix elements calculated as function of the particle-particle interaction strength.

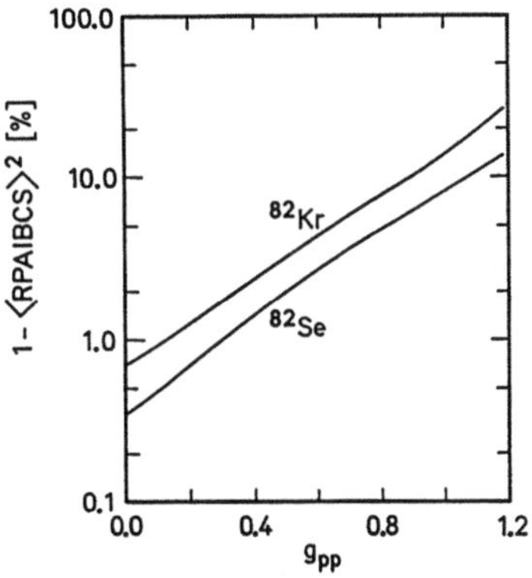

Fig. 4 Probability of RPA-type correlations in the ground state of the parent nucleus ^{82}Se and the daughter nucleus ^{82}Kr, as function of the particle-particle interaction strength.

In the $J^\pi = 1^+$ channel, the ph interaction is repulsive and the resulting mixing of particle-hole states in the odd-odd intermediate nucleus leads to the concentration of the GT strength in the giant resonance. On the other hand, the pp interaction for $J^\pi = 1^+$ is attractive. The two contributions of the opposite tendencies being added together enhance elements of the submatrix B in the RPA eq. (8) and consequently enhance the g.s. correlations. The g.s. correlations give rise to the backward-going transition, the second term of eq. (11). When the transition amplitude becomes comparable in magnitude but with the opposite sign, compared to the forward-going amplitude, the two terms cancel each other almost completely. The cancellation occurs at $g_{pp} \sim 1$ in the transitions from 1^+ intermediate states to the final 0_f^+ state, resulting in the very strong suppression of the $\beta\beta$ decay nuclear matrix element.

4. Determination of g_{pp} and $\beta\beta$ decay half-lives

We have seen that the nuclear matrix element of the 2ν mode $\beta\beta$ decay, and consequently the half-life, are sensitive to a small change of the strength of the pp interaction. One must therefore fix g_{pp} carefully in order to obtain a reliable theoretical half-life. We try to fix g_{pp} from β^+ decay of a number of nuclei with A = 60-160. β^+ decays are the same transitions as those which appear in the $\beta\beta$ decay, $0_f^+ \to 1^+$, and are sensitive to g_{pp}. For a nucleus with a large neutron excess, such as ^{76}Ge, β^+ transitions are possible only from correlated components of the ground state. The g.s. correlations are determined by the pp interaction. The β^- transitions also involved in the $\beta\beta$ decay are much less sensitive to the g.s. correlations. Most part of the β^- transitions comes from the excess neutrons or from spin-flip transitions which are allowed even if there are no g.s. correlations. This difference in the sensitivity to g_{pp} between β^- and β^+ is shown in fig. 5. In a

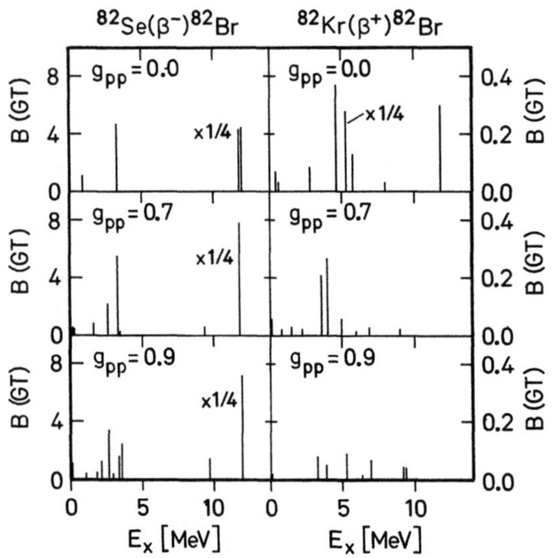

Fig. 5 Calculated Gamow-Teller strength distributions for the β^- and β^+ sectors of $\beta\beta$ decay of $^{82}Se \rightarrow {}^{82}Br \rightarrow {}^{82}Kr$. Note the scale difference between the two sectors. The distributions are shown for three values of the particle-particle interaction strength; $g_{pp} = 0$ (the particle-particle interaction is switched off), $g_{pp} = 0.7$, and $g_{pp} = 0.9$, where the 2ν $\beta\beta$ nuclear matrix element vanishes almost completely.

similar estimation Vogel and Zirnbauer [13] used β^+ decays of only a few semimagic nuclei. However, most of the $\beta\beta$ emitters and their daughters are open-shell nuclei and the RPA is expected to work much better in this type of nuclei (see below).

We have collected β^+-unstable even-even nuclei with A = 60-160 from the Table of Isotopes [23]. For each nucleus the GT strength in the Q_β window is calculated in the QRPA as a function of g_{pp}, and the g_{pp} with which the calculation reproduces the observed B(GT) values is estimated. This procedure is shown in fig. 6 for four examples. The values of g_{pp} thus obtained are summarized in fig. 7. One finds a significant difference between some specific nuclei (such as ^{100}Pd, ^{150}Dy, etc.) and the others. In the former group, a single-particle transition $\pi g_{9/2} \rightarrow \nu g_{7/2}$ or $\pi h_{11/2} \rightarrow \nu h_{9/2}$ dominates the β^+ strength, whereas in the latter various single-particle transitions contribute. From the viewpoint of the random phase assumption β^+

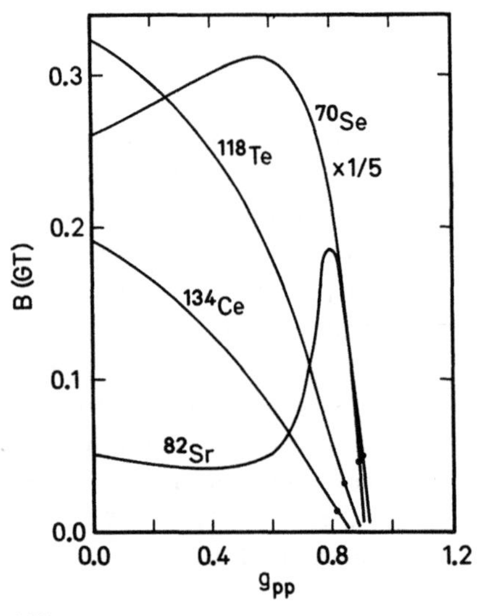

Fig. 6 The β^+ decay Gamow-Teller strength in the Q_β window, calculated as a function of the particle-particle interaction strength. The experimental value is presented by a dot.

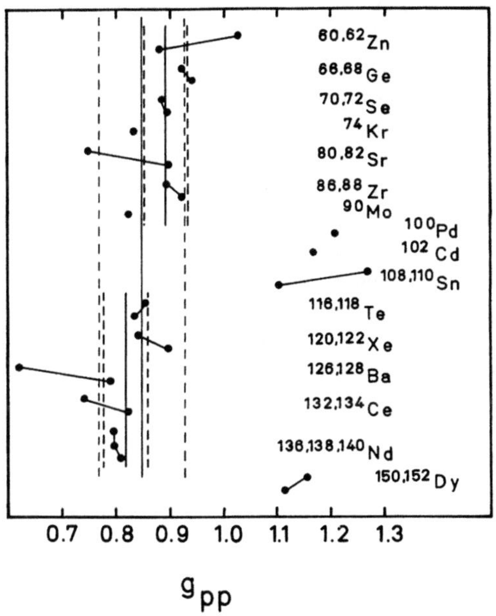

Fig. 7 The values of the particle-particle interaction strength, with which a QRPA calculation reproduces experimental Gamow-Teller strengths of β^+ decay. The vertical solid lines indicate average values in different mass regions, dashed lines correspond to the standard deviation.

transitions of the latter group are expected to be well described by the QRPA model. We therefore exclude the specific nuclei from the analysis of g_{pp}. We then obtain for the remaining 23 nuclei the average and the standard deviation, $g_{pp} = 0.85 \pm 0.08$ ('wide range'). In the distribution of g_{pp} for these 23 nuclei we find a tendency that the values for nuclei with A = 60-90 are generally larger than those of heavier nuclei with A = 116-140. For the two mass regions ('narrow range'), we obtain $g_{pp} = 0.89 \pm 0.04$ and $g_{pp} = 0.82 \pm 0.04$, respectively. It should be noted that the nuclear matrix element of 2ν $\beta\beta$ decay goes through zero at a g_{pp} lying within the range of one standard deviation from the above determined average values for the corresponding mass region.

We now calculate half-lives of 2ν mode $\beta\beta$ decay. We show two numbers, the lower limit of the half-life in the 1σ range of g_{pp}, and the average in the same range. The half-lives thus calculated for the 'wide' and 'narrow' ranges are listed in table 1 and also shown in fig. 8. The agreement with the experimental half-lives is remarkable. In particular for ^{130}Te, the large discrepancies of the earlier calculations of up to two orders of magnitude are removed.

Thus, a solution to the long-standing problem is found in the nuclear structure calculation. The success is based on the fine tuning of the particle-particle interaction strength. This interaction is not a special one, but is involved in any nucleon-nucleon interaction. Therefore, it is automatically taken into account in, for instance, a shell-model calculation. The QRPA approach, although being simplified compared to such a sophisticated model, just because of the simplification gives insight into and clarifies the role of the particle-particle interaction for the suppression of the 2ν mode nuclear matrix elements.

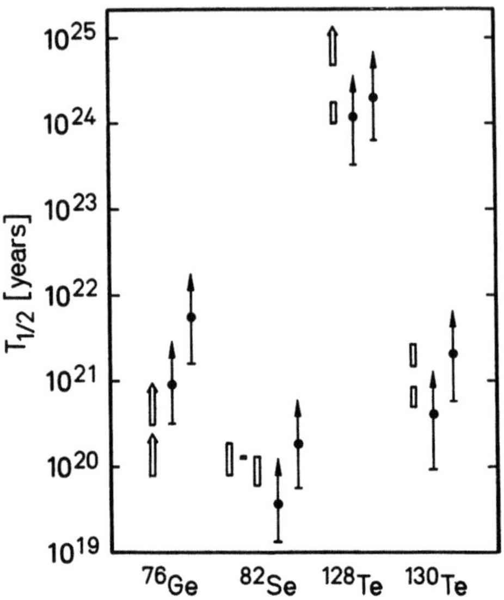

Fig. 8 Comparison of experimental [5-9] half-lives of 2ν $\beta\beta$ decay with the calculations presented in this work (see also [15]). The experimental values are given by open squares or open arrows for lower limits. The theoretical half-lives are presented by arrows. The arrow starts from the lower limit and the circle on it denotes the average. The two arrows for each nucleus correspond to the 'wide' (left) and 'narrow' (right) ranges of the particle-particle interaction strength, which are obtained from the analysis of single β^+ decays.

Table 1. Comparison of calculated and experimental half-lives of 2ν $\beta\beta$ decay (see text for the explanation of the theoretical values)

	^{76}Ge 10^{21}y	^{82}Se 10^{20}y	^{128}Te 10^{24}y	^{130}Te 10^{21}y
Calculation				
$g_{pp} = 0$	0.064	0.027	0.019	0.0069
wide range				
lower limit	0.31	0.13	0.33	0.093
average	0.91	0.37	1.2	0.41
narrow range				
lower limit	1.6	0.57	0.63	0.57
average	5.5	1.9	2.0	2.0
Experiment				
	>0.08$^{a)}$	1.30±0.05$^{d)}$	>5$^{d)}$	1.5-2.75$^{d)*)}$
	>0.3$^{b)}$	1.0±0.4$^{e)}$	1.4±0.4$^{e)}$	0.7±0.2$^{e)*)}$
		$1.1^{+0.8}_{-0.3}$$^{c)}$		

Direct measurements: a) ref. [5], b) ref. [6], c) ref. [7],
Geochemical measurements: d) ref. [8], e) ref. [9]

*) (Mainly 2ν $\beta\beta$ decay mode according to phase space arguments)

5. Conclusion

The neutrino is the only fermion that could have the Majorana self-conjugate character, $\nu = \bar{\nu}$. The 0ν mode $\beta\beta$ decay is at present most sensitive to the existence of

Majorana neutrinos. Observation of this process implies, in addition to Majorana neutrinos, finite masses of the neutrinos and possible admixture of right-handed weak interactions, which are beyond the standard electroweak theory but are both predicted by the grand unified theories based on the SO(10) gauge group.

Extraction of the neutrino mass and the mixing parameters of right-handed currents from 0ν $\beta\beta$ experiments requires a calculation of nuclear matrix elements. A test of nuclear matrix element calculations is provided by the 2ν $\beta\beta$ decay, which can occur within the standard theory. It is shown that the large discrepancy between theoretical and experimental half-lives, which existed for years, is resolved by the recent quasiparticle RPA calculations, which include particle-particle forces. The particle-particle interaction acting between a pair of proton and neutron particles (holes) plays a fundamental role. It enhances correlations in the ground states of even-even nuclei. The ground-state correlations give rise to a transition which destroys a particle-hole pair in the ground state, and the transition amplitude interferes with that of a particle-hole pair creation. The destructive interference in the amplitudes of transitions from 1^+ states of the intermediate odd-odd nucleus to the ground state of the daughter nucleus results in the suppression of the nuclear matrix element.

The strength of the particle-particle interaction is critical for the prediction of 2ν $\beta\beta$ half-lives. We fixed it to Gamow-Teller strengths observed in β^+ decays of nuclei. We showed that the 2ν $\beta\beta$ half-lives calculated with the thus fixed interaction strength are consistent with the existing experimental data on 2ν $\beta\beta$ decay.

Finally, we wish to mention that in 0ν $\beta\beta$ decay <u>no</u> passing of the matrix element through zero is observed (we refer to the paper by A. Faessler in this volume and to [24]). In contrast, deduced limits on the neutrino mass are rather stable against the degree of sophistication of the calculation of nuclear matrix elements, so that the search for 0ν $\beta\beta$ decay remains at present the most powerful tool for investigating the mass of the electron neutrino.

References

[1] H. Primakoff and S.P. Rosen, Ann. Rev. Nucl. Part. Sci. 31 (1981) 145; W.C. Haxton and G.J. Stephenson, Jr., Prog. Part. Nucl. Phys. 12 (1984) 409; M. Doi, T. Kotani and E. Takasugi, Prog. Theor. Phys. Suppl. 83 (1985) 1; J.D. Vergados, Phys. Rep. 133 (1986) 1.

[2] S.L. Glashow, Nucl. Phys. 22 (1961) 579; S. Weinberg, Phys. Rev. Lett. 19 (1967) 1264; A. Salam, Proc. of the 8th Nobel Symp., ed. N. Svartholm (Almqvist and Wiksell, Stockholm) p. 367.

[3] J. Schechter and J.W.F. Valle, Phys. Rev. D25 (1982) 2951; J.F. Nieves, Phys. Lett. 147B (1984) 375; E. Takasugi, Phys. Lett. 149B (1984) 372.

[4] P. Langacker, Phys. Rep. 72 (1981) 185.

[5] D.O. Caldwell, R.M. Eisberg, D.M. Grumm, D.L. Hale, M.S. Witherell, F.S. Goulding, D.A. Landis, N.W. Madden, D.F. Malone, R.H. Pehl and A.R. Smith, Phys. Rev. D33 (1986) 2737.

[6] F.T. Avignone III, R.L. Brodzinski, J.C. Evans, Jr., W.K. Hensley, H.S. Miley and J.H. Reeves, Phys. Rev. C34 (1986) 666.

[7] S.R. Elliott, A.A. Hahn and M.K. Moe, Phys. Rev. Lett. 59 (1987) 2020.

[8] T. Kirsten, E. Heusser, D. Kaether, J. Oehm, E. Pernicka and H. Richter, Proc. Int. Symp. on Nuclear Beta Decays and Neutrino, eds. T. Kotani, H. Ejiri and E. Takasugi (World Scientific, Singapore, 1986) p. 81.

[9] O.K. Manuel, Proc. Int. Symp. on Nuclear Beta Decays and Neutrino, eds. T. Kotani, H. Ejiri and E. Takasugi (World Scientific, Singapore, 1986) p. 71.
[10] W.C. Haxton, G.J. Stephenson, Jr. and D. Strottman, Phys. Rev. D25 (1982) 2360.
[11] H.V. Klapdor and K. Grotz, Phys. Lett. 142B (1984) 323;
K. Grotz and H.V. Klapdor, Nucl. Phys. A460 (1986) 395.
[12] P. Vogel and P. Fisher, Phys. Rev. C32 (1985) 1362.
[13] P. Vogel and M.R. Zirnbauer, Phys. Rev. Lett. 57 (1986) 3148.
[14] O. Civitarese, A. Faessler and T. Tomoda, Phys. Lett. 194B (1987) 11.
[15] K. Muto and H.V. Klapdor, Phys. Lett. 201B (1988) 420.
[16] J.A. Halbleib and R.A. Sorensen, Nucl. Phys. A98 (1967) 542.
[17] K. Muto, E. Bender and H.V. Klapdor, submitted to Phys. Lett. B.
[18] A. Bohr and B.R. Mottelson, Nuclear Structure (Benjamin, New York, 1969) Vol. I.
[19] R. Vinh Mau, Mesons in Nuclei, Vol. I, eds. M. Rho and D.H. Wilkinson (North-Holland, Amsterdam, 1979) p. 151;
M. Lacombe, B. Loiseau, J.M. Richard, R. Vinh Mau, J. Cote, P. Pires and R. de Tourreil, Phys. Rev. C21 (1985) 1.
[20] N. Anantaraman, H. Toki and G.F. Bertsch, Nucl. Phys. A398 (1983) 269.
[21] A.H. Wapstra and G. Audi, Nucl. Phys. A432 (1985) 1.
[22] R. Madey, B.S. Flanders, B.D. Anderson, A.R. Baldwin, J.W. Watson, S.M. Austin, C.C. Foster, H.V. Klapdor and K. Grotz, preprint, 1987.
[23] Table of Isotopes, 7th ed., eds. C.M. Lederer and V.S. Shirley (Wiley, New York, 1978).
[24] T. Tomoda and A. Faessler, Phys. Lett. 199B (1987) 475.

Recent Progress in Ultralow Background Ge Detector Searches for the $\beta\beta$-Decay of ^{76}Ge, Dark Matter Candidates, and Solar Axions

F.T. Avignone, III[1], *R.L. Brodzinski*[2], *H.S. Miley*[2], *and J.H. Reeves*[2]

[1]Department of Physics and Astronomy, University of South Carolina, Columbia, SC 29208, USA
[2]Pacific Northwest Laboratory, Richland, WA 99352, USA

Recent results from a number of ^{76}Ge $\beta\beta$-decay experiments and the different approaches in nuclear structure calculations which impact their interpretation are discussed. An update of current efforts to improve the sensitivities of these experiments is given. A brief review of the recent publications on searches for dark matter candidates and solar axions is also given.

1. Double-Beta Decay

1.1 Introduction

Double-beta decay ($\beta\beta$-decay) is a second order weak process which converts two neutrons in a nucleus into two protons. Figure 1 shows the diagrams of three commonly considered decay modes. Figure 2 depicts the decay scheme and spectral shapes of the sum energy of the two electrons in these decays.

The existence of $\beta\beta$-decay was first suggested by Wigner as reported by GOEPPERT-MAYER in 1935 [1], but was not directly observed in a laboratory experiment until recently. On August 30, 1987, Michael Moe announced the results of his experiment at the American Chemical Society meeting in New Orleans. The 2ν $\beta\beta$-decay half-life of ^{82}Se was determined to be $(1.1^{+0.8}_{-0.3})\times10^{20}$ years using the University of California (Irvine) Time Projection Chamber (TPC). The most interesting connection between 0ν $\beta\beta$-decay and fundamental particle theory is that it can be engendered by Majorana neutrino mass, or explicit right-handed neutrino couplings to hadrons, or both. It was shown recently that these latter mechanisms will produce a vanishing amplitude unless at least one neutrino eigenstate has a non-zero Majorana mass.

The TPC measurements of the electron spectrum from the $\beta\beta$-decay of ^{82}Se [2] can in principle distinguish between 2ν and 0ν decay; however, the energy resolution achievable with a source large enough to obtain statistically significant data will be relatively poor (7-10%). The ^{76}Ge experiments have the distinct advantage of excellent energy resolution ($\Delta E/E \sim 1.8\times10^{-3}$) at 2041 keV, the $\beta\beta$-decay energy. In addition, the source is the 7.78% abundant ^{76}Ge in the detector itself and increases in sample volume will not necessarily affect the energy resolution. Radioactive impurities in the detector and cryostat are more easily identifiable and are less prone to be confused with $\beta\beta$-decay.

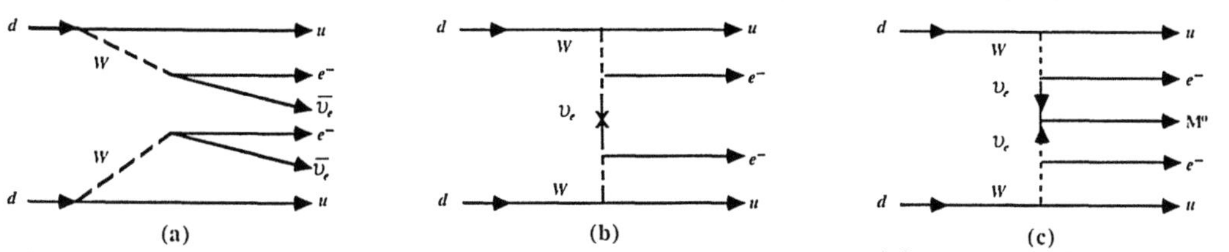

Fig. 1. Diagrams commonly associated with $\beta\beta$-decay modes: (a) ordinary lepton conserving 2ν $\beta\beta$-decay, (b) 0ν $\beta\beta$-decay with no other particles, and (c) 0ν $\beta\beta$-decay with the emission of a goldstone boson (majoron)

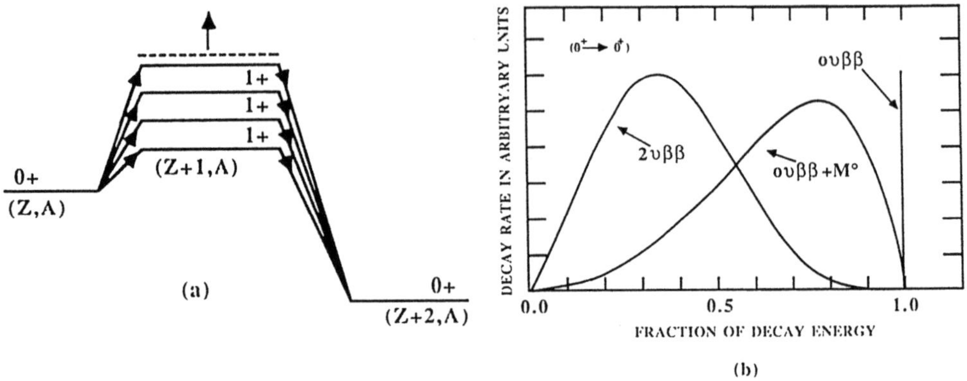

Fig. 2. (a) Schematic decay diagram depicting $\beta\beta$-decay through 1+ states of the intermediate nucleus. (b) Theoretical shapes of the spectra of the total electron energy from the three decay modes shown in the diagrams of Fig. 1

1.2 General Theoretical Considerations

The parameters x, y and z, are introduced with the conventions of [3] and with a priori assumption of CP conservation. Accordingly,

$$x \equiv m_e^{-1} \left|\sum_{k=1}^{2n} \lambda_k^{CP} U_k^L U_k^R (m_\nu)k\right|, \tag{1}$$

where λ_k^{CP} is the CP eigenvalue, and the U transformations diagonalize the mass matrix. The physical quantity to which 0ν $\beta\beta$-decay is sensitive is $\langle m_\nu\rangle_L \equiv x\, m_e$. A quantity $\langle m_\nu\rangle_R$ can be defined analogous to (1). The two parameters associated with right-handed neutrino couplings are:

$$y \equiv \eta_{RR} \left|\sum_{k=1}^{2n} \lambda_k^{CP} U_k^L U_k^R (m_\nu)k\right|, \quad \text{and} \quad z \equiv \eta_{RL} \left|\sum_{k=1}^{2n} \lambda_k^{CP} U_k^L U_k^R (m_\nu)k\right|. \tag{2}$$

The $\beta\beta$-decay rate, for the 0ν mode, can be expressed in the following general form:

$$\omega(yr^{-1}) = \alpha_1(yr^{-1})\{x^2+\alpha_2 y^2+\alpha_3 z^2+\alpha_4 xy+\alpha_5 xz+\alpha_6 yz\}. \tag{3}$$

The quantities $\{\alpha_i\}$ depend on the nuclear and atomic wave functions and relevant matrix elements. The various values of these parameters, from recent calculations discussed below, are given in Table I.

TABLE I. Numerical values for the parameters α_i in (3) for the $\beta\beta$-decay of ^{76}Ge.

	ref. 3	ref. 5	ref. 12	ref. 16
α_1	1.08×10^{-13}	1.12×10^{-13}	2.18×10^{-13}	1.21×10^{-13}
α_2	1.44	0.92	1.04	1.43
α_3	0.65	1.14	7016	30500
α_4	-0.45	-0.33	-0.36	-0.25
α_5	-0.38	-6.68	82.0	86.0
α_6	-1.34	-0.68	-0.82	-0.47

α_1 has dimensions of yr^{-1}; α_2 through α_6 are dimensionless

1.3 Right-Handed Currents and Majorana Neutrino Mass

Recently, KAYSER, PETCOV and ROSEN [4] showed that in the context of any gauge theory, a non-zero amplitude of 0ν $\beta\beta$-decay unambiguously implies that at least one neutrino eigenstate must have a non-vanishing Majorana mass. This is a generalization of an earlier proof by DOI, KOTANI and TAKASUGI and the black box proof of SCHECHTER and VALLE [5]. The most rigorous argument hinges on the demonstration that the diagonalization of the most general possible mass matrix results in $\langle\bar{\nu}_R|\nu_L\rangle=0$; accordingly,

$$\sum_i U^R_{ej}\; U^L_{ej} = 0. \tag{4}$$

Using the general form for the neutrino propagator, one can show that the R-L amplitude for 0ν $\beta\beta$-decay, driven by ν exchange, is proportional to

$$\sum_j I(m_j)\; U^R_{ej}\; U^L_{ej}, \tag{5}$$

where m_j is the mass of the jth neutrino eigenstate and $I(m_j)$ is the usual integral over momentum exchange. In the case that all neutrino mass eigenstates are degenerate, including the case in which they are zero, $I(m_j)$ will factor out, and the amplitude corresponding to R-L coupling vanishes. If $\beta\beta$-decay with the emission of two electrons only is observed, it will constitute unambiguous evidence that at least one neutrino eigenstate has a non-zero Majorana mass.

1.4 Neutrinoless Double-Beta Decay with the Emission of Majorons

Neutrinoless (0ν) $\beta\beta$-decay might also be engendered by a process resulting in the emission of a Goldstone boson. GELMINI and RONCADELLI [6] and, independently but subsequently, GEORGI, GLASHOW and NUSSINOV [7] proposed a model which contains a complex Higgs triplet in addition to the usual doublet. A non-zero vacuum expectation value of the triplet field spontaneously breaks B-L global symmetry, giving mass to light Majorana neutrinos and resulting in a real massless Goldstone boson, the majoron.

Neutrinoless $\beta\beta$-decay with the emission of a majoron was first discussed by GEORGI, GLASHOW, and NUSSINOV [7], and subsequently by VERGADOS [8], HAXTON and STEPHENSON [3], and DOI, KOTANI and TAKASUGI [5]. Using the results given in ref. 7, the ratio of the rate of 0ν $\beta\beta$-decay with the emission of a majoron, $\Gamma(M)$, to that driven by Majorana ν-mass, $\Gamma(m_\nu)$, can be expressed as follows:

$$\Gamma(M)/\Gamma(m_\nu) = 1.21 \times 10^{-3}\, g^2_{ee} \left[\frac{m_e}{m_\nu}\right]^2 \epsilon^2\, R(\epsilon), \quad (6)$$

where

$$R(\epsilon) = \left\{ \frac{\epsilon^4+14\epsilon^3+84\epsilon^2+210\epsilon+210}{\epsilon^4+10\epsilon^3+40\epsilon^2+60\epsilon+30} \right\}. \quad (7)$$

In (6), g_{ee} is the coupling constant of majorons to electron-neutrinos, and ϵ is the $\beta\beta$-decay energy in units of electron rest mass energy. The numerical coefficient in (6) differs from those derived from the results given in ref. 5 and ref. 8 by factors of 8 and 4, respectively. The formulae from different articles should be used with great care in the analysis of data because the root of these discrepancies is not clearly understood. However, DOI and coworkers [9] have addressed these differences.

1.5 Earlier Nuclear Structure Calculations

An overview of the relevant nuclear structure calculations was given earlier [10]. The recent developments concerning possible suppression of both 2ν and 0ν $\beta\beta$-decay will be dealt with briefly.

The most extensive microscopic nuclear structure calculations were the weak-coupling shell model calculations of HAXTON, STEPHENSON and STROTTMAN [11], who treated ^{76}Ge, ^{82}Se, ^{128}Te and ^{130}Te. In the cases of ^{76}Ge and ^{82}Se, the valence space included the $1g_{9/2}$, $1f_{5/2}$, $2p_{3/2}$ and $2p_{1/2}$ levels. The assumed closed core was ^{56}Ni so that the ^{76}Ge ground state consisted of 4p-,6n-holes, while the ^{76}Se ground state consisted of 6p-,8n-holes. The wave functions were constructed with all possible combinations of proton- and neutron-holes. The results exhibited strong coherence in the density matrix, resulting in large Gamow-Teller matrix elements.

The Osaka group included the p-wave effect [5], as well as a weak magnetism correction to the nuclear current. They found these effects to have approximately equal contributions. The Osaka group used the nuclear matrix elements of Haxton, Stephenson and Strottman so that the differences in their results are attributable to other effects.

Subsequently, TOMODA, FAESSLER, SCHMID and GRÜMMER [12], discussed neutrinoless $\beta\beta$-decay by describing the initial and final nuclear states by angular-momentum- and particle-number-projected Hartree-Fock-Bogoliubov wave functions.

Nuclear structure calculations of the $\beta\beta$-decays of ^{128,130}Te, ^{82}Se and ^{76}Ge were published by GROTZ and KLAPDOR [13]. In this work, right-handed couplings were neglected and the nuclear matrix elements calculated in the framework of particle number-projected BCS wave functions with a two body interaction including pairing, double Gamow-Teller, and quadrupole-quadrupole terms.

1.6 Recent Quasiparticle Random Phase Approximation Calculations

In a recent paper, VOGEL and ZIRNBAUER [14] demonstrated that the Gamow-Teller matrix element, in the quasiparticle random phase approximation (QRPA), is sensitive to particle-particle interactions in the spin-isospin polarization force. They claim that this was neglected in previous calculations, and that values of the relevant coupling constant g_{pp}, consistent with β+ ft values in ^{94}Ru, ^{96}Pd, ^{148}Dy, ^{150}Er, and ^{152}Yb, can lead to severe suppression of the Gamow-Teller matrix element for 2ν $\beta\beta$-decay in a number of nuclei. Their two body matrix elements were calculated using a δ-function interaction of the form $V = \sigma_1 \cdot \sigma_2\, \tau_1 \cdot \tau_2\, \delta(\vec{r}_1 - \vec{r}_2)$. In the interaction Hamiltonian, these potentials are weighted by empirical particle-hole and particle-particle coupling constants and the appropriate products of occupation and un-occupation probabilities calculated within the model.

A similar calculation was reported by CIVITARESE, FAESSLER and TOMODA [15]. In their work, however, a more realistic interaction was used, namely the renormalized Bonn one-boson exchange potential. This potential was renormalized by solving the

Bethe-Goldstone equation and parameterizing with g_{pair}, g_{ph}, and g_{pp}. Again there were values for $g_{pp}/g_{pair} \cong 1$ for which 2ν $\beta\beta$-decay was suppressed in the decays of ^{76}Ge, ^{82}Se, ^{128}Te and ^{130}Te. Muto and Klapdor (see K. Muto and H. V. Klapdor in this volume) have very recently completed QRPA calculations of the 2ν $\beta\beta$-decay of ^{76}Ge, ^{82}Se, and $^{128,130}Te$. They fix $g_{pp}=0.85 \pm 0.08$ using a large number of experimental $\beta+$ strengths. They also predict a suppression of 2ν $\beta\beta$-decay.

1.7 QRPA Analyses Extended to Neutrinoless Double-Beta Decay

Very recently there have been two articles addressing 0ν $\beta\beta$-decay, arriving at different conclusions [16,17]. In both cases, the calculations were performed in the Quasiparticle Random Phase Approximation; however, different nucleon-nucleon interactions were used. TOMADA and FAESSLER [16] used realistic effective interactions with a G-matrix derived from the Bonn potential. ENGEL, VOGEL, and ZIRNBAUER [17], on the other hand, accounted for short range correlations by the use of a parametrized function $\rho=1-e^{-\gamma r^2}(1-\gamma r^2)$ multiplying the spin and isospin operators in the Gamow-Teller and Fermi $\beta\beta$-decay matrix elements $M^{0\nu}_{GT}$ and $M^{0\nu}_{F}$, respectively. They used a δ-function interaction for the nucleon-nucleon two body interaction of the nuclear model itself.

The results based on the approach of ref. 17 were significantly affected by the inclusion of short range correlations, while the inclusion of these effects by use of G-matrix of ref. 16 did not significantly impact the calculated decay rates. The calculations of ref. 17 were done with the approximation $y = z \equiv 0$. For comparison if one assumes $\langle m_\nu \rangle = 1$ eV and $x = y \equiv 0$ for both cases, then $T^{0\nu}_{1/2}(^{76}Ge) = 2.9\times10^{24}$ years [16], while $T^{0\nu}_{1/2}(^{76}Ge) = 1.2\times10^{26}$ years [17]. This factor of 41 difference partly results from the different residual interactions used and partly from the different values of the particle-particle interaction parameter; $g_{pp}/g_{pair} = 1.0$ [16] and $g_{pp}/g_{pair} = 1.47$ [17]. If one uses the graphs of $T^{2\nu}_{1/2}$ versus g_{pp}, taken from [15] and shown in Fig. 3, with the experimental half-life of ^{82}Se from the recent report by the UCI group [2], one obtains $g_{pp}/g_{pair} = 1.00\pm0.04$. For the calculations given in ref. 16, the value 1.0 appears to be appropriate.

The retardation of 0ν $\beta\beta$-decay is unique to the results of ref. 17; however, this issue is of crucial importance, and it must be concluded that at this point a clear case is made that a mechanism has been identified which might possibly result in suppressions of 0ν $\beta\beta$-decay.

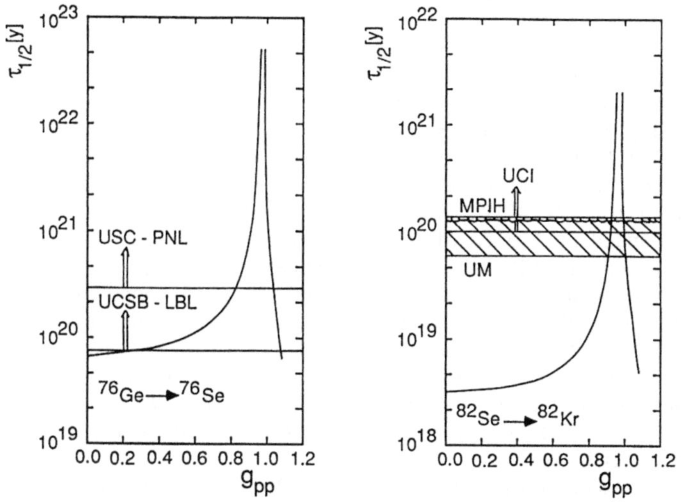

Fig. 3. Calculated 2ν $\beta\beta$-decay half-lives of ^{76}Ge and ^{82}Se with QRPA [15] and g_{pp} normalized with g_{pair}.

In summary, recent developments suggest the possible suppression of the 0ν and 2ν $\beta\beta$-decay rates. The results of the direct measurement of the 2ν $\beta\beta$-decay half-life of ^{82}Se by the UCI group shows excellent consistency between the choice of g_{pp}/g_{pair} by TOMODA et al. [16] and that obtained by intersecting the experimental half-life and theoretical predictions.

2. ^{76}Ge Double-Beta Decay Experiments

2.1 Introduction

The ingenious introduction of Ge detectors into the field of $\beta\beta$-decay more than twenty years ago was an innovation of FIORINI and his colleagues [18]. The first limit on the half-life for 0ν $\beta\beta$-decay of ^{76}Ge was $T_{1/2} > 2\times10^{20}$yr, followed by several more experiments [19,20] which improved that limit by an order of magnitude. These efforts were paralleled by the low background studies of WOGMAN and BRODZINSKI [21]. In 1979, AVIGNONE and GREENWOOD suggested the use of a NaI(Tl), Compton suppression live shield for the search for 0ν $\beta\beta$-decay of ^{76}Ge [22] and numerically calculated ideal suppression factors for the background sources and levels observed by the Milano group [20]. In 1983, the USC and PNL groups joined efforts and set a new limit on 0ν $\beta\beta$-decay of ^{76}Ge of $T_{1/2} > 1.7\times10^{22}$ yr using a Compton-suppressed, low background intrinsic Ge detector in a laboratory on the Earth's surface. Shortly thereafter, BELLOTTI et al. improved the Milano limit to $T_{1/2} > 5\times10^{22}$yr with a singles experiment [23]. During the same year, BOEHM and coworkers at Cal Tech produced a very low background Ge detector, housed in a copper cryostat, and operated in a 4π anticosmic-ray veto detector above ground [24]. This was a small detector without Compton suppression, and the limit set was only $T_{1/2} > 1.0\times10^{22}$yr. The background in the region of the $\beta\beta$-decay energy (~2041 keV) was clearly lower than previous attempts and showed great promise for their planned larger detector. Other efforts by SIMPSON et al. [25], EJIRI et al. [26], LECCIA et al. [27], and CALDWELL et al. [28] have all served to pioneer this important class of experiments. The UCSB-LBL detector [28] is the largest volume and lowest background Compton-suppressed Ge detector and has produced the best limit to date on 0ν $\beta\beta$-decay of ^{76}Ge ($T_{1/2} > 5\times10^{23}$yr).

2.2 The Milano (Mont Blanc Tunnel) Experiment

The details of this experiment and its history were given in 1986 in an article by BELLOTTI et al. [29].

The apparatus consisted of two Ge detectors in cryostats of specially selected materials. Their Ge-1 was a detector with a volume of 143 cm^3, corresponding to a fiducial volume of 117 cm^3. The end cap was made of titanium, and the cold finger of oxygen-free high-conductivity (OFHC) copper with a sapphire insulator in the detector holder in place of the usual boron nitride insulator. This detector was operated for a total of 2.361 yr in several shielding configurations. In one, it was surrounded by a 2.5-cm-thick band of Te to search for the $0^+\to2^+$, 0ν $\beta\beta$-decays of ^{128}Te and ^{130}Te. In another, the detector was surrounded by 2.5 cm of OFHC copper and 25 cm of lead. The energy resolutions at 2041 keV and at 1482 keV were 2.4 keV and 2.0 keV, respectively. The Ge-2 detector had an active volume of 148 cm^3 which corresponds to a fiducial volume of 138 cm^3 by Monte Carlo calculation. The end cap was manufactured from OFHC copper 0.5 mm thick. The detector was operated for 0.844 yr with 3.5 cm of triple-distilled mercury, 4 cm of OFHC copper, and 25 cm of low-activity lead shielding. This detector was operated for a second period of 0.917 years with 5.0 cm of copper substituted for the mercury. The energy resolutions for this detector were 2.5 keV at 2041 keV, and 2.1 keV at 1482 keV.

The spectra were subjected to maximum likelihood analyses with the following results: detector-1, 0ν $\beta\beta$-decay mode, $T_{1/2}(0^+\to0^+) > 1.2\times10^{23}$yr (68%CL); $> 5.4\times10^{22}$yr (90%CL).

For detector-2, their analysis yielded $T_{1/2}^{0\nu}(0^+\to 0^+) > 2.3x10^{23}$yr (68%CL) and > $1.0x10^{23}$yr (90%CL).

In this case the authors added the individual likelihood functions and obtained an overall limit:

$$T_{1/2}^{0\nu}(0^+\to 0^+) > 3.3x10^{23}\text{yr (68\%CL)}. \quad (8)$$

This experiment has produced a large body of very useful data, however, it has been discontinued in favor of a ^{136}Xe $\beta\beta$-decay experiment.

2.3 The University of California, Santa Barbara-Lawrence Berkeley Laboratory (UCSB-LBL) Experiment

The UCSB-LBL ^{76}Ge $\beta\beta$-decay experiment incorporates the largest volume of germanium detectors at this time. (See ref. 28 for earlier results.) The most recent data derive from eight Ge detectors operated for different periods of time. The crystals average about 160 cm^3 of fiducial volume, or about 0.9 kg of natural Ge, operating inside of a 15-cm-thick NaI(Tl) shield. (See Fig. 4.)

The apparatus is located 200 meters underground in the powerhouse of the Oroville, California dam. This location experiences approximately $7x10^6$ muons per m^2 per year, whereas the Kamioka site has ~$2x10^4$, and the Homestake mine and the Frejus Tunnel ~$1x10^3$, all in the same units. All Ge detectors which have been on the surface for a year or more will suffer significant cosmogenic production of ^{56,57,58}Co, ^{65}Zn, and ^{68}Ge [10]. These are due mainly to energetic neutron reactions on ^{70}Ge in the detectors. The ^{56}Co and ^{58}Co isotopes are relatively short-lived but ^{57}Co, ^{65}Zn and ^{68}Ge have half-lives of 272, 244 and 271 days, respectively. These isotopes appear to be decaying in the UCSB-LBL experiment.

The main interfering component thus far appears to be the internal β+ spectrum of ^{68}Ga, the equilibrium daughter of ^{68}Ge. The β+ end-point energy is 1.90 MeV, which sums with the 0.511 annihilation gamma rays to produce a continuum up to 2.92 MeV. This component also appears to be decaying.

Some of the background comes from primordial radioactivities as clearly demonstrated by γ-ray lines at 185.72, 911.07, 1001.03, 1460.75, and 2614.47 keV and the other expected lines associated with the decays of ^{235}U, ^{228}Ac, ^{234m}Pa, ^{40}K, and ^{208}Tl. The last published background rate at 2041 keV for this experiment was 1.4 counts/keV/kg/yr, where the mass is that of the natural germanium in the fiducial

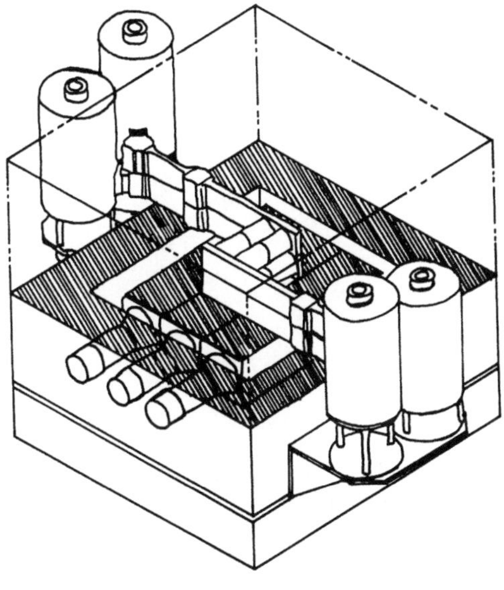

Fig. 4. The UCSB-LBL multi-crystal Ge detector and NaI(Tl) shield

olume of the detector. Even after the ^{68}Ge decays for several more half-lives, the rimordial radioactivity will dominate and any improvement on the half-life will ncrease only as the square root of the counting time.

The data from this detector were subjected to a maximum likelihood analysis ielding, $T^{0\nu}_{1/2} > 5x10^{23}$yr [28]. This limit is about twice as long as the recently ublished "world limit" [30], which included some UCSB-LBL data.

This detector has also been used to search for heavy neutrinos, 0ν $\beta\beta$-decay with he emission of majorons, cold dark matter candidates, and other exotica.

.4 Other Compton-Suppressed Ge detectors

.4.1 The Frejus Tunnel Experiment

he Frejus Tunnel ^{76}Ge $\beta\beta$-decay experiment is a collaboration between French and panish Groups [31,32]. The system consists of four intrinsic Ge detectors with a otal active volume of 417 cm^3 in a single cylindrical vacuum chamber 160 mm in iameter by 120 mm. The assembly is placed in a 4π annulus of hexagonal NaI(Tl) etectors, as shown in Fig. 5.

The NaI(Tl) scintillators are 13.5 cm across and 20.4 cm long. Fourteen crystals, anufactured to low background specifications, are wrapped in Teflon®. They have uartz optical windows and are contained in low-activity stainless steel shells 0.5 mm hick. Five more are packaged in aluminum and form the floor of the annulus. The etector's location in a laboratory in the Frejus tunnel experiences about 1x10^3 muons er m^2 per year.

The most interesting result to date is the limit $T^{0\nu}_{1/2}$ $(0^+ \rightarrow 2^+) \gtrsim 2.1x10^{22}$yr from an xperiment requiring coincidences between the Ge detector and the NaI(Tl) annulus. A mall bump was observed a few keV above the expected energy of 1482 keV for the decay o the first excited state of ^{76}Se at 559 keV.

A special low-background 200-cm^3 detector is being built for this experiment. A ollaboration between PNL-USC and the University of Zaragoza, to better understand ackground in ^{76}Ge $\beta\beta$-decay experiments, will include searches for cold dark matter andidates, solar axions, and $(0^+ \rightarrow 2^+)$, 0ν $\beta\beta$-decay.

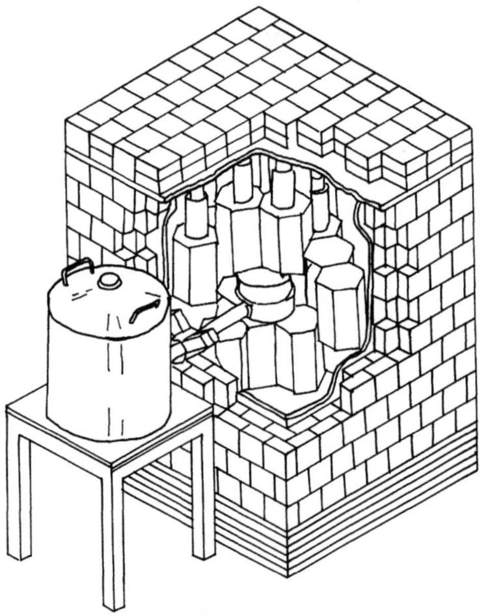

Fig. 5. The Frejus Tunnel, Compton-suppressed Ge spectrometer

®Registered Trademark of E.I. duPont de Nemours and Co., Inc., Wilmington, Delaware

2.4.2 The Osaka, Kamiokanda Underground Laboratory Experiment

The engineering details of this experiment are given in an article by KAMIKUBOTA et al. [33]. The most recent published results appeared in a 1987 article by EJIRI et al. [26]. The detector, called ELEGANTS, is a Compton-suppressed detector in an underground laboratory. ELEGANTS comprises a 171-cm^3 intrinsic Ge detector placed in a 5-segment, NaI(Tl) annulus 25.4 cm in diameter and 30.48 cm in length. There is a NaI(Tl) crystal at each end of the annulus to form a 4π geometry. The Ge detector is surrounded by 2.7 cm of pure mercury, which is in turn surrounded by 15 cm of OFHC copper and 15 cm of lead. Radon is purged using the boiloff nitrogen gas which is piped to the central cavity. All materials were carefully selected for low radioactivity [33]. Even with these precautions, the γ-ray lines from ^{40}K, ^{208}Tl, ^{212}Pb, ^{214}Pb, ^{214}Bi, ^{228}Ac, ^{234m}Pa, and ^{226}Ra are clearly visible in the spectrum. Turning on the annulus to reject background reduces the counting rate in all energy regions by more than a factor of 10.

The suppressed background was improved to ~4.3x10^{-4} counts/keV/hr in the 2041 keV region which converts to ~4 counts/keV/yr/kg and which can be compared with the rate 1.4 counts/keV/yr/kg of the UCSB-LBL experiment. This apparatus is to be used in a ^{100}Mo search for $\beta\beta$-decay described in ref. 33.

2.4.3 The ITEP-Erevan, Avansk-Mine, ^{76}Ge Enriched Experiment in the USSR

The advantages of ^{76}Ge enrichment by an order of magnitude over the natural abundance of 7.78% are obvious. Consider that the limit which can be set with the current PNL-USC 120 cm^3 fiducial volume detector, which presently has a background of $\lesssim 1$ count/keV/yr/kg, is $T_{1/2}^{0\nu}$ (^{76}Ge) $\gtrsim 2\text{x}10^{23}$yr with Nt $\simeq 4\text{x}10^{23}$yr, where N is the number off ^{76}Ge atoms and t is the counting time in years (~1 yr). If this detector was enriched to 85% ^{76}Ge, the limit that could be set with the same background would be $T_{1/2}^{0\nu} \gtrsim 2.2\text{x}10^{24}$yr.

A very brief report on a 90-cm^3 Ge detector enriched to 85% ^{76}Ge and contained in a NaI(Tl) Compton-suppression shield was presented by Kirpichnikov at the July 1987 Underground Physics workshop at the Baksan Laboratory in the USSR. The configuration of the detector and shielding and a gross spectrum after only 436 hr of operation are shown in Fig. 6.

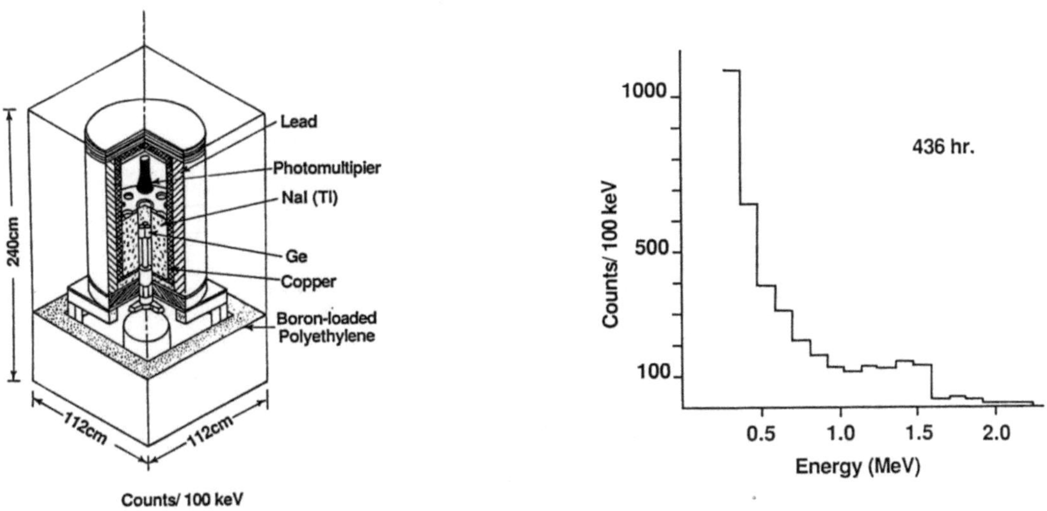

Fig. 6. Shielding configuration and gross spectrum of the 90 cm^3, 85% enriched ^{76}Ge ITEP-Erevan experiment

After 652 hrs of counting, there was one count in the interval 2035-2045 keV or less than 2 counts to a level of confidence of 68%. The quantity Nt is 2.53×10^{23} yr, hence the implied limit is $T^{0\nu}_{1/2}(^{76}Ge) \gtrsim 0.9\times10^{23}$yr. The background is ~2.8 counts/keV/yr/kg, which can be compared to that of the 1987 level of the UCSB-LBL experiment (1.4 counts/keV/yr/kg) and the new PNL-USC level (1.0 counts/keV/yr/kg).

This spectrometer will not be powerful for studying 2ν $\beta\beta$-decay and 0ν $\beta\beta$-decay with majoron emission because the background below ~1500 keV is at least a factor of 22 higher than contemporary experiments. The combination, however, of enrichment and ultralow background singles counting technology would indeed represent a more powerful tool than now exists, and should be encouraged.

2.5 The PNL-USC ^{76}Ge $\beta\beta$-Decay Experiment

Many of the details of the development of this experiment already appear in print [10,30,34]. For purposes of comparison to experiments already discussed, the current mean background level is $<$ 1 count/keV/yr/kg in the 2041 keV region which is somewhat lower than the Compton-suppressed spectrometers and about the same as the lowest background Milano experiment. The background is believed to be dominated by ^{68}Ga in this energy region. The shielding configuration and gross spectra under various conditions are shown in Fig. 7.

The most recent modifications resulted in substantial reductions in both the low- and high-energy regions of the spectrum. The drop at energies below ~500 keV was due to replacement of the indium contact ring with one made of 448-year-old lead. The dramatic decrease above 3000 keV was due to the removal of the solder in proximity to the detector. There was no improvement in the average background in the energy interval between about 800 keV and 2000 keV. This region is now dominated by ^{54}Mn,

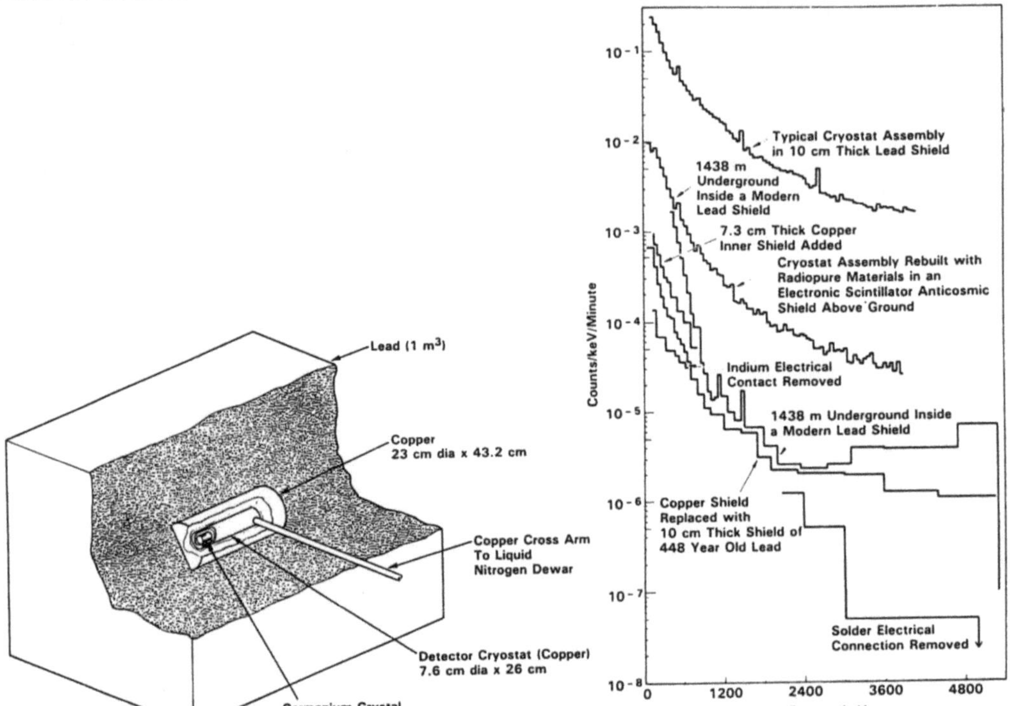

Fig. 7. PNL-USC detector configuration and spectra under various conditions

^{58}Co, ^{65}Zn, ^{68}Ge and other cosmic-ray-produced radionuclides, because the detector was on the surface for almost 8 months for renovation. The earlier experiments suffered significantly lower background from cosmogenic isotopes in the crystal itself because of its long history of containment in neutron moderators, followed by several years in the Homestake mine. The spectrum is remarkably free of γ-ray lines from primordial radioactivity. The striking feature is that there are no obvious lines above the 1461 keV γ-ray from the decay of ^{40}K.

The dominant remaining background is due to cosmogenic radioactivities in the Ge crystal itself: primarily ^{68}Ge, ^{65}Zn, ^{58}Co and ^{56}Co with half-lives of 271, 244, 71, and 78 days, respectively. The ^{68}Ge leads to a β+ and annihilation continuum out to 2920 keV. The ^{65}Zn decays by electron capture and produces a peak at 1124 keV, due to the sum of γ-ray and x-ray energies, and a continuum due to partial escape of the photons.

To mitigate this background, enough Ge ore was recently mined to produce 20 kg of metal after purification and zone refinement. This process was completed in a few days, and the Ge ingots were rushed underground into the Homestake mine to await final purification and fabrication into large, ~200 cm^3, detectors. This process should be completed in a short enough time to produce only a few percent of the saturation level of ^{68}Ge. The ^{65}Zn and Co isotopes should be removed up to the final zone refinement. In addition, the copper cryostat parts have been electroformed to reduce the ^{60}Co induced by energetic cosmic ray neutrons via the reaction ^{63}Cu(n,α)^{60}Co.

Two ~200-cm^3 Ge crystals, low in cosmogenic background, and installed in the electroformed copper cryostat, will be operational by early 1988. It is anticipated that the low background level above 3000 keV, shown in Fig. 7, will extend below 2000 keV, allowing a sensitivity of at least 6 to 10 times that of the present prototype for 0ν $\beta\beta$-decay.

2.6 Other Non Compton-Suppressed ^{76}Ge $\beta\beta$-Decay Experiments

There are two other important single counting experiments which have the potential to make major contributions in the near future. One is the assembly of ~208 cm^3 detectors of SIMPSON et al. [25], which was last reported in the literature in 1984. A large body of data (Nt~6.6x10^{23}yr) was made available in 1986 to assist in producing the "world limit" [30]. These background data were comparable to the other four experiments used in that analysis. New, high-quality results are expected in the not too distant future.

The other experiment is that of BOEHM and coworkers. This detector was originally above ground in a lead shield placed inside a 4π cosmic-ray shield [24] and was then moved to the St. Gotthard tunnel laboratory. There, it had a 21-cm-thick copper shield inside of a lead shield. The background count rate at ~2000 keV was slightly lower than 2 counts/keV/yr/kg. The background below this energy increased rapidly due, in part, to the cosmogenic ^{60}Co activity in the copper. The ^{68}Ga background is also very evident in this detector. A large-volume (~1120 cm^3) version of this detector is already operating and will produce a valuable body of data on 0ν $\beta\beta$-decay of ^{76}Ge in the near future.

3. Searches for Neutrinoless $\beta\beta$-Decay with the Emission of a Majoron

3.1 The PNL-USC Search for 0ν $\beta\beta$-Decay with Majoron Emission

A description of the experiment has been given in Section 2.5 above. The data used in the following analyses were taken in two different time frames. The first period was equivalent to 281 days during which the inner shield was a copper collar more than 7 cm thick, followed by a period of 129 days during which only the very low-energy portion of the spectrum was investigated. The final data of interest were collected in the next 121-day period during which the copper inner shield had been replaced by

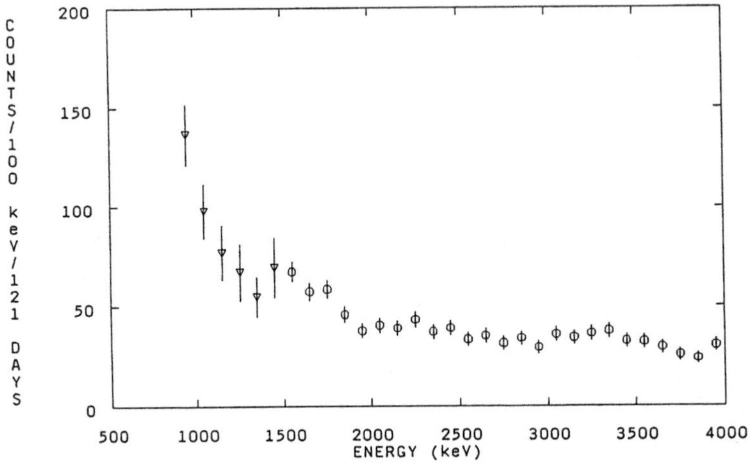

Fig. 8. The combination of the 121 days of data, corrected for ^{40}K and ^{60}Co lines and continua (triangles), and the normalized 402 days of data (circles)

bricks cast from 448-year old lead. This reduced the ^{60}Co, produced by the cosmogenic reaction $^{63}Cu(n,\alpha)^{60}Co$, by a factor of 5.4.

The raw data were combined into 100-keV bins as shown in Fig. 8. The data up to 1440 keV are taken from the final 121-day counting period during which the ^{60}Co was much reduced. The data above this energy are from the full 402 days to gain statistical significance. A good fit to the data above the $\beta\beta$-decay end point energy at 2041 keV is obtained with a straight line. The maximum of the theoretical energy spectrum for the two electrons emitted in 0ν $\beta\beta$-decay with a majoron is located at approximately 1600 keV. A curve of the shape shown in Fig. 2, plus the straight line, was fit to the data between 1600 keV and 3950 keV. A three-parameter likelihood function was calculated, and the variances obtained were added in quadrature to the Poisson statistical fluctuation of the most probable area under the theoretical curve. The resulting number of events which could be attributed to such an effect is 196 ± 28 counts in 402 days. This corresponds to 588 ± 84 events in the entire three-body decay spectrum or 1.46 ± 0.21 events per day. An analysis of the first 281 days of data, taken with the copper inner shield, yields a rate of 1.7 ± 0.3 day^{-1} under the spectral bump in question. The rate during the final 121 days, during which the inner shield was 448-year-old lead, was 1.4 ± 0.4 day^{-1}. These results eliminate cosmogenic radioactivity in the copper shield as a probable cause of the observed effect, while they also place constraints on the half-life of radioactive candidates which might mimic an exotic decay. The number of ^{76}Ge atoms in the fiducial volume is 4.3×10^{23}, and if these events are in fact due to this exotic decay process, the half-life is $(6\pm1)\times10^{20}$ yr. Combining (6) and (7) with the calculations of HAXTON and STEPHENSON [3], the following relation is obtained:

$$g_{ee}^{2} = 3.5\times10^{14}\ yr/T_{1/2}^{0\nu,M}\ yr. \quad \text{Accordingly, } g_{ee} = (8\pm1)\times10^{-4}. \tag{9}$$

Two notes of caution are emphasized. First, although no alternate explanation for the bump was found, the same effect would have to be seen in another Ge crystal before existence of this decay mode can be seriously entertained. Second, the assumption that the background above 2041 keV extends linearly to lower energy may be wrong.

Since these data were taken, the radioactive solder was removed, but the detector resided on the surface for almost eight months, increasing the ^{68}Ge background significantly. The data from this new configuration are shown in Fig. 9. The dashed

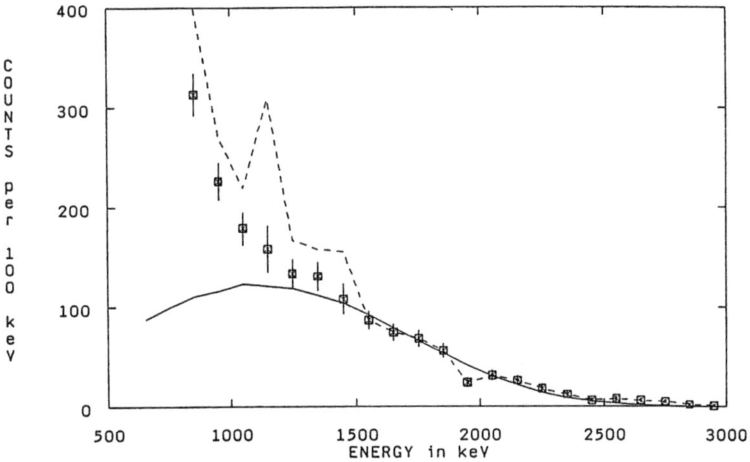

Fig. 9. The most recent 144 days of data taken with the PNL-USC detector. The dashed line is the raw data summed into 100-keV bins. The squares are the data corrected for gamma activity from ^{40}K, ^{60}Co, ^{54}Mn, ^{58}Co, and ^{65}Zn. The solid line is the expected shape of the ^{68}Ga spectrum superimposed on the data at high energies

line represents the raw data summed in 100 keV bins. The squares show the data corrected for the photopeaks and their experimentally measured continua from ^{40}K and ^{60}Co and for the photopeaks only from ^{54}Mn, ^{58}Co, and ^{65}Zn. The solid line represents the expected shape of the ^{68}Ga spectrum in this detector and is arbitrarily superimposed on the data at the high energy end. The amount of ^{68}Ga is less than the amount which could be present based on measurement of the 1077 keV gamma-ray from electron capture in ^{68}Ga. This line, which would be expected to appear at 1087 keV, the sum of the gamma-ray and x-ray energies, is not observed in the spectrum, but would not be expected to be statistically significant at these levels. The current electronic arrangement does not permit direct measurement of the ^{68}Ge x-ray. Therefore, no conclusions can be drawn from these data regarding a contribution from 0ν $\beta\beta$-decay with majoron emission.

Data obtained from detectors made of deep-mined germanium are anticipated to be reasonably free of ^{68}Ga and, hence, offer much lower background in the 2ν and $0\nu(M)$ $\beta\beta$-decay regions of interest.

3.2 The St. Gotthard Laboratory Result

This experiment was discussed in Section 2.6. The 90 cm^3 Ge detector was operated for 9373 hrs underground. There were seven γ-ray lines observed between 1700 and 2700 keV. Six of them (1730, 1764, 1847, 2119, 2204, and 2448 keV) are associated with the ^{238}U chain, and the 2614 keV line is associated with the ^{232}Th chain. The data points and error bars shown in Fig. 10 are traced from the data from ref. 35. These data have been corrected by subtracting the β-continuum from cosmogenically created ^{68}Ga in the crystal and eight gamma-ray lines between 1700 and 2614 keV and their associated continua as modeled by Monte Carlo techniques. For purposes of this discussion, this corrected spectrum is accepted at face value. The authors conclude that the maximum decay rate allowed for this exotic process corresponds to a half-life $> 1.2\times10^{21}$ yr and that this conflicts with the PNL-USC result.

The assumption that the portion of the continuum background spectrum below 1500 keV can be smoothly connected to the portion above ~2040 keV may not be valid. In fact, this assumption would tend to hide the very subject of the search; namely, a continuum rising from zero energy, broadly peaking at ~1600 keV, and decreasing to zero at 2041

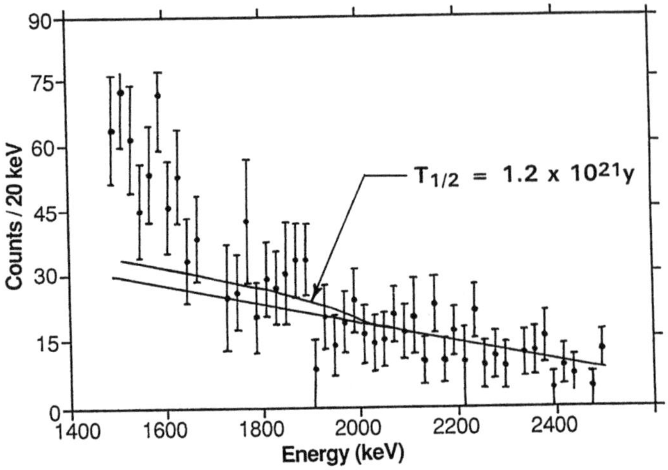

Fig. 10. The corrected data points given by FISHER et al. [35]. The analysis is done by assuming the background below 1500 keV cannot be smoothly connected to that above 2041 keV

keV. The energy distribution corresponding to $T^{0\nu,M}_{1/2}$=1.2x10^{21} yr is shown in Fig. 10,and when plotted above the best fit straight line, is not convincingly eliminated. In fact, the best fit curve corresponds to a half-life of 4.3x10^{20} yr.

However, the St. Gotthard tunnel results do not necessarily support those of the PNL-USC experiment. This is only one of several reasonable assumptions, and the only way to resolve this issue is with additional experiments with significantly improved sensitivities. Finally, it is important that when the PNL-USC data are subjected to an analysis very similar to that discussed in ref. 35, the result is $T^{0\nu,M}_{1/2} > 2\text{x}10^{21}$ yr, which demonstrates that half-lives for this decay mode depend very sensitively on the assumptions made to construct the unknown continuum background.

3.3. The UCSB-LBL Result

The results of the UCSB-LBL search for 0ν $\beta\beta$-decay with the emission of a majoron [36] are not as straightforward as the above cases. The authors do not present a completely corrected spectrum which they then test to determine how much of the decay mode in question it will tolerate. Their results are shown in Fig. 11. The data minus the γ-ray peaks are plotted in 50-keV bins and are denoted by crosses, the height of which indicates the statistical accuracy. The intensities of the γ-ray peaks were used to Monte Carlo model the Compton continua. No systematic errors are shown on the solid line which represents the modeled background and passes very nearly through all of the data points. It is remarkable that the data are so well fit by the modeled continuum, particularly when the exact location of each source in this complicated detector geometry is not known. The ^{68}Ge contribution was determined by integrating under the x-ray from Ga following electron capture. The contribution of Compton continua comprises about 50% of the background, while ^{68}Ga accounts for 20%. Very little can be done to critique this result further or to interpret the data in any other manner because of this very complex set of corrections.

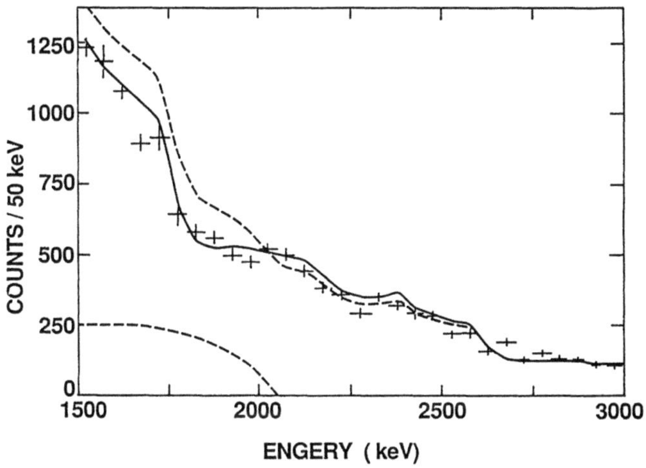

Fig. 11. Data points of the UCSB-LBL spectrum plotted every 50 keV. The solid line is composed of "known" backgrounds, modeled by Monte Carlo, plus a constant term. The dashed line crossing the solid line is the solid line plus the expected contribution from 0ν $\beta\beta$-decay with majoron emission with $T_{1/2} = 6\text{x}10^{20}$ yr. The lower dashed curve is the expected spectrum from the exotic decay itself

4. Applications of Ge Detectors in Astrophysics

Recently two efforts have been published in which the PNL/USC detector has been used to place limits on the masses of weakly interacting cold dark matter candidates and on the coupling constant of Dine-Fishler-Srednicki (DFS) axions to electrons [37,38]. In this section these results will be reviewed, and the reader will be referred to the original articles for a more complete discussion and references to the original literature. Subsequent to the PNL/USC work, the UCSB-LBL group has placed similar limits on dark matter candidates with their experiments and is also examining new approaches.

4.1 Limits on Cold Dark Matter Candidates

In the conventional wisdom of astrophysics there is a significant portion of the matter in the universe which is non-luminous and might be weakly interacting massive particles as suggested by Whitten and which he calls WIMPs. The Ge detector experiment involves coherent scattering of the WIMPs off of Ge nuclei via Z° exchange. The recoiling Ge nuclei generate a signal by creating electron-hole (e-h) pairs.

The measurement of the nuclear recoil due to the scattering of WIMPs requires a detector with a low energy threshold and excellent background rejection. Because of its low band gap (0.69 eV at 77°K) and high efficiency for converting electronic energy loss to e-h pairs (2.96 eV per e-h at 77°K), germanium is probably the best suited material for a semiconducting WIMP detector. In late 1985, the threshold on the PNL/USC detector was reduced to an equivalent incident electron energy of 4 keV, which permitted detection of Ge nuclei-WIMP scattering with recoil energies greater than 15 keV. Ten weeks of low-energy data are shown in Fig. 12. Interpretation of the data requires knowledge of the relative efficiency factor (REF), which is the ratio of e-h pairs produced by an electron, to the number produced by a recoiling Ge nucleus.

The particles that comprise the halo are assumed to have a velocity distribution function, $f(\bar{v})$, with an RMS of 250 km/s and a maximum of 550 km/s while the halo, like the galactic spheroid, slowly rotates with a velocity of 80 km/s [39]. The

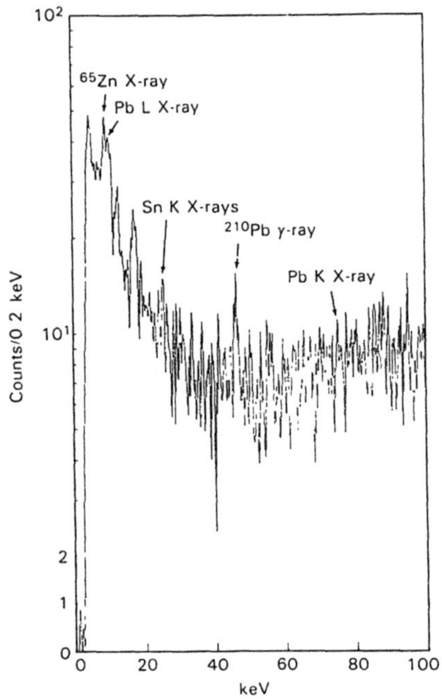

Fig. 12. Ten weeks of data from the low energy portion of the Ge detector spectrum

predicted detection rate of recoils having energy T, $R_p(T)$, was calculated according to,

$$R_p(T) = n_X \Delta T \int \frac{d\sigma}{dT}(v,T) f(\bar{v}) v d^3v, \qquad (10),$$

where ΔT is the range of recoil energies detected in a given channel and n_X is the local density of dark matter. The cross section as a function of recoil energy, $d\sigma/dT$, for an incident WIMP depends upon its mass, m_X, and velocity, v, according to

$$\frac{d\sigma}{dT} \simeq \frac{G_F^2 m_N c^2}{8\pi v^2} [Z(1-4\sin^2\theta_W) - N]^2 \left[1 + (1 - T/E)^2 - \frac{m_N T + m_X^2}{E^2}\right] \qquad (11)$$
$$\bullet \exp(-m_N T R^2/3).$$

In (11), m_N is the mass of the nucleus, Z and N are the number of protons and neutrons respectively, T is the recoil energy, G_F is the weak coupling constant, θ_W is the Weinberg angle and $E = m_X(c^2 + v^2/2)$. When the de Broglie wavelength, corresponding to the momentum transferred in the recoil, is smaller than the nucleus, the assumption of a coherent interaction with a point-like mass is no longer valid, and the finite size of the nucleus must be included in the calculation. The exponential factor in (11) is a nuclear form factor.

The halo model used in the calculation is conservative. If the maximum halo velocity is chosen closer to the local escape velocity of ~750 km/s [40], the lower limit of the excluded mass range decreases. Figure 13 shows the limits obtained for standard as well as a "perverse" halo model in which all of the dark matter particles were assumed to lie on purely circular orbits.

Since the predicted count rate depends upon WIMP density, limits on the density of interacting dark matter particles in the halo can be obtained. Figure 13 also shows these limits for particles with spin independent (s.i.) Z° exchange interactions for the standard isotropic halo model and for the model that assumes that all of the particles are on purely circular orbits. Both models include the sun's motion

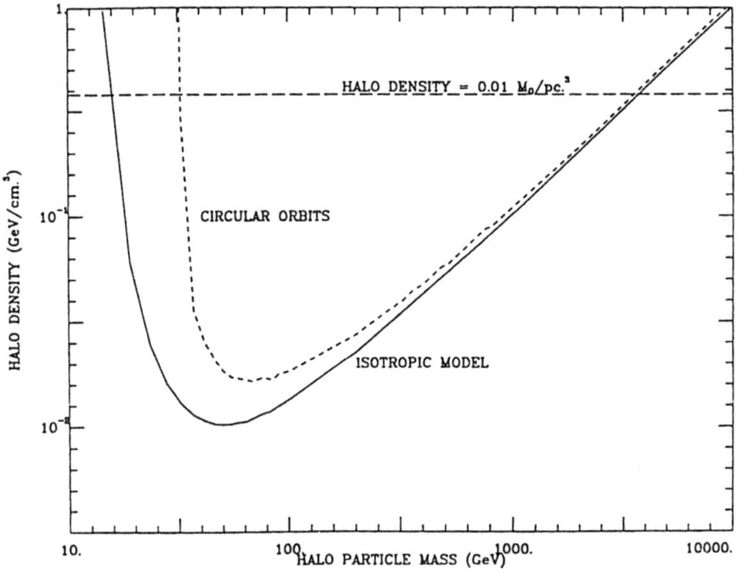

Fig. 13. Maximum halo density consistent with the observed count rate

relative to the galactic halo. Figure 13 can be used for other s.i. vectorial interactions by multiplying the vertical axis by the ratio $(\sigma_{weak}/\sigma_{s.i.})$. For example, for neutral technibaryons this ratio is approximately 0.1.

The most important results are shown in Fig. 14, where the range of mass and cross-section of particles excluded as main components of the halo are presented. The ratio g/g_w is defined as $(\sigma/\sigma_{weak})^{1/2}$ where σ_{weak} is the cross section for standard heavy Dirac (s.i.) or Majorana spin dependent (s.d.) neutrino interactions. The validity of the experimental bound extends to cross sections on the order of 10^{-28} cm^2. The halo

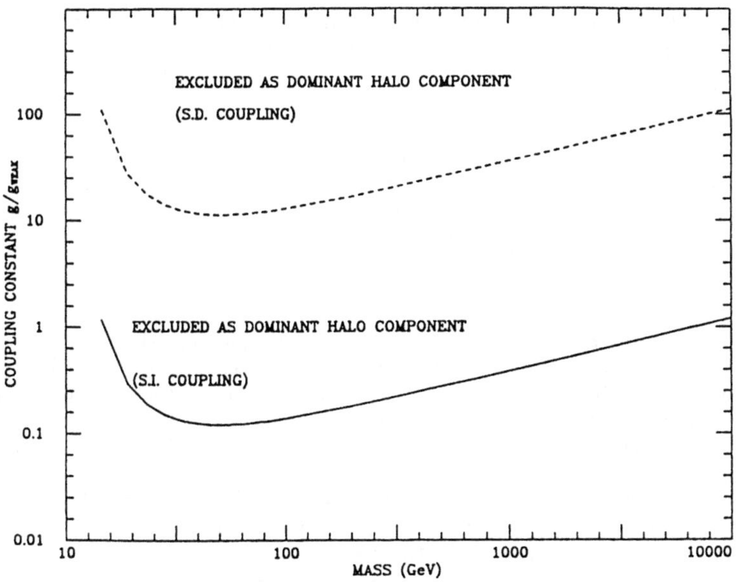

Fig. 14. Excluded regions of the coupling constant-mass plane lie above the curves

cannot be composed of particles that interact with nuclei through s.i. interactions whose coupling constant (normalized to the coupling of massive Dirac neutrinos to baryons) lies above the solid line. Nor can the halo be composed of particles that interact with nuclei through s.d. interactions whose coupling constant (normalized to the coupling of massive Majorana neutrinos to baryons) lies above the dashed line.

4.2 Laboratory Limits on Solar Axions

It is well known that the strong CP problem can be solved by the introduction of a light pseudoscalar axion. Its mass and couplings are inversely proportional to the vacuum expectation value (VEV) which breaks the Peccei-Quinn U(1) symmetry [41]. The CP problem is not sensitive to the magnitude of the particular VEV, but the detectability of the axion is sensitive, through its mass and coupling constants. With few exceptions, the standard axion has been ruled out by a large number of experiments [42].

Atomic enhancements can be exploited when axions interact with bound electrons [43] in a process analogous to the photoelectric effect, which is enhanced by factors of $\sim 10^3$ when the photon energy is near the electron binding energy. This process is called the axioelectric effect. Such effects are expected to be large for solar axions, since their energy should be comparable to atomic energies; the average solar temperature is near 1 keV. In the dipole approximation with axion energies $\omega \ll m_e$ (in natural units $\hbar = c = 1$), we have

$$\sigma_{axioelectric} = \frac{\alpha(axion)}{\alpha(em)} \left(\frac{\omega}{2m_e}\right)^2 \sigma_{photoelectric}, \qquad (12)$$

and

$$\alpha(axion) = (2x_e' m_e/F)^2 \frac{1}{4\pi}, \qquad (13)$$

where $\alpha(em) \cong (137)^{-1}$ and x_e' is a constant of order unity and which Srednicki argues is greater than one in the DFS model. F is defined by the axion-electron interaction Lagrangian

$$L = 2x_e' \frac{m_e}{F} a\bar{e}i\gamma_5 e, \qquad (14)$$

where a is the axion field. Equation (18) includes all Coulomb effects for the nonrelativistic electron. The axion mass can be related to F by

$$m_{axion} \cong 7.2\ \text{eV} \left[\frac{10^7\ \text{GeV}}{F}\right]. \qquad (15)$$

A reliable lower bound may be placed on F by requiring that the solar bremsstrahlung axion luminosity not exceed the photon luminosity. Such a large axion luminosity would imply that the sun is significantly younger than $\sim 4.5 \times 10^9$ years, the age of the oldest known meteorites. This gives

$$F/2x_e' \gtrsim 1.08 \times 10^7\ \text{GeV}. \qquad (16)$$

Motivated by this bound, the axionization cross sections per kg for C, Si, Ge and Pb were calculated with (18) for $F/2x_e' = 10^7$ GeV. Because of its low band gap and high efficiency for converting electronic energy loss to e-h pairs, germanium detectors are probably the best suited particle detectors for DFS axions.

In Fig. 15, the number of events per kg per day for germanium are plotted against the incoming axion energy for $F/2x_e' = 0.5 \times 10^7$ GeV (solid line) and $F/2x_e' = 10^7$ GeV (dashed line). The major contribution to the event rate comes from a narrow band between 1 keV and 10 keV. This is because both the solar axion flux and the axioelectric cross section peak in this region. In the following, the expected rates are compared with the count rates observed in the ultralow background germanium spectrometer described earlier. Three months of data were accumulated, of which six

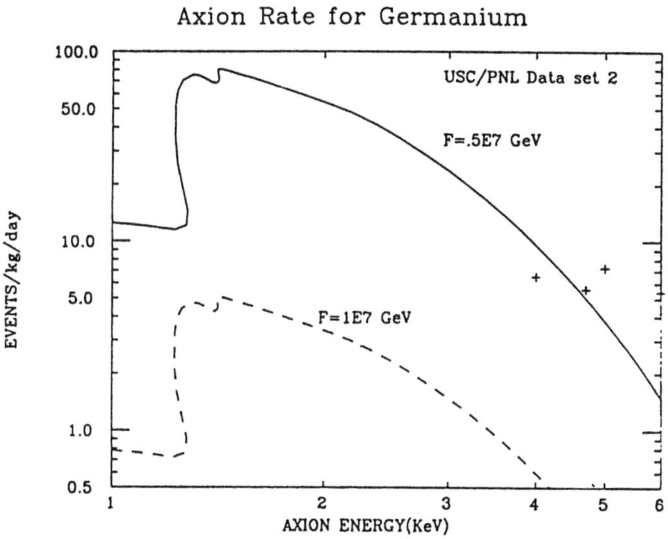

Fig. 15. Solar axion events per kg per day for Ge, $F/2x_e' = 0.5\times10^7$ GeV (solid), 1.0×10^7 GeV (dashed). The crosses are data points

weeks were low in noise. Also plotted in Fig. 15 are some of the experimental points (crosses) for $\omega \cong 4$ keV. The statistical error on these data is estimated at ± 25%. From this, the experimental bound

$$F/2x_e' \gtrsim 0.5 \times 10^7 \text{ GeV} \tag{17}$$

may be deduced. For $2x_e' = 1$ the laboratory bound on the DFS axion mass is

$$m_a \gtrsim 15 \text{ eV}. \tag{18}$$

Bounds similar to (18) are obtained for any light pseudoscalars or light scalars that couple directly to electrons. Familons [44] and singlet majorons [6,7,45], associated with right-handed neutrinos, have couplings similar to (20) where $F/2x_e'$ is replaced by a model dependent coupling constant and F is the large global horizontal symmetry breaking scale. Triplet majorons, associated with left-handed neutrinos, appear if lepton number is a global symmetry spontaneously broken (up to this point the same applies to singlet majorons) at a scale v_T (the vacuum expectation value of a triplet Higgs field) small with respect to the electroweak scale. From the coupling of majorons, M, to a pseudoscalar electron current we have

$$L = 2\sqrt{2}\, G_F v_T m_e M \bar{e} i \gamma_5 e. \tag{19}$$

By comparison with (14), the bound analogous to (17) becomes

$$v_T \lesssim 6.9 \text{ MeV}. \tag{20}$$

There is a problem of self-consistency here, however. A laboratory bound for axions $F/2x_e' \gtrsim 0.5\times10^7$ GeV has been given. Suppose $F/2x_e'$ were indeed 0.5×10^7 GeV. Then the solar axion luminosity, ℓ_a, would be approximately four times the solar photon luminosity, ℓ_γ. To calculate the axion flux, a model of the sun was used which was dominated by QED, weak, and nuclear processes; axions were assumed unimportant for solar dynamics. This might not be the case if $\ell_a \simeq 4\, \ell_\gamma$, and therefore the laboratory bound might be self-inconsistent. However, a bound stronger by a factor of 2 to 3 would be self-consistent. Thus, future improvements in our laboratory bounds are crucial.

Using the axioelectric effect, semiconducting Ge detectors could eventually set limits $F/2x_e' > 10^8$ GeV ($v_T < 0.3$ MeV) or, more exciting, see solar axions or other light bosons. The discovery of these particles would allow us to study physics at energies beyond the reach of accelerators and provide us with a new laboratory tool to study the interior of stars.

The results reported in Sections 4.1 and 4.2 are from work in collaboration with S. P. Ahlen, A. K. Drukier, G. Gelmini, D. N. Spergel, S. Dimopoulos, G. D. Starkman, and B. W. Lynn. The details and more complete reference to the literature on the subject are given in ref. [25] and [26].

5. Summary and Conclusions Concerning $\beta\beta$-Decay

There have been many recent exciting developments in $\beta\beta$-decay, both theoretical and experimental. A question has evolved concerning whether or not particle-particle interactions in nuclear structure calculations actually suppress 2ν and 0ν $\beta\beta$-decay, and if they do, then how much. The first direct observation of 2ν $\beta\beta$-decay, by ELLIOTT, HAHN and MOE [2] was reported in September 1987 with the result $T_{1/2}^{2\nu}(^{82}Se) = (1.1^{+0.8}_{-0.3})\times10^{20}$years which is only a factor of 1.7 longer than shell model predictions [3].

This clearly implies that the aforementioned suppression is not serious in this case.

Several Ge detector experiments have produced interesting new limits on the 0ν $\beta\beta$-decay half life of ^{76}Ge. These were, in order of their publication, $T_{1/2} > 3.3\times10^{23}$ yr [29], $T_{1/2} > 3.1\times10^{23}$ yr [30], and $T_{1/2} > 5\times10^{23}$ yr [28]. In addition, preliminary results from the first enriched ^{76}Ge detector were announced in 1987. A combination of vast improvements in low background techniques and enrichment could result in extremely tight limits on the Majorana mass of the neutrino (~0.03 eV).

The controversy which arose over the announcement of an anomalous bump in the spectrum of the PNL-USC experiment [28,46,35], has led to a deeper understanding of the continuum background of several of the Ge $\beta\beta$-decay experiments, which has resulted in increased sensitivity in searches for the 2ν $\beta\beta$-decay of ^{76}Ge, as well as 0ν $\beta\beta$-decay with the emission of a majoron.

Accelerated progress in $\beta\beta$-decay experiments has led to improvement of the sensitivity of low-background Ge detectors for a host of other applications.

The most dramatic improvement would be use of ~85% ^{76}Ge enriched material in an ultralow-background experiment which, though worthwhile, is extremely costly. In searches for 2ν $\beta\beta$-decay of ^{76}Ge, the technique of deep mining Ge, protecting it from energetic cosmic ray neutrons during the preparation of the detector, and then returning it underground, is expected to produce a new generation of sensitivity.

This subject is experiencing an exciting stage of rapid development, and the levels of sensitivity which would allow direct observation of 2ν $\beta\beta$-decay of ^{76}Ge should be achieved in the near future.

First generation low-background Ge detector searches for cold dark matter candidates for the non-luminous mass of the galactic halo have been performed. Low-energy threshold Ge detectors could decrease the lower detectable mass-limit of WIMPs to 5-10 GeV [37]. Larger arrays of lower background Ge detectors could improve the laboratory bounds on solar axions to $F/2x_e' > 10^8$ GeV.

References

1. M. Goeppert-Mayer: Phys. Rev. 48, 512 (1935)
2. S.R. Elliott, A.A. Hahn, M.K. Moe: Phys. Rev. Lett. 59, 2020 (1987)
3. W.C. Haxton, G.J. Stephenson, Jr.: Prog. Part. Nucl. Phys. 12, 409 (1984)
4. For a brief report on the theorem of Kayser, Petkov and Rosen see B. Kayser: Proc. XXIII Int. Conf. on High Energy Physics, ed. S.C. Loken, (World Scientific, Singapore) p.945 (1987); see also J. Schechter, J.W.F. Valle: Phys. Rev. D25, 2951 (1982) and Ref. 8
5. M. Doi, T. Kotani, E. Takasugi: Prog. Theor. Phys. Suppl. 83, 1 (1985)
6. G.B. Gelmini, M. Roncadelli: Phys. Rev. Lett. 99B, 411 (1981)
7. H.M. Georgi, S.L. Glashow, S. Nussinov: Nucl. Phys. B193, 297 (1981)
8. J.D. Vergados: Phys. Lett. 109B, 96 (1982)
9. M. Doi, T. Kotani, E. Takasugi: Osaka preprint OS-GE-97-07 (1987)
10. F.T. Avignone, R.L. Brodzinski, J.C. Evans, Jr., W.K. Hensley, H.S. Miley, J.H. Reeves: Phys. Rev. C34, 666 (1986)
11. W.C. Haxton, G.J. Stephenson, Jr., D. Strottman: Phys. Rev. D25, 2360 (1981)
12. T. Tomoda, A. Faessler, K.W. Schmid, F. Grummer: Nucl. Phys. A452, 591 (1986)
13. K. Grotz, H. V. Klapdor: Phys. Lett. 153B, 1 (1985);
14. P. Vogel, M.R. Zirnbauer: Phys. Rev. Lett. 57, 3148 (1986)
15. O. Civitarese, A. Faessler, T. Tomoda: Phys. Lett. B194, 11 (1987)
16. K. Grotz, H.V. Klapdor: Phys. Lett. 153B, 1 (1985)
17. J. Engel, P. Vogel, M.R. Zirnbauer: CAL TECH Preprint MAP-95, June 1987
18. E. Fiorini, A. Pullia, G. Bertolini, F. Cappellani, G. Restelli: Phys. Lett. 25B, 602 (1967)
19. E. Fiorini, A. Pullia, G. Bertolini, F. Cappellani, G. Restelli: Lett. Nuovo Cimento 3, 149 (1970)
20. E. Fiorini, A. Pullia, G. Bertolini, F. Cappellani, G. Restilli: Lett. Nuovo Cimento 13A, 747 (1973)
21. N.A. Wogman and R.L. Brodzinski: Nucl. Instrm. Methods 109, 277 (1973); IEEE Trans. Nucl. Sci. NS-20, 73 (1973); R.L. Brodzinski: Radiochem. Radioanal. Letters 13, 375 (1973)
22. F.T. Avignone, III, Z.D. Greenwood: Nucl. Instr. and Methd. 160, 493 (1979)
23. E. Bellotti, O. Cremonesi, E. Fiorini, C. Liguori, A. Pullia, P.P. Sverzellati, L. Zanotti: Phys. Lett. B121, 72 (1983)
24. A. Forster, H. Kwon, J.K. Markey, F. Boehm, H.E. Henrikson: Phys. Lett. 138B, 301 (1984)
25. J.J. Simpson, P. Jagam, J.L. Campbell, H.L. Malm, B.C. Robertson: Phys. Lett. 53, 141 (1984)
26. H. Ejiri, N. Kamikubota, Y. Nagai, T. Nakamura, K. Okada, T. Shibata, T. Shima, N. Takashasi, W. Watanabe: J. Phys. G13, 839 (1987)
27. F. Leccia, Ph. Hubert, D. Dassie, P. Mennrath, M.M. Villard, A. Morales, J. Morales, R. Nunez-Lagos: Lett. Nuovo Cimento 78A, 50 (1983)
28. D.O. Caldwell, R.M. Eisberg, D.M. Grumm, D.L. Hale, M.S. Witherell, F.S. Goulding, D.A. Landis, N.W. Madden, D.F. Malone, R.H. Pehl, A.R. Smith: Phys. Rev. Lett. 59, 419 (1987)
29. E. Bellotti, O. Cremonesi, E. Fiorini, C. Liguori, A. Pullia, P.P. Sverzellati, L. Zanotti: Lett. Nuovo Cimento 95A, 1 (1986)
30. F.T. Avignone, III, R.L. Brodzinski, H.S. Miley, J.H. Reeves: Phys. Rev. D35, 1713 (1987)
31. Ph. Hubert, F. Leccia, D. Dassie, P. Mennrath, M.M. Villard, A. Morales, J. Morales, R. Nuñez-Lagos, J.A. Villar: Lett. Nuovo Cimento 85A, 19 (1985)
32. A. Morales, J. Morales, R. Nuñez-Lagos, J. Puimedón, J.A. Villar, D. Dassie, Ph. Hubert, F. Leccia, P. Mennrath, M.M. Villard, J. Chevallier, B. Hass: Proc. XXIII Int. Conf. on High Energy Physics, ed. S.C. Loken, World Scientific, Singapore (1987) p.948

33. N. Kamikubota, H. Ejiri, T. Shibata, Y, Nagai, K. Okada, T. Watanabe, T. Irie, Y. Itoh, T. Nakamura, N. Takahasi: Nucl. Instr. and Methd. A245, 379 (1986)
34. R.L. Brodzinski, D.P. Brown, J.C. Evans, Jr., W.K. Hensley, J.H. Reeves, N.A. Wogman, F.T. Avignone, III, H.S. Miley: Nucl. Instr. and Methd. A239, 207 (1985)
35. P. Fisher, F. Boehm, E. Bovet, J.-P. Egger, K. Gabathuler, H. Henrikson, J.-L. Vuilleumier: Phys. Lett. B192, 460 (1987)
36. D.O. Caldwell, R.M. Eisberg, D.M. Grumm, D.L. Hale, M.S. Witherell, F.S. Goulding, D.A. Landis, N.N. Madden, D.F. Malone, R.H. Pehl, A.R. Smith: Phys. Rev. D33, 2737 (1986)
37. F.T. Avignone, III, R.L. Brodzinski, S. Dimopoulos, G.D. Starkman, A.K. Drukier, D.N. Spergel, G. Gelmini, and B.W. Lynn: Phys. Rev. D35, 2752 (1987)
38. S.P. Ahlen, F.T. Avignone, III, R.L. Brodzinski, A.K. Drukier, G. Gelmini, and D.N. Spergel: Phys. Lett. B195, 603 (1987)
39. J. Bahcall and S. Casertano: Kinematics and Density of the Galactic Spheroid, IAS preprint 11/85
40. J. Caldwell and J.P. Ostriker: Astrophys. J. 251, 61 (1981)
41. R.D. Peccei and H.R. Quinn: Phys. Rev. Lett. 3B, 1440 (1977); Phys. Rev. D16, 1791 (1977); S. Weinberg: Phys. Rev. Lett. 40, 223 (1978); F. Wilczek, Phys. Rev. Lett. 40, 279 (1978)
42. A. Zehnder: SIN Report No. PR-83-03 (1983)
43. S. Dimopoulos, G.D. Starkman, and B.W. Lynn: Mod. Phys. Lett. 1, 491 (1986)
44. D.B. Reiss: Phys. Lett. B115, 217 (1982); F. Wilczek: Phys. Rev. Lett. 49, 1549 (1982); G.B. Gelmini, S. Nussinov and T. Yanagida: Nucl. Phys. B219, 31 (1983)

A Direct Laboratory Measurement of Two-Neutrino Double Beta Decay in ^{82}Se

S.R. Elliott, A.A. Hahn, and M.K. Moe

Physics Department, University of California, Irvine, CA 92717, USA

An experiment employing an enriched ^{82}Se source in a time projection chamber has yielded an observation of two-neutrino double beta decay at a half life of $(1.1^{+0.8}_{-0.3}) \times 10^{20}$ years at 68% confidence.

1. Introduction

It was here at the Max-Planck-Institut für Kernphysik that T. Kirsten and his coworkers in 1967-8 found the first solid evidence that double beta decay actually occurs, KIRSTEN et. al. [1,2]. An unmistakable excess of ^{130}Xe in a geologically old sample of Tellurium ore was attributed to the double beta decay of ^{130}Te. Similar evidence for the decay of ^{82}Se to ^{82}Kr was reported by the Max-Planck group in 1969, KIRSTEN & MÜLLER [3]. For 20 years, in spite of many experimental efforts to observe double beta decay directly, the geochemical technique provided the only measurements of double beta decay half lives.

The geochemical results, together with lower limits on the half life for ^{82}Se neutrinoless double beta decay ($\beta\beta(0\nu)$) from direct counting, forced the conclusion that double beta decay proceded predominantly by the two-neutrino mode ($\beta\beta(2\nu)$). In recent years a disparity between the geochemical values and $\beta\beta(2\nu)$ half lives derived from theoretical matrix elements caused some people to question the gas retention and dating of billion-year-old ores. Others questioned the theory, which should after all, successfully predict the $\beta\beta(2\nu)$ half lives if valid conclusions about neutrino mass and right handed currents are to be drawn from applications of similar theoretical matrix elements to $\beta\beta(0\nu)$ data.

Therefore, a direct laboratory measurement of $\beta\beta(2\nu)$ was needed. The two-neutrino mode results in a nondescript energy spectrum with a broad peak at less than half of the full transition energy. A time projection chamber was selected for the search because the kinematics of the reconstructed electron tracks provide more information than just the indistinctive spectrum. This additional information is extremely helpful in distinguishing $\beta\beta(2\nu)$ from the ever present background.

2. The Apparatus

A sectional view of the TPC is shown in Fig. 1. The central vertical plane contains the double beta decay source which consists of 14 g of selenium enriched to 97% in ^{82}Se. The total thickness of the source, including its thin aluminized Mylar envelope, is 7.5 mg/cm^2. The source plane also serves as the central electrode of the TPC. A gas filling of 92.5% helium and 7.5% propane at one atmosphere gives good electron tracks with little scattering. In the 700 G magnetic field these tracks are helices traced out by ionization electrons formed by the passing beta particles. The ionization electrons drift away from the source on each side until they pass through a grid and approach crossed anode and cathode wires where their X and Y coordinates are recorded. The Z coordinate corresponds to the arrival time. Thus, the helices can be reconstructed in three dimensions with 5 mm spatial resolution in all three directions.

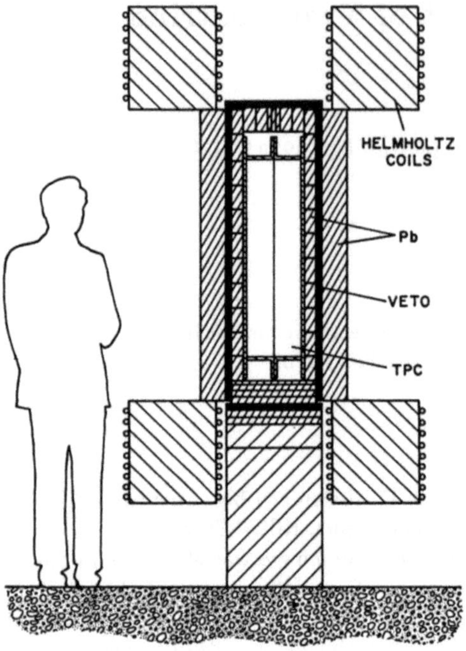

Figure 1: A section through the TPC.

Also shown in the figure is the 10 to 15 cm thick lead shield surrounding the TPC, and a 4π cosmic-ray veto made of six wire chambers operated with argon and carbon dioxide. Large Helmholtz coils keep the magnetic field uniform to 1% over the 80 cm diameter sensitive region of the 20-cm-thick TPC.

The system has been described in greater detail elsewhere MOE [4] and ELLIOTT [5].

3. The Data

The sum spectrum (sum of the kinetic energies of the two electrons) is shown in Fig. 2. The histogram represents 7960 hours (nearly a year) of live time and includes all negatron pairs having a common origin on the selenium source and having the kinetic energies of both particles greater than 150 keV.

The figure includes the theoretical $\beta\beta(2\nu)$ spectrum for a 150 keV singles threshold, shown as a smooth curve normalized to a half life of 1.1×10^{20} y. The signal-to-background ratio is greatest in the region between 1.3 and 2.0 MeV, and we have selected the events in this region for analysis. The kinematic information available in these tracks has been exploited to test for a $\beta\beta(2\nu)$ signal, as described in Section 6.

The data points in Fig. 2 show a preliminary spectrum from a new TPC (TPC2) which has been running for a few months, but normalized to the same live time as TPC1. More will be said about TPC2 in Section 7.

4. Background Events

A tracking chamber in a magnetic field eliminates many background processes, but a few remain. These include any mechanism capable of causing two negative electrons with double-beta-like energies to emerge from a common point on the source within the spatial and temporal resolution of the device.

The most serious such mechanism is ordinary beta decay to an excited level of a daughter nucleus which then relaxes to its ground state by internal conversion. These two electrons, the beta

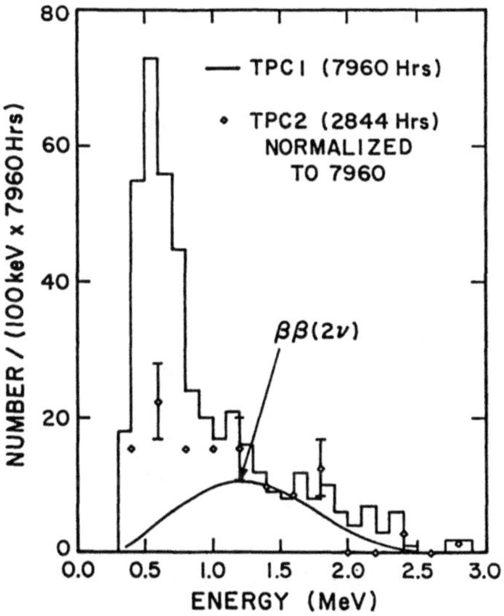

Figure 2: The measured sum-energy spectrum for negatron pairs observed in TPC1 (histogram) and TPC2 (data points), and the theoretical $\beta\beta(2\nu)$ spectrum, all with a threshold of 150 keV imposed on the individual electrons. The signal-to-background ratio is greatest for TPC1 in the energy range from 1.3 to 2.0 MeV.

particle and the conversion electron, look very much like the products of double beta decay. The nuclides which do this most frequently in the TPC are ^{214}Pb and ^{214}Bi in the uranium series, and ^{208}Tl in the thorium series. All are descendants of radon which, as an inert gas, provides them with a high degree of mobility. As far as we are able to determine, none of these nuclides is present at a significant level in the selenium, but all are attracted to the outer surface of the Mylar envelope by the electric field of the chamber acting on the predominantly positive daughters of radon decaying in the chamber gas. The radon diffuses out of materials containing radium, whether they be in the chamber or in the gas supply system.

^{214}Pb is perhaps the worst with a very high internal conversion coefficient of 25%. Fortunately it has a Q-value of only about 1 MeV which confines it to the lower part of the spectrum. It can be seen rising as a great excess in the histogram of Fig. 2 below about 800 keV. It is not a problem in the 1.3 to 2.0 MeV region selected for analysis.

^{214}Bi has a much more dangerous Q-value of 3.2 MeV, but fortunately has an internal conversion coefficient of only 1.8%. It is also a bit lucky that the daughter nucleus ejects an energetic alpha particle after only 164 μs, and the selenium source is thin enough that most of these alpha particles show up in the chamber gas. Consequently, ^{214}Bi makes a negligible contribution to the data.

Because radon daughters are in equilibrium in the chamber, if the decay rate of one is known, as by the alpha particle identification, the rates of the others can be inferred. From the easily recognized beta-alpha cascade from ^{214}Bi, for example, we are confident in our identification of the large low-energy excess as ^{214}Pb. Similarly a beta-alpha cascade in the thorium series from ^{212}Bi allows us to estimate the background from ^{208}Tl in the 1.3 to 2.0 MeV region as 2.8±0.7 events

It is noteworthy in retrospect that without the alpha-particle incentive for making the source thin we probably would have made it a good deal thicker, and been inundated with the other type of serious background which is Möller scattering. This is the process whereby a beta particle or Compton electron knocks out an orbital electron from say, an atom of selenium, so again we have two electrons emerging from the source. The opening angle between these electrons for Möller scattering is uniquely determined by their sum energy, and is usually in the neighborhood of 60 or 70 degrees. The angle can be altered by scattering, however, so a cut at the correct angle is not 100% effective in eliminating this background.

Most of the time, electrons from the source do not Möller scatter, so there is a stream of lone electrons (not members of pairs) originating in the source. From the lone electrons and the known Möller scattering cross section, the rate of Möller scattering can be calculated. In the 1.3 to 2.0 MeV range the Möller contribution is 9.3±2.6 events. (See Table I.)

5. Detector Bias

There is very little bias in the energy spectrum of electrons over the 150 keV threshold. For example, the detector reproduces the ^{214}Bi beta spectrum quite faithfully between 0.2 and 3.0 MeV.

The opening angle distribution for pairs of electrons is another matter. The two electrons from double beta decay are nearly as likely to emerge on the same side of the source as on opposite sides, but the same-side events are usually more difficult to identify because of intertwining of the tracks. The software is not as efficient at saving same-side events, and the result is a bias toward the large opening angles typical of opposite-side events.

The opening angle bias is further complicated by the tendency to lose electrons that come off nearly parallel to the magnetic field, and by the tendency of the source to scatter electrons toward the normal. A Monte Carlo calculation allowed us to see the general effect that each of these factors had on the shape of the bias function, but the final result was an approximation dependent on simplifying assumptions. It was necessary to have an accurate knowledge of the bias before we could make use of the opening angles in testing the data for the presence of double beta decay. A solution was found by sacrificing the same-side events and using a trick to force the opposite-side events to reveal the bias associated with this subset. Each side-A electron was paired with each side-B electron other than its own mate, thereby breaking any angular correlation in the real pairs. The opening angle distribution of these false pairs was taken to be representative of an isotropic primary distribution. This "bias function" is shown in Fig. 3.

As expected, the negative cosines (large opening angles) predominate, since only opposite-side events are included. When the distribution of real opposite-side events is divided by this function the true opening angle distribution emerges. It has the 1 - cos(θ) form characteristic of double beta decay, and gave the first clue that $\beta\beta(2\nu)$ was present in the data.

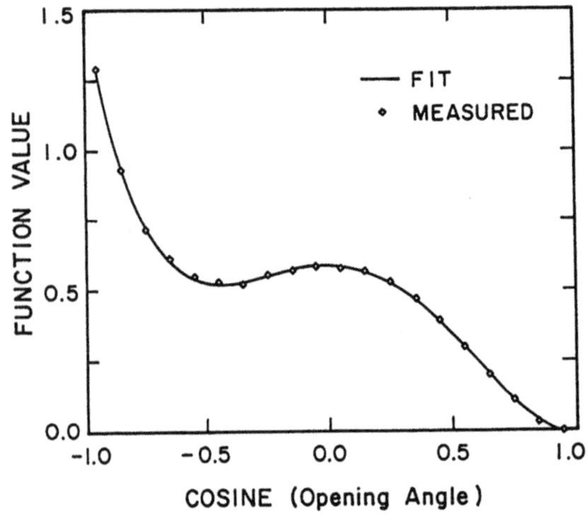

Figure 3: The measured bias function used to correct the opening angle distribution for the distortions imposed by the experiment.

6. Data Analysis

Three distributions characteristic of data were used in the analysis. These are the sum energy spectrum, the singles energy spectrum (kinetic energies of the electrons taken singly,) and the opening angle distribution. These were treated independently, and correlations between them were not exploited. There is additional information in the energies, such as the distribution of ratios of the lower to the higher energy in each electron pair, but this spectrum is not very distinctive for double beta decay visa vis the background, and so was not used.

The analysis was done for the 46 opposite-side events in the sum energy range 1.3 to 2.0 MeV. The two significant background processes in this range, as discussed in Section 4, are beta decay of ^{208}Tl with internal conversion, and Möller scattering. The energy spectra and opening angle distributions for these processes have been measured in the TPC, and for $\beta\beta(2\nu)$ the theoretical distributions have been calculated DOI et. al. [6]. Thus, the energy and opening angle distributions can be constructed for any arbitrary mix of double beta decay and the two background components. Since the distributions are quite distinctive for these three contributions one would expect a rather unique mix would give the best fit to the data in a maximum likelihood procedure. The resulting probability contours are shown in Fig. 4.

The greatest probability corresponds to 78%, or 35.9 of the 46 events being double beta decay. The associated proportions of ^{208}Tl and Möller scattering are shown in Table I, and correspond closely to the values calculated from beta-alpha cascades and lone electrons.

The analysis was repeated for the energy range 0.8 to 2.5 MeV and resulted in a very similar conclusion for the $\beta\beta(2\nu)$ rate. A more complete discussion of these procedures can be found in ELLIOTT et. al. [7].

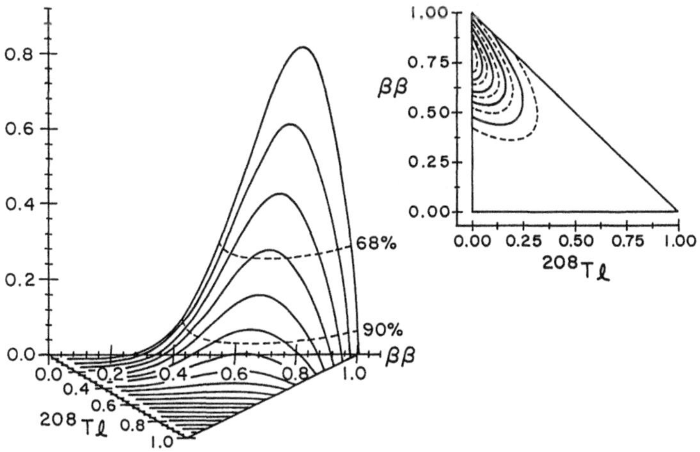

Figure 4: Mesh and contour plots for the maximum-likelihood analysis results. The vertical scale in the mesh plot is the relative probability of that particular combination of the three contributions giving the observed data. The probability is shown as a function of the fractions of ^{208}Tl and $\beta\beta(2\nu)$. The fraction due to Möller scattering can be deduced by the requirement that the sum of the three fractions be unity. The 68% and 90% curves indicate the regions which contain those percentages of the volume under the surface. The coutour plot is the same function projected on the base plane.

Table I: Comparison of maximum likelihood and independently calculated components of the data.

	Möller scattering	^{208}Tl	$\beta\beta(2\nu)$
Estimate from $\beta+\alpha$ cascades and lone electrons	9.3±2.6	2.8±0.7	x
Maximum likelihood	10.1	0	35.9

7. The New TPC

TPC1 contained several materials known by our measurements with a NaI crystal or the TPC itself, to be slightly radioactive. These included ribbon cable, resistors, solder, header strips, and most notably the TPC wire (beryllium-copper.) TPC2 was designed to locate all of these components, except the wire, outside the lead shield. This was done by continuing the TPC wire through the shield wall, sandwiched between Mylar sheets, and attaching all electronic components externally. The new wire is of stainless steel, a bulk sample of which was tested for radioactivity before being drawn. This wire is at least a factor of 10 cleaner than the beryllium-copper. The minimum shield thickness was increased from 10 to 15 cm.

An immediate benefit of these changes was an order of magnitude reduction in the trigger rate. Another indication of reduced activity is the lower rate of lone electrons from the source as shown in Fig. 5, although this reduction is only about a factor of two. The Möller scattering background is expected to drop by the same factor.

It is clear from Fig. 1 that the TPC2 background below 800 keV is well down from the old level. This improvement is a result of a reduction of radon, and consequently of ^{214}Pb. Some of the radon had been coming from components inside TPC1, and some had been from the helium cylinders which apparently contain a little radium. The worst cylinders have been eliminated.

With TPC2 we are saving all events that are followed by a second trigger within one millisecond, providing that the second trigger does not coincide with a cosmic ray veto pulse. This sample includes most of the ^{214}Bi beta decays with their telltale ^{214}Po alpha particles. When a

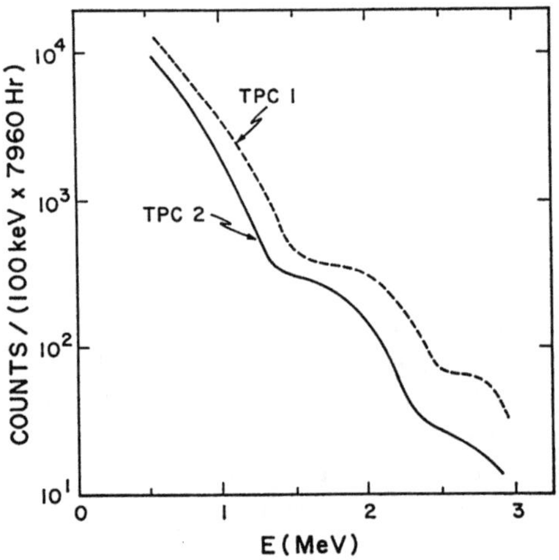

Figure 5: The spectra of lone electrons from the selenium source for the old and new TPC's. From these spectra and the Möller scattering cross sections, one can calculate the background from Möller scattering.

low-energy $\beta\beta(2\nu)$ candidate appears, one can look ahead in time for several 20-minute ^{214}Bi half lives and see if a ^{214}Bi decay appears at the same location on the source. Since ^{214}Bi is the daughter of ^{214}Pb this technique allows the rejection of most of the remaining ^{214}Pb events.

The event rate in the 1.3 to 2.0 MeV range is about the same in TPC2 as before, which is expected if, indeed, most of the TPC1 events in this region were double beta decays.

Above 2.0 MeV the rate in TPC2 appears to be down, but statistics are still poor there. A count has appeared at 2.8 MeV in TPC2 which, when added to the counts of TPC1, make a cluster of five events near this energy. These events do not have the appearance of Möller scattering, and their origin is not understood.

8. Conclusions

The energy spectra and opening angle distribution for the data in the region of highest signal to background conform closely to the expected shapes for $\beta\beta(2\nu)$, as reflected in the sharpness of the maximum likelihood contours. The background fraction from the maximum likelihood fit agrees with calculations based on the lone electron spectrum and observed beta-alpha cascades. The rate deduced for $\beta\beta(2\nu)$ is independent of the energy range chosen for analysis. And finally, the signal seems to be holding up in the new TPC while backgrounds have been reduced.

We believe that these observations constitute strong evidence for $\beta\beta(2\nu)$. The 35.9 events from the maximum likelihood fit correspond to a ^{82}Se half life of $(1.1^{+0.8}_{-0.3}) \times 10^{20}$ years at 68% confidence, in good agreement with geochemical measurements KIRSTEN [8] and MANUEL [9]. TPC2 continues to run in an attempt to reduce the uncertainty in the $\beta\beta(2\nu)$ half life and extend the search for neutrinoless decay in ^{82}Se.

This project is supported by the U.S. Department of Energy.

References:

1. T. Kirsten, W. Gentner, and O. Schaeffer: Z. Physik 202, 273 (1967)

2. T. Kirsten, et. al.: Phys. Rev. Lett. 20, 1300 (1968)

3. T. Kirsten and H. W. Müller: Earth and Planetary Science Letters 6, 271 (1969)

4. M. K. Moe, A. A. Hahn, and H. E. Brown: In *The Time Projection Chamber*, ed. by J. A. MacDonald, AIP Conference Proceedings No. 108 (American Institute of Physics, New York, 1984), p. 37.

5. S. R. Elliott, A. A. Hahn, and M. K. Moe: Phys. Rev. Lett. 56, 2582 (1986).

6. M. Doi, T. Kotani, and E. Takasugi: Prog. Theor. Phys. Suppl. 83, 1 (1985).

7. S. R. Elliott, A. A. Hahn, and M. K. Moe: Phys. Rev. Lett. 59, 2020 (1987).

8. T. Kirsten: In *Proceedings of the International Symposium on Nuclear Beta Decay and Neutrinos, Osaka, Japan,* ed. by T. Kotani, H. Ejiri, and E. Takasugi (World Scientific, Singapore, 1986), p. 81.

9. O. K. Manuel: In *Proceedings of the International Symposium on Nuclear Beta Decay and Neutrinos, Osaka, Japan,* ed. by T. Kotani, H. Ejiri, and E. Takasugi (World Scientific, Singapore, 1986), p. 71.

Indication of Neutrinoless Double Beta Decay of ^{76}Ge to the First Excited State of ^{76}Se. Results of the Frejus Experiment

J. Busto[1], *J. Chevallier*[2], *D. Dassié*[1], *Ph. Hubert*[1], *A. Larrea*[3], *P. Larrieu*[1], *F. Leccia*[1], *P. Mennrath*[1], *A. Morales*[3], *J. Morales*[3], *R. Nunez-Lagos*[3], *J. Puimedon*[3], *J.A. Villar*[3], *and M.M. Villard*[1]

[1]Centre d'Etudes Nucléaires de Bordeaux-Gradignan, F-33170 Gradignan, France
[2]Centre de Recherches Nucléaires, F-67000 Strasbourg, France
[3]Departamento de Fisica Nuclear, Universidad de Zaragoza, Zaragoza, Spain

I. Introduction

In the present paper we report on the experimental results of a search for the neutrinoless double beta decay of ^{76}Ge, whose decay scheme is shown in fig. 1.

As can be seen, the $\beta\beta$ decay is energetically possible not only to the ground state ($J^{\pi}=0^{+}_{1}$) of ^{76}Se, but also to the first excited state ($J^{\pi}=2^{+}_{1}$) at 559.1 keV excitation energy and the second excited state ($J^{\pi}=0^{+}_{2}$) at 1122.3 keV. The energy available for the $\beta\beta$ decay of ^{76}Ge, $Q_{\beta\beta}$ = 2041.4 ± 0.5 keV is taken from the atomic mass difference determinations of ELLIS et al. [1].

The main goal of the present experiment is to look for the decay to the first excited state of ^{76}Se using a coincidence technique among several Ge and NaI detectors. A coincidence event should therefore correspond (see fig. 1) to a 1482.3 ± 0.5 keV two electron kinetic energy deposition in the Ge detectors and a 559.1 ± 0.1 keV γ-ray energy deposition in the NaI detectors. After 6207 h. of statistical time the experimental data show a coincidence signal at the 2.5σ level whose energies of 1483.6 ± 0.5 keV and 559 ± 15 keV agree quite well with the expected

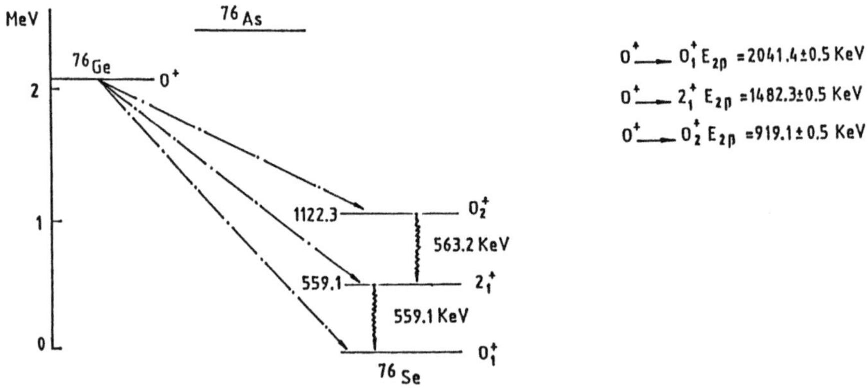

Fig. 1. Nuclear scheme for the double beta decay in ^{76}Ge. The energies are based on the $Q_{\beta\beta}$ = 2041.4 ± 0.5 keV measured by ELLIS et al. [1].

values given above. A rather detailed analysis of the experimental results and of the possible origin of this cascade suggests that we may be seeing a neutrinoless double beta effect.

II. Experimental set-up

The experimental set-up consists mainly of three different parts, a central detector containing four Ge crystals (≃ 100 cm^3 each) in the same cryostat, a halo of 19 NaI scintillators in a 4π geometry and a passive shielding of 2.5 cm Cu and 10 cm Pb, all this being located in the Frejus underground facility. More details can be found in ref. 2. It should be noted that all the materials needed for the experiment were selected using low-background criteria already published in the literature, and that no specific preliminary measurements have been done. Therefore, the low background perfermance could be improved rather easily. Of paramount importance is the energy resolution of the detectors (FWHM). The intantaneous energy resolutions of the Ge crystals range from 1.9 to 2.2 keV on the 1332 keV ^{60}Co line, depending on the crystal used. The overall energy resolution (summed over the 4 Ge-detectors and after 6207 h. of statistics) is 2.4 keV at 1500 keV, that is in the vicinity of the $\beta\beta(0^+ \rightarrow 2^+_1)$ energy. The instantaneous energy resolution of one NaI scintillators is around 8 % on the 661 keV ^{137}Cs line. The overall energy resolution (summed over the 19 NaI scintillators and after 6207 h. of statistics) is 11 % at 580 keV, (close to the $2^+_1 \rightarrow 0^+_1$ γ-ray energy in ^{76}Se).

Finally, as already quoted above, the experiment has been running for 6207 h. of active time in the Frejus underground laboratory whose great depth (≃ 4500 mWe) strongly suppresses any effect of the cosmic rays.

III. Experimental result

The final result of the experiment consists of a two-dimensional matrix whose axes are the Ge and NaI energies and with the multiplicity requirement that only one Ge counter and only one NaI counter have fired within the ≃ 60 ns time window. From this matrix different selections can be done, the most important one being a cut on the NaI energy at 559 keV with a window width of 70 keV (±1.4σ). The corresponding partial Ge spectrum around ≃ 1.5 MeV is shown in Fig. 2.

In such conditions, if the $\beta\beta(0\nu)$ $0^+ \rightarrow 2^+_1$ decay is possible and if we adopt the ELLIS $Q_{\beta\beta}$ = 2041.4 ± 0.5 keV, then a sharp peak should appear at the energy of 1482.3 ± 0.5 keV. Clearly, no such a peak shows up at this energy and a lower limit of $T_{1/2} > 6.10^{22}$y can be set. However, as shown in fig. 2, an accumulation of counts is present at a nearby energy of 1483.6 keV, that is 1.3 keV above the ELLIS value. Therefore, two important questions arise immediately :i) is this accumulation of events really a "peak" ? and ii) if yes, does it correspond to neutrinoless ββ decay ?

An answer to the first question can be found using the Poisson distribution. With an energy bin of 4 channels (2.8σ) and 30 channels on both sides of the "peak"

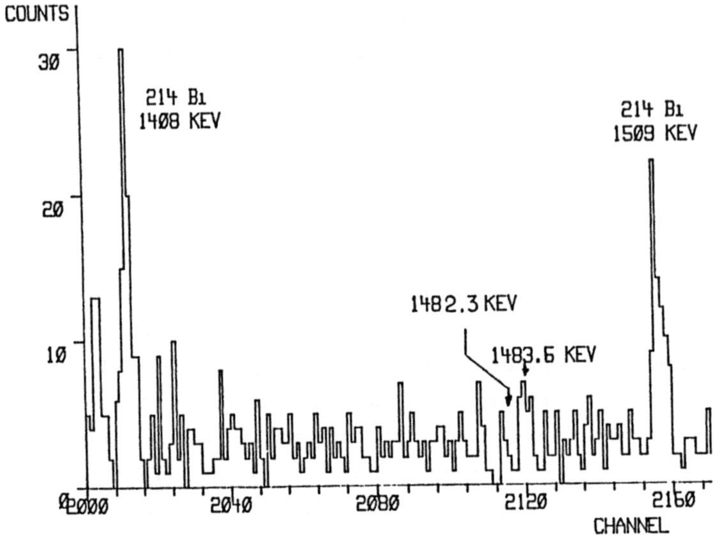

Fig. 2. Partial Ge spectrum obtained in coincidence with the NaI halo. The NaI window is centered at 560 keV and has width of 70 keV. The energy calibration is strictly 0.7 keV/channel (0.0 keV intercept).

there is a 0.1 % probability to find such an accumulation of events due to statistical fluctuations. A maximum likelihood method leads to the result :

E_{Ge} = 1483.6 ± 0.5 keV

Background = (6.2 ± 0.6)10^{-4} counts/keV * h

Area = 13.3 ± 5.4 counts

Using a likelihood ratio test [3] we found a probability of 0.5 % for no peak hypothesis.

The intensity of the 1483.6 keV peak has been followed when the energy of the NaI window is varied. As shown in fig. 3, if present when the NaI window centered at 560 keV, the 1483.6 peak clearly disappears when the NaI window is shifted either to 520 keV or 580 keV. Such an analysis gives 556 ± 20 keV for the energy deposition in the NaI detectors, in agreement with the expected 559.1 0.1 keV γ-ray energy.

A two-dimensional analysis of the data has also been done both with the maximum likelihood method and the least-squares method, leading to approximatively the same results, i.e., E_{Ge} = 1483.6 ± 0.5 keV, E_{NaI} 559 ± 15 keV, Area = 13 ± 5 counts and a probability ≤ 1 % for the no peak hypothesis.

At this point it should be stressed that, if assigned to the double beta decay, our measurement of the kinetic energy of the two electrons, 1483.6 ± 0.5 keV, is in agreement within two standard deviations, with the updated $Q_{\beta\beta}$ value of 1482.3 ± 0.5 keV [1].

The answer to the second question depends to the knowledge of our background. Let us define as a criterion for the existence of a peak, the relation : area > 2σ,

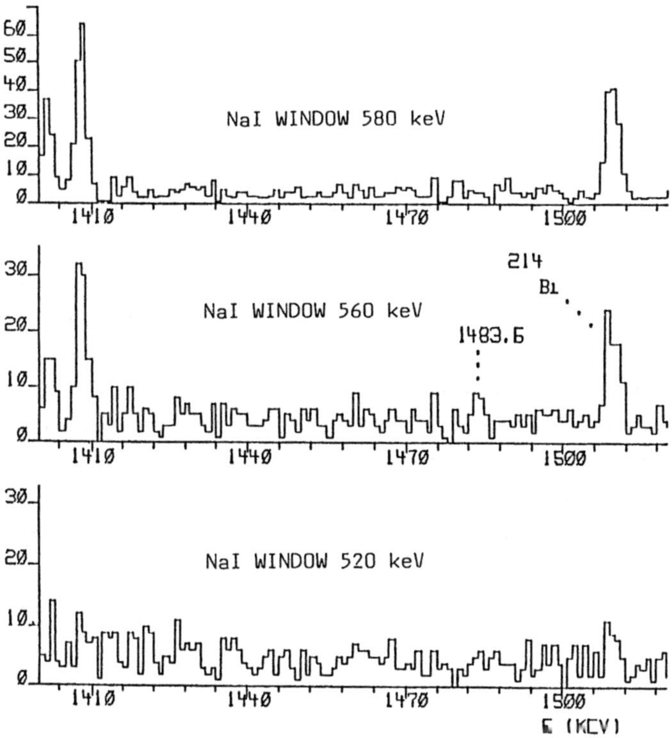

Fig. 3. Partial Ge spectra (calibration 1 keV/channel) in coincidence with the NaI halo and corresponding to 3 differents NaI windows. (window width = 70 keV)

where σ is the statistical fluctuation of the background. With the above criterion, all the coincidence cascades in the full Ge-NaI matrix have been identified as coming from the "natural" background (Th and U series). Only the 1483.6 keV - 559 keV cascade is left unidentified. Since the background comes only from the natural radioactivity, we studied carefully the decay schemes of all the natural isotopes, with the result that no similar cascade has been found. Beside this analysis, we ran in our set-up several sources (^{228}Th, natural U, Monazite sand rich in Ra and Th isotopes) and no γ-γ cascade has been observed with the above energies.

A triple γ-cascade in ^{76}Ge with following energies : 1482.5 keV - 545 keV - 563 keV, and de-exciting a state at 2591 keV could conceivably have simulated our signal. However this possibility has been completely discarded since i)we found no way to populate this excited state through (n,n'), (α,α') or (γ,γ') reactions, ii) the probability to detect a triple cascade in our set-up is very small ($\simeq 10^{-4}$ as compare to 20 % for a double cascade) and iii) other cascades either in the ^{76}Ge isotope or the more enriched ones should have been present in our data.

Finally, we checked that the time distribution of the events is in agreement with a Poisson distribution. Also, the geometrical distribution of the events among the Ge and NaI detectors shows that almost all the detectors, and not only few of them, have contributed to the coincidence data. These two distributions allow one to reject the possibility of some artifact in the electronics.

IV. Summary and conclusions

We have presented in this paper the results of a ^{76}Ge neutrinoless double beta decay experiment which has been running for approximately one year in the Frejus underground facility. We have observed a coincidence cascade of $\simeq$ 13 ± 5 counts between a 1483.6 ± 0.5 keV Ge energy deposition and a 559 ± 15 keV NaI energy deposition, in agreement with the expected energies of 1482.3 ± 0.5 keV and 559.1 ± 0.1 keV. No other unidentified cascade was seen in the full Ge-NaI matrix and no explaination within our background analysis has been found. All these features suggest the possibility of neutrinoless double beta decay of ^{76}Ge to the first excited state of ^{76}Se with a half-life $T_{1/2} \simeq 10^{22}$ y. Such a half-life based on a signal at the 2.5σ level does not appear, at the present time, in contradiction with the 1σ level limits given by several authors [4]. However, due to some lack of statistics, the present experimental result should encourage further studies of the $\beta\beta(0\nu)$ decay of ^{76}Ge with major improvements in the background level and in the energy resolution of the detectors .

References :

1. R.J. Ellis et al., Nucl. Phys. A435 (1985) 34
2. A. Morales et al., Weak and Electromagnetic Interactions in Nuclei, Proc. of the Int. Symp., Heidelberg, July 1986, Ed. H.V. Klapdor, Springer-Verlag.
3. W.T. Eadie et al., in Statistical Methods in Experimental Physics, North Holland Publishing Company, 1971, Amsterdam.
4. See ref. 2, and also Proc. of the 12th Int. Conf. on Neutrino Physics and Astrophysics, Sendaï, Japan, 1986.

Search for Double Beta Decay at the Gotthard Underground Lab

J.-L. Vuilleumier

Institut de Physique, A.L. Breguet 1,
CH-2000 Neuchâtel, Switzerland

The Gotthard underground lab is located in the Gotthard road tunnel at a depth of 3000 m.w.e. It is used by the Caltech-Neuchâtel- SIN collaboration to search for double beta decay. First ^{76}Ge was studied with a 90 cc Ge detector [1]. This part of the experimental program is finished, and results will be presented below. This is followed by a second experiment, conceptually very similar but using a much larger detector consisting of 8 Ge crystals of 140 cc active volume each. The data taking time is not yet very long, and very preliminary results only are available. A third experiment is in preparation. It will use a TPC [2] filled with 60% enriched ^{136}Xe.

1. The 90 cc Ge Detector. This detector was made from carefully selected low activity materials. Copper was chosen for the cryostat. The detector was shielded against local activities by 20 cm of Cu and 15 of Pb. Data were accumulated during a total of 9373 h (0.51 kg·y). The spectrum above 1.5 Mev is shown in fig. 1. The continuous background in this region is rather low and, in spite of the small detector size, these data are particularly sensitive to double beta decay with Majoron emission ($\chi^0\beta\beta$), which peaks around 1.5 MeV. The background is thought to come primarily from three sources:

i) γ activity from the ^{232}Th chain (one line at 2614 keV) and ^{238}U chain (six more lines between 1.5 and 2.5 MeV). The peaks were removed, and the associated

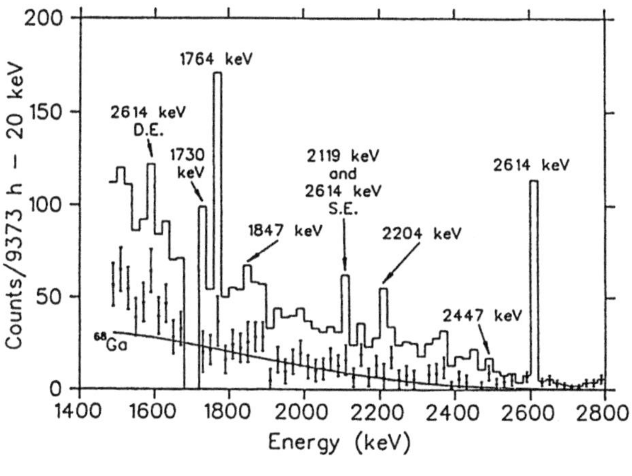

Fig. 1

continuous background was calculated assuming that the ^{238}U lines come from ^{226}Ra gas around the cryostat, and that the ^{232}Th is on the surface of the Ge crystal. This calculated background, obviously an underestimate, was removed.

ii) Positrons and annihilation γ rays from the β^+ decay of ^{68}Ga in the Ge crystal. ^{68}Ga is produced by electron capture in ^{68}Ge, itself the result of the cosmogenic activation $^{70}Ge(n,3n)^{68}Ge$ when the crystal was above ground. The rate was determined from the intensity of the ^{68}Ga X-rays at 9 keV. The background was then calculated (fig. 1) and, like background i), subtracted from the data. The remaining spectrum is shown in fig. 2.

iii) γ rays from ^{60}Co, presumably produced by cosmogenic activation of the Cu cryostat and Cu shielding via the reaction $^{63}Cu(n,\alpha)^{60}Co$. The 1173 and 1332 keV lines are clearly visible in our spectrum. From their intensity we calculate the continuous background from sum events above 1.5 MeV. This background accounts for about one third of the remainig counts between 1.5 and 2.5 MeV. It decreases smoothly with energy, almost linearly.

The spectrum of fig. 2 still contains background iii) and whatever is left of background i). These have a rather smooth energy dependance, as would any additional background we can think of. We thus assume that the remaining background has a linear shape multiplied by a power law (the detector response is well approximated by a power law E^{-b}). The fit of the data with this three parameter function is shown in fig. 2. The χ^2 (36.69 for 45 degrees of freedom) is excellent, and there is no indication of $\chi^0\beta\beta$ decay. To get a lower limit on the half-life fits were performed adding the expected $\chi^0\beta\beta$ spectrum for different decay rates. We find $T_{1/2}(\chi^0\beta\beta) > 1.2\cdot10^{21}$ y (90% C.L.), in good agreement with the UCSB-LBL limit of $T_{1/2}(\chi^0\beta\beta) > 1.4\cdot10^{21}$ y [3]. This is however conflicting with the positive PNL-USC result $T_{1/2}(\chi^0\beta\beta) = (6\pm1)\cdot10^{20}$ y [4], obtained in an experiment very similar to ours, with comparable background level. The corresponding expected spectrum for our set up is shown in fig. 2 (dotted line).

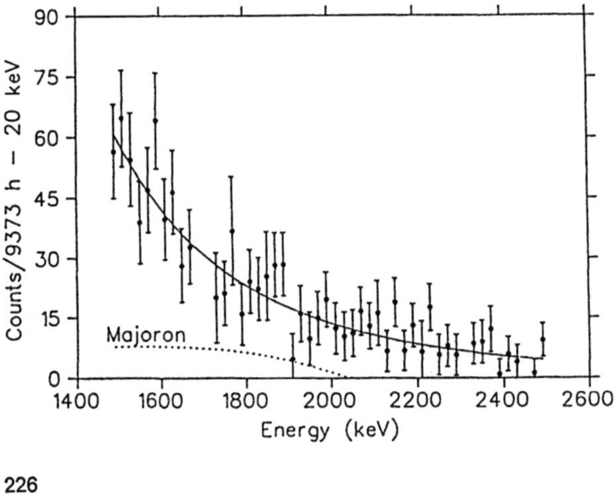

Fig.2

2. The 8x140 cc Ge Detector. This large detector array consisting of 8 crystal of 140 cc active volume in one single cryostat has also been built with clean materials. About all components have been tested with the 90 cc detector. The energy resolution is satisfactory, varying from 1.9 keV FWHM at 1332 keV for the best crystal, to 2.1 keV for the worst one. The background, presently 3 counts$\cdot$keV$^{-1}\cdot$y^{-1} at the $0\nu\beta\beta$ energy (2041 keV), is reasonably low. We expect it to drop further since it it is to a significant extent due to ^{68}Ga, which is not produced anymore under ground and which decays away with a 275 days half-life. Such a drop was observed in the 90 cc detector. The data sample (0.81 kg$\cdot$y^{-1}) is still modest and the limits we obtain for $0\nu\beta\beta$ decay are not yet very constraining compared to those of the UCSB-LBL collaboration. From background fluctuations we find: $T_{1/2}(0+ \rightarrow 0+) > 1.1\cdot10^{23}$ y and $T_{1/2}(0+ \rightarrow 2+) > 0.4\cdot10^{23}$ y at the 68% confidence level. Much better results will come out whithin a year. Background studies are in progress. They will allow to improve our results on $\chi^0\beta\beta$ and $2\nu\beta\beta$ decay.

3. The Xe TPC. An independent experiment to study the double beta decay of ^{136}Xe has been initiated to complement and improve on the ^{76}Ge data. Given the large uncertainty afflicting the nuclear matrix elements [5, 6, 7] and the large differences from nucleus to nucleus it appears important to look for double beta decay in more than one nucleus. The ^{136}Xe experiment can be performed with a large target, particularly since Xe enriched to about 60% ^{136}Xe is available commercially (the natural abundance is 8.9%). Also the end point energy (2479 keV) is quite high, giving a large phase space. The detector is being built and tested at Caltech [2]. It is a TPC of 208 l active volume which can be filled with Xe at 5 to 10 atm. It has good tracking capability in spite of multiple scattering. Electron tracks can be identified by the increased ionization at the end of the track. A double beta decay candidate will look like one single continuous track with increased ionization at both ends. This signature will allow a high background suppression. This should compensate, in the case of $0\nu\beta\beta$ decay, for the poorer energy resolution (ideally 5 % instead of 0.1 % in the case of Ge). For $2\nu\beta\beta$ decay the background should be low enough so that this decay should be detectable, even if the most pessimistic estimates of the matrix elements are correct. This is important since the measurement of this decay would, to a point, allow to check the calculation of the nuclear matrix elements. In Ge the detection of $2\nu\beta\beta$ is marginal under the same assumptions. The sensitivity to $\chi^0\beta\beta$ decay should also be excellent.

The tests of the TPC have shown that the spatial resolution already is sufficient. The energy resolution however is not yet satisfactory. It is hoped that this problem will be solved soon and that the TPC can be installed at the Gotthard in 1988.

References

[1] P. Fisher et al., Phys. Lett. B 192(1987)460

[2] M.Z. Iqbal et al., Nucl. Inst. Meth. A259(1987)459

[3] D.O. Caldwell et al., Phys. Rev. Lett. 59(1987)419

[4] F.T. Avignone et al., Talk American Physical Society (Salt Lake City, UT, January 1987), AIP Conf. Proc. (AIP New York), to be published; F.T. Avignone, these proceedings.

[5] J. Engel, P. Vogel and M.R. Zirnbauer, Caltech preprint MAP-95 (1987), submitted to Phys. Rev. C; M.T. Zirnbauer, these proceedings

[6] A. Faessler, these proceedings

[7] Muto, these proceedings

Part IV

Solar and Cosmological Neutrinos

The Present Status of the Gallium Solar Neutrino Detector GALLEX

W. Hampel

Max-Planck-Institut für Kernphysik, P.O. Box 103980, D-6900 Heidelberg, Fed. Rep. of Germany

GALLEX Collaboration*

The present status of the Gallium Solar Neutrino Experiment to be performed by the International GALLEX Collaboration is described. This experiment is now in its implementation phase. After a short introduction and an outline of the basic experimental procedure, details will be given on the following topics:

- Gallium acquisition
- Gran Sasso Underground Laboratory
- Germanium extraction and conversion
- ^{71}Ge counting
- ^{51}Cr source experiment
- Time table of the experiment

Introduction

The Homestake Chlorine Solar Neutrino Detector of Raymond Davis Jr. and collaborators [1] has now been in operation since almost two decades. It is well known that there is a persistent discrepancy between the signal observed with this radiochemical detector and the theoretical prediction. This discrepancy is usually called the "solar neutrino problem". The measured production rate of 2.1 ± 0.3 SNU [1] is about a factor of four lower than the rate (7.9 SNU) predicted by the Standard Solar Model (SSM) of Bahcall and Ulrich [2] (see Table 1 and Fig. 1). A second detector, the KAMIOKANDE water Cerenkov detector, is now close to confirm the result of the Cl detector [3]. This detector (using a completely different experimental technique) does not observe the high energy tail ($\geq$ 10 MeV) of the ^{8}B neutrinos at the level predicted by the SSM. However, as long as the KAMIOKANDE detector has no evidence for a signal above zero, one even cannot be completely certain that the signal observed by the Cl detector is in fact caused by solar neutrinos.

***GALLEX Collaboration**

M. Breitenbach, W. Hampel, G. Heusser, J. Kiko, T. Kirsten, A. Lenzing, E. Pernicka, R. Plaga, B. Povh, S. Richter, C. Schlosser, M. Schneller, H. Völk, R. Wink (**MPIK Heidelberg**)
R. v. Ammon, K. Ebert, E. Henrich, G. König, L. Stieglitz, J. Unk (**KfK Karlsruhe**)
R. Mössbauer, A. Urban (**TU München**)
M. Cribier, G. Dupont, B. Pichard, J. Rich, M. Spiro, D. Vignaud (**CEN Saclay**)
G. Berthomieu, E. Schatzman (**University of Nice**)
E. Bellotti, O. Cremonesi, E. Fiorini, C. Liguori, S. Ragazzi, L. Zanotti (**University of Milano**)
S. d'Angelo, R. Bernabei, L. Paoluzi, R. Santonico (**University of Rome**)
I. Dostrovsky (**WIS Rehovot**)
G. Friedlander, R.L. Hahn, J.K. Rowley, R.W. Stoenner, J. Weneser (**Brookhaven Nat. Lab.**)

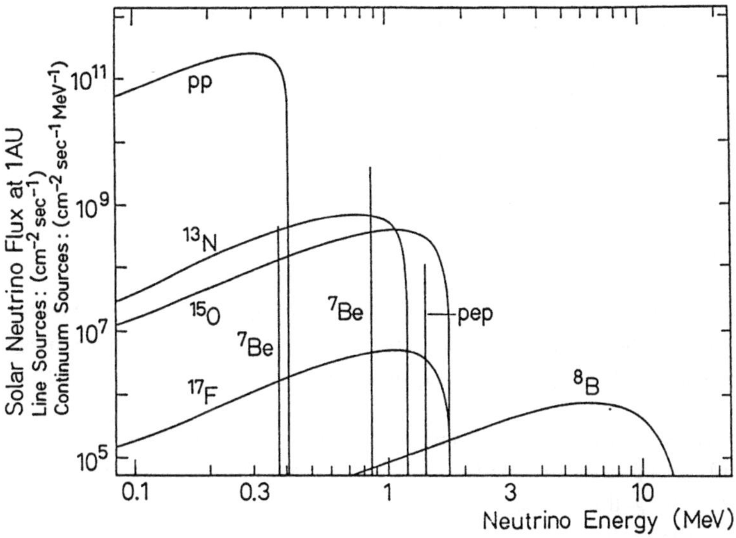

Fig. 1: Solar neutrino energy spectrum according to the Standard Solar Model [2]

Table 1: Neutrino fluxes according to the Standard Solar Model and production rates for the ^{37}Cl and ^{71}Ga solar neutrino detectors (from Bahcall and Ulrich [2]).

Neutrino source and energy [MeV]		Flux at earth $[10^{10}cm^{-2}sec^{-1}]$	Production rate [SNU†] ^{37}Cl experiment	^{71}Ga experiment
pp	($\leq$ 0.42)	6.0	-	70.8
pe^-p	(1.44)	0.014	0.2	3.0
^{7}Be	(0.38, 0.86)	0.47	1.1	34.3
^{8}B	(< 15)	0.00058	6.1	14.0
^{3}He p	($\leq$ 18.77)	$7.6 \cdot 10^{-7}$	0.03	0.06
^{13}N	($\leq$ 1.20)	0.061	0.1	3.8
^{15}O	($\leq$ 1.73)	0.052	0.3	6.1
^{17}F	($\leq$ 1.74)	0.00052	0.003	0.06
Total		6.60	7.9	132

†1 SNU ≡ 1 Solar Neutrino Unit ≡ 1 neutrino capture per sec and 10^{36} target atoms

It has long been recognized that an experiment capable to observe the neutrinos from the primary fusion reaction in the sun, the pp neutrinos (see Fig. 1) is urgently needed. Such an experiment could provide

- a test of the SSM in view of the solar neutrino problem;
- the experimental verification that hydrogen burning via the pp chain is indeed the principal energy source of the sun;
- information on neutrino oscillations with squared mass differences ($\geq 10^{-12}$ eV^2) much smaller than those accessible in terrestrial experiments.

The only experimental approach in this direction which is close to realization is a radiochemical detector based on gallium. Two experiments using this technique are in preparation: an experiment with 60 tons of Ga metal to be installed in the Baksan Underground Laboratory by the Academy of Sciences of the USSR [4], and an experiment with 30 tons of Ga in form of gallium chloride solution which will be performed by the International GALLEX Collaboration. Here we shall report on the present status of the latter.

Outline of the basic experimental procedure

The gallium detector is based upon the neutrino capture reaction $^{71}Ga(\nu_e,e^-)^{71}Ge$. The energy threshold (233.2 ± 0.5 keV) is well below the maximum energy of the pp neutrinos (420 keV) (see Fig. 1). ^{71}Ge decays back to to ^{71}Ga by electron capture with a halflife of 11.43 ± 0.03 d.

Table 1 lists the neutrino fluxes for the different solar neutrino sources according to the SSM [2] along with the corresponding capture rates for the chlorine and gallium neutrino detectors. The total rate of 132 SNU for the gallium detector is dominated by 74 SNU from the pp and pe^-p neutrinos resulting from the basic fusion reaction in the pp chain.

The feasibility of a gallium solar neutrino detector based on $GaCl_3$ solution has been studied in a pilot experiment using 1.3 tons of gallium [5,6]. The experimental procedure for a full scale experiment as developed in this pilot experiment is as follows. 30 tons of gallium in form of a concentrated $GaCl_3$-HCl solution will be exposed to solar neutrinos. Natural gallium has two isotopes, ^{71}Ga (39.6 %) and ^{69}Ga (60.4 %). ^{69}Ga is useless for solar neutrino detection, but it can serve as a monitor for the interfering (p,n) background reaction (see below). In $GaCl_3$-HCl solution, the neutrino induced ^{71}Ge atoms (as well as the inactive Ge carrier atoms added to the solution at the beginning of a run) form the volatile compound $GeCl_4$, which at the end of an exposure is simply swept out of the solution by means of a gas stream (air). This gas stream is then passed through a gas scrubber where the $GeCl_4$ is absorbed in water. The $GeCl_4$ is finally converted to GeH_4, which together with xenon is introduced into a proportional counter in order to determine the number of ^{71}Ge atoms by observing their radioactive decay.

Status of the experiment

Funding of the GALLEX detector was assured at the end of 1985. The detector is now in its implementation phase. In the following we shall describe the present status of the experiment in detail.

Gallium acquisition

The acquisition of 30 tons of gallium in form of 105 tons of gallium chloride ($GaCl_3$) solution requires special measures (apart from the question of funding). This is because such an amount of gallium corresponds to roughly one half of the world's annual production; it is therefore not readily available on the market. As a consequence, a demand of this magnitude must lead to an expansion of production capacities. In addition, because of background reasons (see below), the solution must be extremely pure with respect to the radioactive elements uranium (<50 ng/g), thorium (<1 ng/g) and radium (<1.9×10^{-5} Bq/g). After funding was assured a contract for delivery of 30 tons gallium in form of gallium chloride solution was signed with the Sulzer Company (Winterthur, Switzerland). Since gallium is naturally concentrated in the Bayer liquor used to produce alumina from bauxite, Sulzer built a new gallium production plant on the premises of the Kaiser Aluminum Plant in Gramercy, Lousiana (USA). This gallium plant is now (end of 1987) nearly completed and will start production in spring 1988. It has been demonstrated that test solutions (produced from the original Bayer liquor by the chemical procedure to be used in the production process) do indeed meet the requirements listed above.

Gran Sasso Underground Laboratory

The most important side reaction for ^{71}Ge production in $GaCl_3$ solution by sources other than solar neutrinos is the $^{71}Ga(p,n)$ reaction (threshold energy 1.02 MeV), the protons being generated in the $GaCl_3$ solution as secondaries from (α,p) and (n,p) reactions and from cosmic ray muon interactions. As mentioned above, the (α,p) background puts severe constraints on the concentration of the α-emitting elements U, Th and Ra in the solution. In order to escape the muon induced background, the experiment has to be performed in an underground position, shielded by at least one kilometer of rock. The GALLEX experiment has been approved for installation in hall A of the Gran Sasso Underground Laboratory (GSUL). This underground facility, shielded by about 3500 meters of water equivalent, has been built by the Italian INFN (Istituto Nazionale di Fisica Nucleare) in the middle of a high-way tunnel through the Abruzzese mountains, about 150 km east of Rome.

Within hall A, the gallium experiment will be accommodated in two buildings. The GALLEX main building (12x10x9 m^3) will house the process tank containing the 105 tons $GaCl_3$ solution, a spare tank (70 m^3 each) and the equipment for the Ge extraction. ^{71}Ge counting will be performed in the Low-Level counting laboratory (10x10x6 m^3). Fig. 2 is a view of the south wing of hall A showing the GALLEX main building and (in the background) the Low-Level counting building.

Germanium extraction and conversion

The basic chemical procedures for Ge extraction out of large quantities of $GaCl_3$ solution have been developed during the pilot experiment [5,6]. The tanks to be used in the experiment will be two identical plastic tanks reinforced with glass fiber backing and with an inner teflon liner (PVDF). The second tank is needed for safety reasons in the event of a tank leak. It also can take up the $GaCl_3$ solution if an inspection of the process tank is necessary. The process tank will be equipped with a reentrant tube to hold a ^{51}Cr neutrino source (see below). There are provisions to install a neutron shield (25 cm polyethylen) around the process tank in order to reduce the background production of ^{71}Ge by fast neutrons originating from the rock walls. However, preliminary neutron flux measurements in the GSUL indicate that such a fast neutron shield might not be required. Fig. 3 shows a cross section through hall A and the GALLEX main building with the process tank.

The extraction is performed by circulating air through the process tank and the absorption columms at a rate of about 100 m^3 per hour. Germanium is removed from the solution as volatile germanium chloride ($GeCl_4$) and then absorbed in the water of the absorption columns. A series of columns serves to concentrate the germanium into smaller and smaller volumes. The

Fig. 2: Gran Sasso hall A: View of the GALLEX main building and (behind it) the Low-Level counting building (see text).

final concentration step is an extraction into CCl_4 and a back-extraction into tritium-free water. This step also serves to separate the germanium from any tritium in the tank or the absorber system, since tritium would contribute to the background in ^{71}Ge counting.

The last part of the chemical procedure is the conversion of $GeCl_4$ to the gas GeH_4 by means of the reducing reagent sodium borohydride ($NaBH_4$). The GeH_4 is dried and purified by gas chromatography. Its volume is then measured in order to determine the overall yield of the chemical procedures. Finally, it is placed together with xenon into a proportional counter.

^{71}Ge counting

The ^{71}Ge decay rates expected in a full scale gallium experiment are below 1 per day. The measurement of such low decay rates is an extreme low-level task and can only be achieved with miniaturized proportional counters. ^{71}Ge decays by K (87.7 %), L (10.3 %), and M (2.0 %)

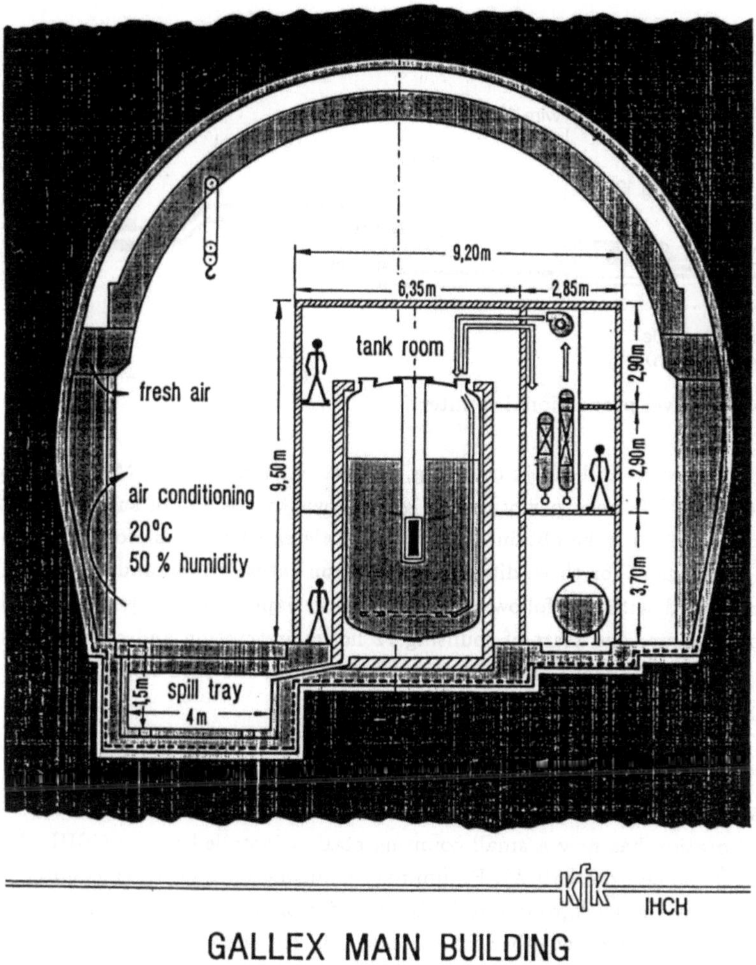

Fig. 3: Cross section through hall A, the GALLEX main building and the process tank.

electron capture. Neglecting the M events, the energy deposition from Auger electrons and X-rays emitted in the decay results in a spectrum with two peaks: an L peak at 1.2 keV and a K peak at 10.4 keV. All provisions typical for low-level counting have to be applied in order to reach the extremely low background rates required: miniaturization (see Fig. 4), use of ultrapure (with respect to radioactive contamination) materials, anticoincidence shield with sodium iodide and plastic scintillation detectors, and a heavy passive shield of lead and iron. The remaining background in the L and K peak energy windows is of the order of 1 count per day. It is mainly due to (a) beta particles from natural radioactivity in the counter construction materials, (b) Compton electrons caused by external γ-rays, and (c) electronic noise pulses. Most of these background events, fortunately, produce pulse shapes distinct from ^{71}Ge decay events. The computer-controlled counting system to be used for the experiment is therefore equipped with a transient digitizer in order to record the whole shape of each preamplifier output pulse.

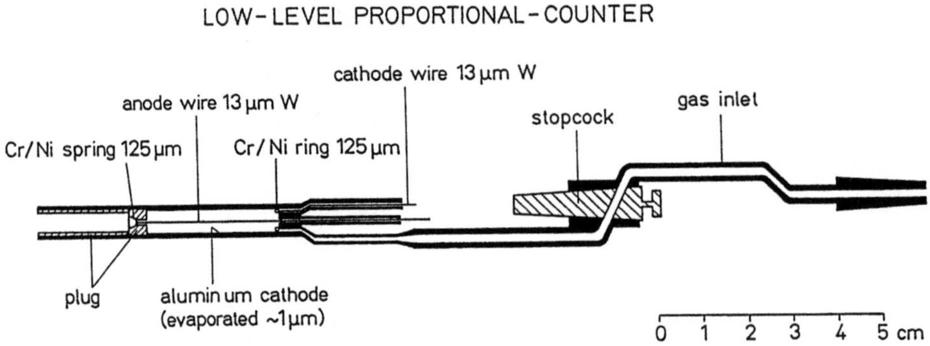

Fig. 4: Miniaturized low-level proportional counter

In order to judge our present abilities in counting, a series of Monte Carlo (MC) simulations of the full scale experiment has been performed. Fig. 5 shows the results of such simulations. The absolute 1σ-error (in SNU) to be obtained in the full scale experiment is plotted versus the running time of the experiment for three different production rates (10, 40 and 90 SNU). The calculations were performed with the following assumptions: exposure time per run 21 days, delay between end of exposure and start of counting 12 hours, extraction and conversion yield 92%, counting efficiency 24% (L peak) and 38% (K peak), background rate 0.22 counts per day (L peak) and 0.11 cpd (K peak).

The background data used as input in the MC calculations have all been obtained in the Low-Level Counting Laboratory at Heidelberg (shielding about 15 meters of water equivalent). The GALLEX Collaboration has now a small counting station installed at the GSUL (bypass No. 12, about 500 m distance from hall A). Preliminary counting data on three counters filled with Ar-CH_4 indicate that the background at GSUL is a factor 3 to 4 lower than that at the Heidelberg Counting Lab. However, whether this reduction factor is applicable also for counters filled with GeH_4 or Xe-GeH_4 mixtures remains to be demonstrated.

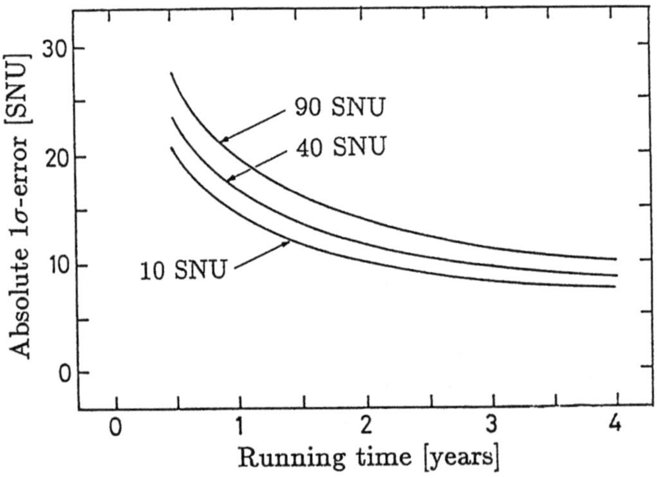

Fig. 5: Results of Monte Carlo simulations of the full scale experiment (see text).

^{51}Cr source experiment

A calibration of the gallium detector by means of an artificial ^{51}Cr neutrino source provides the most direct test of the whole experimental procedure. ^{51}Cr decays by electron capture (halflife 27.7 days) and emits monoenergetic neutrinos with energies of 746 keV (81%), 751 keV (9%), 426 keV (9%) and 431 keV (1%). The last two neutrino lines are accompanied by emission of a 320 keV γ-ray. ^{51}Cr can be produced by neutron activation of natural chromium through the reaction $^{50}Cr(n,\gamma)^{51}Cr$ (natural isotopic abundance of ^{50}Cr 4.35%). A source strength of ≥ 600 kCi is required in order to perform a sensible source experiment. This is because the neutrino signal from the source has to compete with the "background" from solar neutrinos.

A 21 day test irradiation with 12.4 kg Cr powder (natural isotopic composition) has been performed at the 35 MW Siloe reactor in Grenoble. From this test irradiation it could be concluded that a 600 kCi ^{51}Cr neutrino source is indeed feasible [7]. Determination of the obtained source strength is relatively simple by measuring the 320 keV γ-ray activity of an aliquot sample. This sample is taken after the Cr powder is homogenized through mixing in order to smooth out activity differences caused by neutron flux gradients during irradiation.

The source is then put into a sealed container surrounded by 10 cm lead or tungsten. The thickness of this shield is entirely determined by the γ-ray activity of activated impurities in the chromium powder (mainly ^{110m}Ag and ^{124}Sb). Active cooling of the source is not required. After transport to the GSUL, the source is placed for about two months in the reentrant tube of the process tank (see Fig. 3). During that time there will be about five Ge extractions from the $GaCl_3$ solution. In order to achieve a precision of 10% in the final result, a series of four

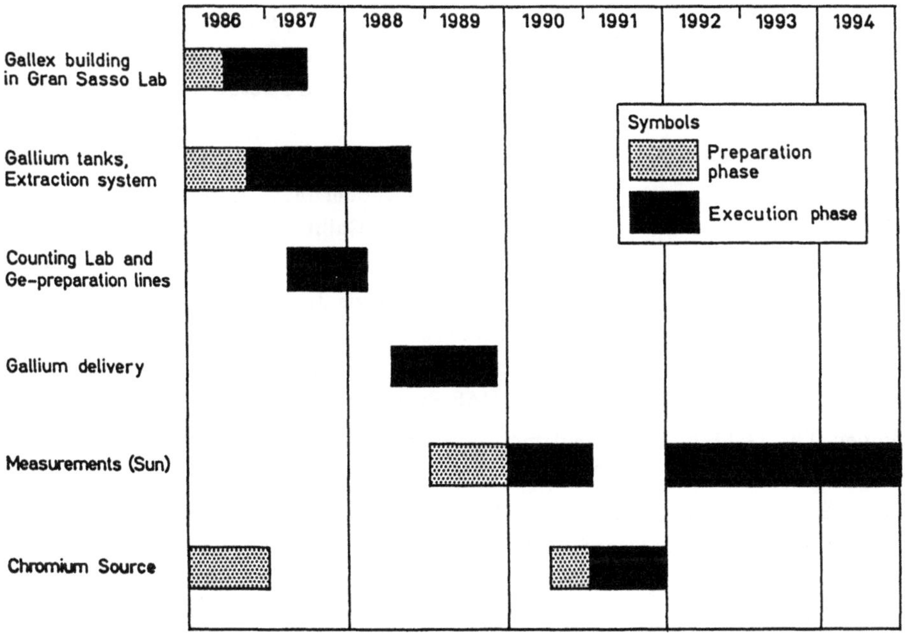

Fig. 6: Time schedule for GALLEX.

source experiments is necessary. Since in chromium of natural isotopic composition about 77% of the neutrons are captured by isotopes other than ^{50}Cr, the idea of using chromium enriched in ^{50}Cr is presently investigated.

Time table of the experiment

The time table of the experiment is mainly determined by the gallium delivery (see Fig. 6). According to the contract with the gallium supplier, the total of 30 tons of gallium will be available for the experiment at the end of 1989. By that time, all installations at the GSUL required for the experiment should have been completed and tested. However, test runs with less than the full amount of gallium will start already in early 1989. The measurement on solar neutrinos will last for about four years, eventually interrupted for a period of 6 to 12 months in order to perform the ^{51}Cr source experiment. Thus, a preliminary result from GALLEX can be expected around 1992, for the final result we will have to wait until 1994.

References

1. R. Davis Jr.: Report on Status of Solar Neutrino Experiments, Paper presented to the Seventh Workshop on Grand Unification, ICOBAN '86, April 16-18, 1986, Toyama, Japan.
2. J.N. Bahcall, R.K. Ulrich: Solar Models, Neutrino Experiments and Helioseismology, Preprint IASSNS-AST 87/1, IAS Princeton, September 1987. Submitted to Reviews of Modern Physics.
3. K. Hirata, T. Kajita, M. Koshiba, M. Nakahata, Y. Oyama, N. Sato, A. Suzuki, M. Takita, Y. Totsuka, T. Kifune, T. Suda, K. Takahashi, T. Tanimori, E.W. Beier, L.R. Feldscher, S.B. Kim, A.K. Mann, F.M. Newcomer, R. Van Berg, W. Zhang, B.G. Cortez: Search for ^8B Solar Neutrinos at KAMIOKANDE-II, Preprint UT-ICEPP-87-04, UPR-0144E, July 1987. Submitted to the 1987 International Symposium on Lepton and Photon Interactions at High Energies, Hamburg, FRG, July 1987.
4. I.R. Barabanov, E.P. Veretenkin, V.N. Gavrin, S.N. Danshin, L.A. Eroshkina, G.T. Zatsepin, Yu.I. Zakharov, S.A. Klimova, Yu.B. Klimov, T.V. Knodel, A.V. Kopylov, I.V. Orekhov, A.A. Tikhonov, M.I. Churmaeva: Pilot installation of the Gallium-Germanium solar neutrino telescope, AIP Conf. Proc. 126, M.L. Cherry, W.A. Fowler, K. Lande (Eds.), 175 (1985).
5. Pilot Experiment Collaboration: B.T. Cleveland, R. Davis Jr., G. Friedlander, S. Katcoff, L.P. Remsberg, J.K. Rowley, J. Weneser (Brookhaven National Laboratory); W. Hampel, G. Heusser, M. Hübner, J. Kiko, T. Kirsten, E. Pernicka, R. Schlotz (MPIK Heidelberg); I. Dostrovsky (WIS Rehovot), J.N. Bahcall (IAS Princeton).
6. W. Hampel: The Gallium Solar Neutrino Detector, AIP Conf. Proc. 126, M.L. Cherry, W.A. Fowler, K. Lande (Eds.), 162 (1985).
7. M. Cribier, B. Pichard, J. Rich, M.Spiro, D. Vignaud, A. Besson, A. Bevilacqua, F. Caperan, G. Dupont, P. Sire, J. Gorry, W. Hampel, T. Kirsten: Study of a High Intensity 746 keV Neutrino Source for the Calibration of Solar Neutrino Detectors, Submitted to Nuclear Instruments and Methods in Physics Research, July 1987.

The Sudbury Neutrino Observatory

D. Sinclair (for the SNO Collaboration [1])
Nuclear Physics Laboratory, Keble Road, Oxford, United Kingdom

1 Introduction

Neutrinos are expected to be produced copiously in astrophysical conditions such as the energy generating reactions in the sun, and in the collapsing cores of supernovae. Their weak interaction with matter allows them to pass out of these regions carrying, in their energy spectra, direct information about the processes occuring under these conditions which is completely inaccessible to other forms of observation. The recent detection of neutrinos from the supernova SN1987A marks the first experimental observation in neutrino astronomy. However, the neutrinos from the sun have been sought for almost 20 years [2] and the flux is clearly much less than the Standard Solar Model (SSM) predicts [3]. Two classes of solution to this problem have been suggested; either our model of the sun is deficient in some way so that the neutrinos are not produced in the expected quantities, or the neutrinos change in some way in transit to the earth so they are not detected. Both possibilities have been discussed in the literature for many years but it has been realized only recently that there is a mechanism [4] whereby the electron neutrinos might be converted into other neutrino types with high efficiency. This mechanism requires that at least some neutrino types have mass and that there is some mixing of neutrino states. The observation of this effect would have profound implications for fundamental physics.

To resolve the Solar Neutrino Problem requires new data and hence new tools. The gallium experiment will provide crucial information about the low energy part of the spectrum. In this talk I will discuss a heavy-water Čerenkov detector which would characterize completely the high energy solar neutrino flux.

2 Neutrino detection in heavy water

An outline of the proposed detector is shown in Figure 1. The neutrinos interact in the D_2O producing energetic electrons which, in turn, emit

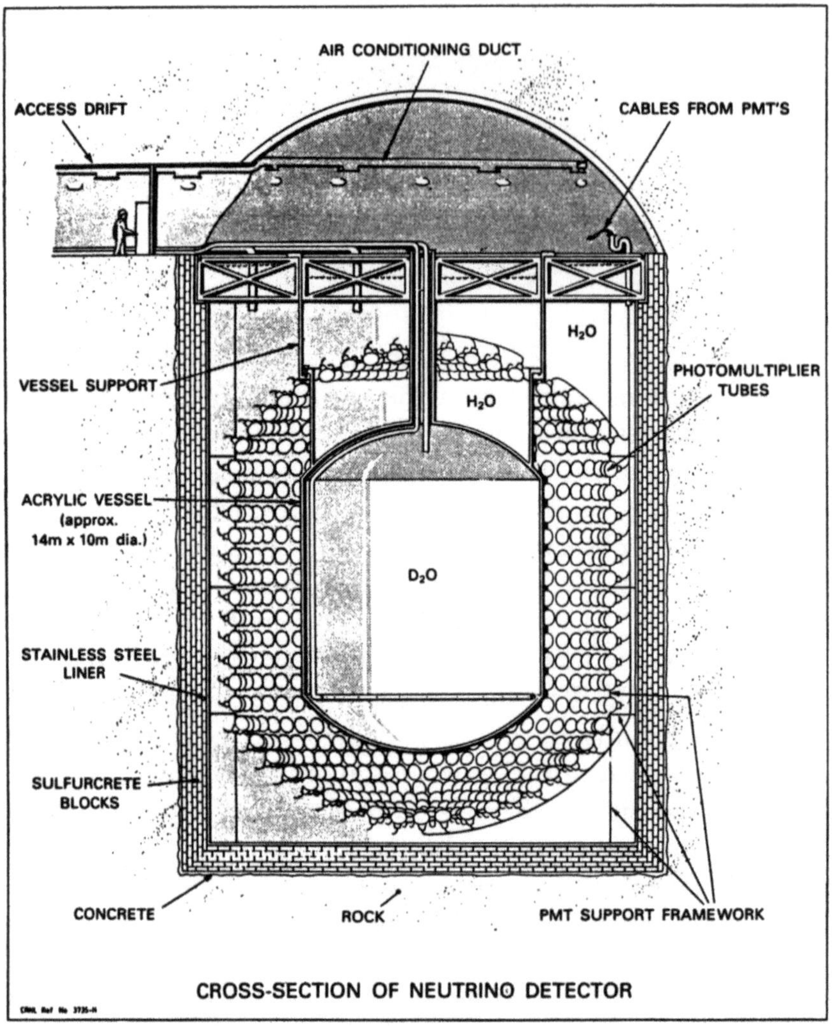

Figure 1: Outline of the proposed heavy-water neutrino detector.

Čerenkov light. This light passes through the acrylic vessel containing the D_2O to photomultiplier tubes. The pattern of signals in these tubes allows the neutrino interaction to be characterized. The design is determined largely by consideration of the radioactive backgrounds. The D_2O is shielded from radioactivity in the rock by one metre of low activity, borated Sulfurcrete [5] and by 4 metres of ultra-pure light water. The photomultiplier tubes must be separated from the D_2O by 2.5 metres of H_2O to prevent γ-rays from radioactivity in their glass envelopes from reaching the D_2O and dissociating deuterons. The present design requires 1955 phototubes of 50 cm diameter to give 40% coverage of the sensitive volume.

High energy solar neutrinos would be identified using three reactions in the Electron-neutrinos would be detected by their charged-current (CC) interaction with deuterons to form two protons together with an electron which carries off most of the free energy.

$$\nu_e + d \rightarrow p + p + e^-$$

By measuring the energy of the electrons the neutrino spectrum could be deduced. All neutrino types can scatter inelastically from deuterons via the neutral-current (NC) interaction causing the deuteron to break up into a neutron and a proton.

$$\nu_x + d \rightarrow n + p + \nu_x$$

This reaction would be detected by observing the γ-rays from the subsequent capture of the free neutron. *The relative rates for these two reactions would establish whether conversion of neutrinos from electron type to some other type is responsible for the solar neutrino problem, independent of any solar model* [6]. The neutron capture rate would be identified by running first with pure D_2O in which the capture probability is 24%, and then with 0.25% NaCl which increases the capture probability to 83%. The difference between these data sets will allow the CC and NC components to be extracted. The third reaction to be studied is the elastic scattering of neutrinos from electrons.

$$\nu_x + e \rightarrow \nu_x + e$$

This gives a good measure of the direction of the neutrino flux and would confirm that the observed neutinos came form the sun. The cross section for ν_e scattering is about 6 times larger than for $\nu_{\mu,\tau}$ but the sensitivity the $\nu_{\mu,\tau}$ might allow their flux to be determined if the ν_e flux is attenuated, for example by oscillations.

3 Background Considerations

In order to realize a practical solar neutrino detector it is vitally important to understand the sources of background. Much of the work of our collaboration has been focused on this problem.

For signals above 7 MeV the main background is expected to come from neutron capture γ-rays in the rock. The Sulfurcrete and light water shield reduce this background to a level two orders of magnitude below the signal expected form the Standard Solar Model.

Below 7 MeV the background arises from the radioactivity in the components used to construct the detector and these must be made extremely clean if the threshold for the CC reaction is to be lowered and if the signal from the NC reaction is to be observed. The problem is particularly acute for the NC reaction because not only must the neutron capture signal be identified but also the photodisintegration rate must be small by comparison with the neutrino disintegration rate. The most serious activities are the high energy β and γ-decays found well down the ^{232}Th and ^{238}U decay chains. Thus it is necessary to study both the Th and U concentrations, and the possible disequilibra in the chains.

Most of the sensitive volume of the detector is filled with water and thus the lowest levels of activity must be achieved for both light and heavy water. It is proposed to purify the water by passing it through Mn coated acrylic fibres [7]. It has been demonstrated that in small quantities ^{224}Ra and ^{228}Th and be removed to levels of about 5 disintegrations per day per tonne (equivalent to a ^{232}Th concentration of $10^{-14} g/g$). Brine can also be purified by this technique.

Measurements have been made of the Th and U content in acrylics. UV-transmitting, dye-free material is laser trimmed and vapourized in Suprasil quartz tubes in an inert atmosphere. The residues are dissolved in HNO_3 and analysed by mass spectroscopy or α-counting. The best results to date obtained with mass spectroscopy gave $0.23 \times 10^{-12} g/g$ but there were some difficulties with the measurement (there were traces of residue which did not dissolve in HNO_3) and we assume levels one order of magnitude higher than these.

The glass of the photomultiplier tubes also contributes a significant background because, although they are far from the fiducial volume, the probability of chance coincidence between two decays reconstructing as a high energy event in the fiducial volume is not negligible. Direct γ-counting on the PMT materials shows that each tube contains about 4 mg of both ^{232}Th and ^{238}U. There is also a significant flux of high energy γ-rays from $(\alpha, p\gamma)$ and $(\alpha, n\gamma)$ reactions in the glass. A programme aimed at finding cleaner materials from which to make the glass is in hand.

The expected background rates are summarized in Table 1, while Fig. 2 shows the spectra expected after collecting data for one year. The potential photodisintegration background to the NC reaction requires the

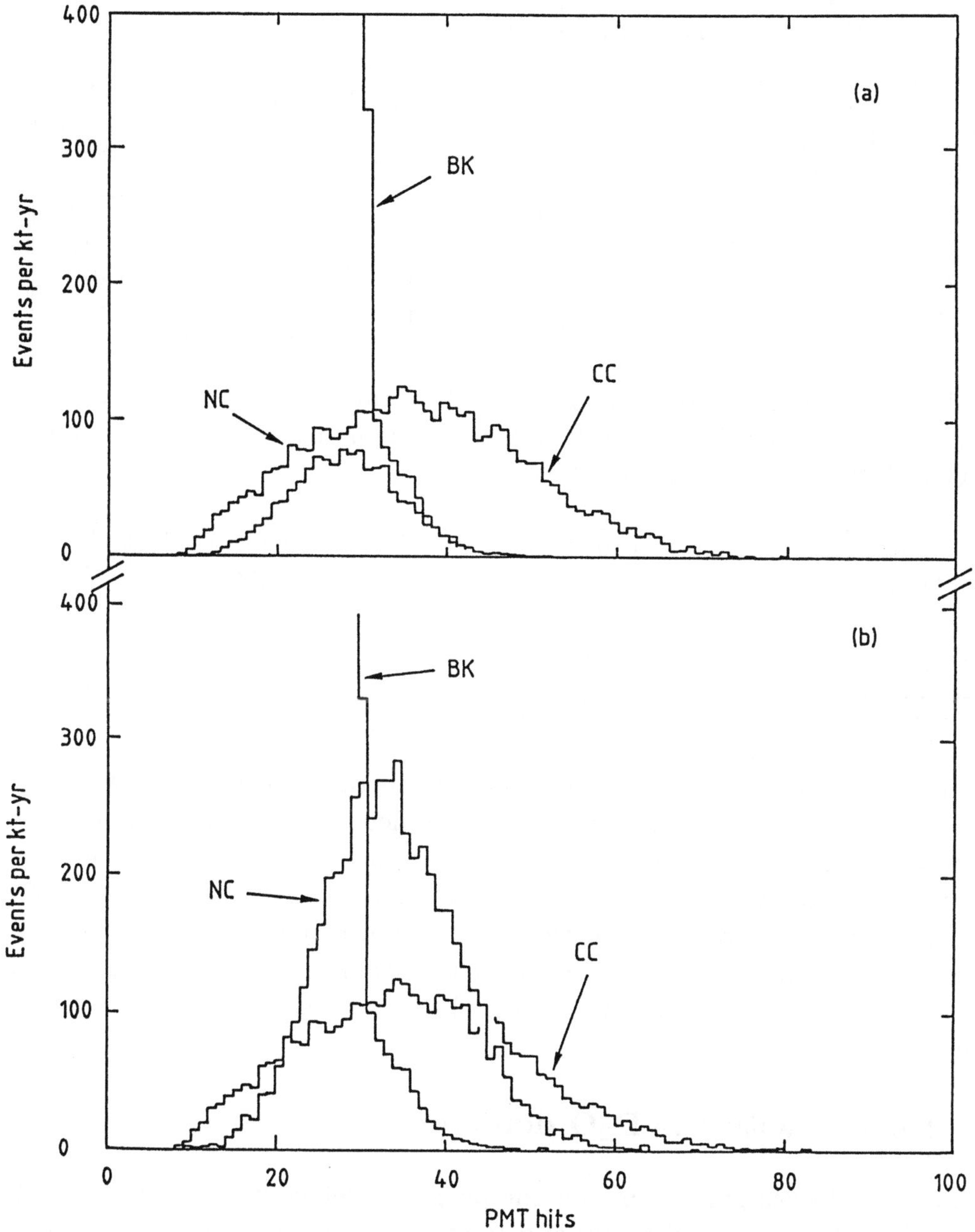

Figure 2: Spectra for the charged-current (CC) and neutral-current (NC) reactions together with the background (BK) expected for one year's data taking (a) with pure D_2O and (b) with 0.25% NaCl added to enhance the neutron capture. The ^8B ν_e flux is assumed to be $2 \times 10^6 cm^{-2}s^{-1}$, the upper limit from the Cl experiment, while the total ν flux is taken to be $6 \times 10^6 cm^{-2}s^{-1}$ as predicted by the SSM.

Table 1: Background event rates (events per year) from various sources above the indicated threshold. The charged-currect event rate for a ^{8}B ν_e flux of $2 \times 10^6 cm^{-2} s^{-1}$ is shown for comparison.

Threshold PMTs	Threshold Energy	PMTs	H_2O	Acrylic	D_2O	Rock	Total	CC Signal
24	4.44	7262	1501	36729	1731	159	47827	3090
28	5.19	650	270	4459	88	227	5680	2718
32	5.94	32	101	87	0	190	344	2269
36	6.69	0	34	87	0	88	208	1805
40	7.45	0	0	0	0	55	55	1425

Table 2: The neutron production and capture rates leading to high energy γ-rays are given for the SSM ν-flux and for the main background sources. D_2O with 2.5 tonnes of NaCℓ is assumed.

Source	Neutron Production Rate	Neutron Capture Rate
SSM ν flux	6570	5450
γ's in D_2O	1095	909
γ's in Acrylic	2920	700
γ's in PMTs	730	175

highest levels of purity in materials; Table 2 lists the expected rates with the above purity assumptions. In this table the neutron leakage from surface-generated and volume-generated neutrons is included, as is the effect of capture on residual hydrogen in the D_2O. The rate of surface-generated background can be determined from the spatial distribution so the major worry is the neutrons generated by activity within the D_2O itself. However, at the present estimates of activity it is clear that a definitive measurement can be made.

4 Physics with the D_2O detector

The main physics programme for the proposed detector will be to identify the solar ^{8}B ν_e and total ν flux and to search for distortions in the energy spectrum which might indicate neutrino oscillations. These measurements will determine unambiguously whether the solar neutrino problem is due to neutrino oscillations. These aspects of the programme have been published recently [8]. I wish to concentrate on a few aspects of the programme which have not been published.

If the SNP is not due to neutrino oscillations then some aspect of the SSM must be incorrect. One of the ways of reducing the ^{8}B production is to introduce some mixing of the solar material. A number of authors have noted that this would significantly increase the rate for the ^{3}He(p,$\nu_e e^+$)^{4}He (hep) reaction. If we use the most recent calculation of the hep cross section [9] together with the SSM we predict an event rate of about 100 per year of which about 20 would be above the ^{8}B endpoint where there is very little background. Thus the flux should be measurable.

The dramatic appearance of supernova Shelton 1987A in February of this year has highlighted the importance of neutrino detection in the understanding of the physics of stellar collapse. The uniquely sensitive and versitile detector proposed would play an important role in studying any galactic supernovae. Table 3 shows the number of events of different types that we would expect for a supernova at the centre of our galaxy. In addition to offering information on the dynamics of the supernova because of its high ν_e sensitivity, the sensitivity to the mu and tau type neutrinos might allow their masses to be identified. For a supernova at 10 kpc a mass of 50 eV would give a delay in the neutral current events of 3 seconds with respect to the $\bar{\nu}_e$ which would be measurable even with a 10 second burst time. Thus the detector would be sensitive to masses far beyond the limits of terrestrial measurements and in a range where the neutrino masses may lie if they are to form dark matter.

Table 3: Events from stellar collapse at 10 kpc.

	Reaction	Target medium	Events in ν_e burst kt^{-1}	Events in cooling phase kt^{-1}
I	$\nu_e + d \longrightarrow p + p + e^-$	D_2O	10	33
II	$\nu_x + e \longrightarrow \nu_x + e$	D_2O/H_2O	1	16
III	$\nu_x + d \longrightarrow p + n + \nu_x$	D_2O	6	760
IV	$\bar{\nu}_e + d \longrightarrow n + n + e^+$	D_2O	0	20
V	$\bar{\nu}_e + p \longrightarrow n + e^+$	H_2O	0	120

5 References

1. Members and *Institutions* of the collaboration are: G.T. Ewan, H.C. Evans, H.W. Lee, J.R. Leslie, H.B. Mak, W. McLatchie, B.C. Robertson, P. Skensved, *Queen's University Kingston.* R.C. Allen, G. Buehler, H.H. Chen, P.J. Doe, *University of California at Irvine.* D. Sinclair, N.W. Tanner, *University of Oxford.* J.D. Anglin, M. Bercovitch, W.F.

Davidson, C.K. Hargrove, H. Mes, R.S. Storey, *National Research Council of Canada*. E.D. Earle, *Chalk River Nuclear Laboratories*. P. Jagam, J.J. Simpson, *University of Guelph*. E.D. Hallman, *Laurentian University*. A.B. McDonald, *Princeton University*. A.L. Carter, D. Kessler, *Carleton University*.

2. R. Davis Jr. et al. Phys. Rev. Lett. 20, 1205 (1968).
   J.K. Rowley *et al.*, in *Solar Neutrino Astronomy*, ed. M.L. Cherry, W.A. Fowler and K. Lande (AIP Conf. Proc. No. 126, New York), p. 1.

3. J.N. Bahcall, Rev. Mod. Phys. 50, 881 (1978).
   J.N. Bahcall *et al.* Rev. Mod. phys. 54, 767 (1982).
   J.N. Bahcall and R.K. Ulrich, to be published.

4. S.P. Mikheyev and A. Yu. Smirnov, Nuovo Cimento C9, 17 (1986).

5. Sulfurcrete is a form of concrete in which the portland cement is replaced by sulfur plus a plasticizer, produced by Sulfurcrete Products Inc, Calgary, Alberta.

6. H.H. Chen, Phys. Rev. Lett. 55,1534 (1985).

7. W.S. Moore and L.M. Cook, Nature 253, 262 (1975).

8. G. Aardsma *et al.*, Phys. lett. B194, 321 (1987).

9. P.E. Tegnér and Chr. Bargholtz, Ap. J. 272, 311 (1983).

Resonant ν-Oscillations in Matter: Measuring of Neutrino Masses and Mixing

S.P. Mikheyev and A.Yu. Smirnov

Institute for Nuclear Research, Academy of Sciences of the USSR, 60th October Anniversary prospect 7a, SU-117312 Moscow, USSR

1. Introduction

Resonant neutrino oscillations in matter [1-3] open unique possibilities in searching for neutrino masses and mixing. To the present moment (October'87) there are the following data which refer to the matter oscillations:

i) Measuring of average ^{37}Ar - production rate in Cl - Ar experiment with solar neutrinos [4].

ii) searching for annual modulations of ^{37}Ar production rate [5].

iii) KAMIOKANDE-II limit on ^{8}B - solar neutrino flux [6].

iv) Registration of upgoing and downgoing atmospheric neutrino events by IMB - collaboration [7].

v) Possible observations of ν - burst related to supernova 1987A in LMC [8].

With these data we are at the beginning of the way, which may lead to the establishing of massiveness of neutrinos and moreover to the measuring of Δm^2 and neutrino mixing ($\sin^2 2\theta$). Otherwise matter oscillations experiments result in exlusion of very large and very important Δm^2 vs. $\sin^2 2\theta$ region.

2. Neutrino Matter Oscillations Experiment - Neutrino Optics

The idea of experiments is very simple (Fig.1). Matter filter distorts initial ν - spectrum $F_o(E)$:

$$F(E) = P(E, \Delta m^2, \sin^2 2\theta, \rho(r)) \cdot F_o(E). \qquad (1)$$

The distortion (suppression) factor P depends on $\Delta m^2 = m_1^2 - m_2^2$, θ (2ν - case), density profile and neutrino energy. Measuring the spectrum at the exit ($F(E)$) one can extract for known $\rho(r)$ the information on Δm^2 and $\sin^2 2\theta$.

The key points which determine the shape of P are the following.

1) Matter effects originate from elastic forward neutrino scattering on components of medium [1]. This scattering reduces to the refraction indexes appearance for the waves, which describe the propagation of the mixed components. If the oscillating neutrinos (for definiteness ν_e and

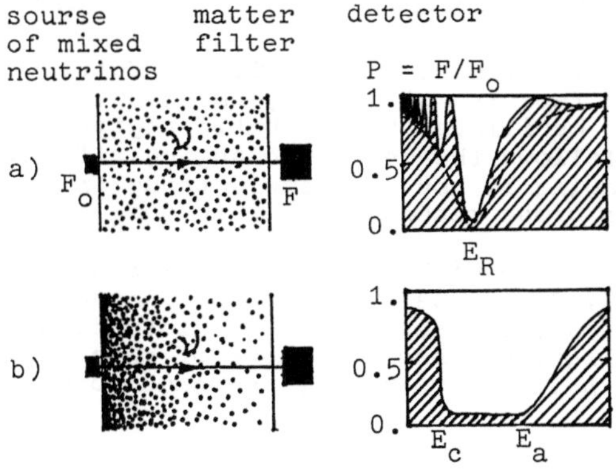

Fig.1 Resonant ν - oscillations experiments. The distortion of initial ν - spectrum by filter a) with a constant density and b) with monotonious decreasing of density

ν_μ) scatter in different way, then an additional phase difference $\Delta\varphi_m$ appears between them during the propagation:

$$\Delta\varphi_m = (n_e - n_\mu)\cdot k\cdot x. \qquad (2)$$

Here n_e (n_μ) is the refraction index for ν_e (ν_μ), k is the moment of neutrino and x is the distance. It is this additional phase difference leads to the matter resonant effects.

The distance in matter with ρ = const on which $\Delta\varphi_m$ equals to 2π is called the refraction length:

$$l_o = 2\pi/(k\cdot(n_e - n_\mu)) = 2\pi(\sum_i \Delta f^i(0)\cdot N_i/k)^{-1} \sim m_N/(G_F\rho). \qquad (3)$$

Here $\Delta f^i(0) = f_e^i - f_\mu^i$, f_α^i is the amplitude of ν_α scattering on i - component of medium, and N_i is the concentration of i - component, G_F is the Fermi constant and m_N is nucleon mass.

If the size of matter layer (d) is much smaller, than l_o: $d \ll l_o$, then matter effects are small: $P \simeq P_{vac}$, P_{vac} is the vacuum oscillation factor. Strong effects appear, when

$$\int_0^d dx\ \rho(x) \gtrsim l_o\bar{\rho} \approx m_N/G_F.$$

2) Matter changes mixing ($\theta \to \theta_m(\rho)$) and oscillations length ($l_\nu \to l_m(\rho)$). The dependence of mixing parameter $\sin^2 2\theta_m$ on ρ or E has a resonant form [2] (Fig.1a and 6). The resonant condition ($l_\nu = l_o \cos\theta$) determines the resonant density (energy):

$$(E/\Delta m^2)\rho = (m_N \cos2\theta)/(2\sqrt{2}\,G_F) \simeq m_N/(2\sqrt{2}\,G_F) \qquad (4)$$

(last equality for small θ).

Consider filter with constant density. The distortion factor will be the oscillating function which insribes in resonant curve $\sin^2 2\theta_m(E)$ (Fig.1a and Fig.6). Quasiperiod of P(E) depends on the size of filter. Measuring $F(E)/F_o(E)$, one can restore resonant curve. Then the position of resonance that is E_R determines Δm^2 according to (4); and the width of resonance: $\Delta E_R = E_R\cdot tg2\theta$ gives mixing angle.

3) Dynamics of oscillations in matter with varying density is governed by adiabacity condition. In the most important form it looks like [2]

$$2\Delta r_R > l_m^R \qquad (5)$$

where

$$2\Delta r_R = 2\ (d\rho/dr)^{-1}\Delta\rho_R \qquad (6)$$

is the spatial width of resonance ($\Delta\rho_R = \rho_R\cdot tg2\theta$) and

$$l_m = l_\nu/\sin2\theta \qquad (7)$$

is the oscillation length in resonance. Introduce adiabacity parameter:

$$\text{æ}_R = 2\Delta r_R/l_m^R = 2\ (\Delta m^2\cdot\sin^2 2\theta\cdot h)/(4\pi E\cdot\cos2\theta) \qquad (8)$$

$h = (d(\ln\rho)/dr)^{-1}$, so that for $\text{æ}_R > 1$ adiabacity condition is fulfilled

Condition (5) gives upper limit on rapidity of density changing. If density changes slowly, then neutrino oscillates in adiabatic regime. The (averaged) flavour of ν - state follows the density changing. If density changes quickly, then this following is not complete. Strong transformations take place, when neutrinos cross resonant layer ($\rho = \rho_R$) adiabatically.

The most interesting case is the profile with monotonious decreasing of density (the Sun, collapsing star). Neutrinos are produced in the layer with maximal density ρ_c. In this case the dependence of distortion factor (averaged over period of oscillations) has the shape of bath (Fig. 1b). The position of left edge of bath is determined by resonant condition at maximal density

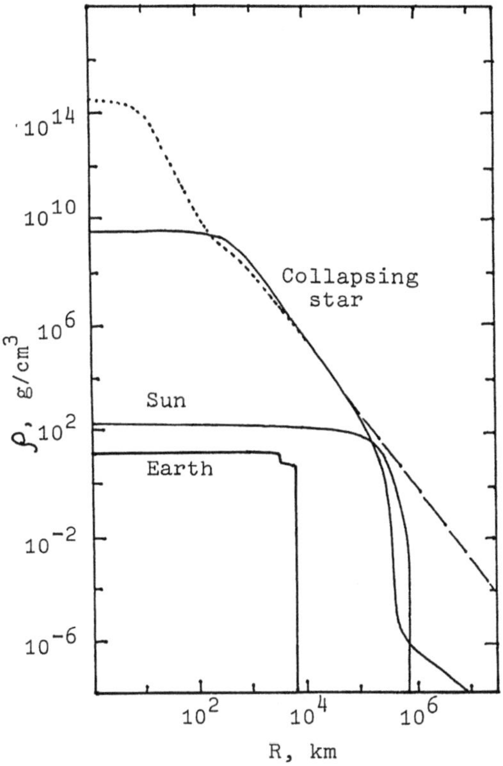

Fig.2 Density profiles of different matter filters; for collapsing stars the density distributions are shown: for progenitor with compact profile (solid line), for progenitor without sharp border between mantle and envelope (dashed line) and for the star on ν - opacity stage (dotted line)

$$(E/\Delta m^2)_c \simeq m_N/(2\sqrt{2}\, G_F \rho_c) \quad (9)$$

(the point of ν - production). The energy of right edge is fixed by adiabacity condition: ($\varkappa_R = 1$)

$$(E/\Delta m^2)_a \simeq (h(E)\cdot \sin^2 2\theta)/(2\pi \cos 2\theta). \quad (10)$$

In the interval $(E/\Delta m^2)_c - (E/\Delta m^2)_a$:

$$P = \sin^2\theta. \quad (11)$$

If there is no averaging over period, then P(E) has oscillations on the edges of bath; bottom of the bath, corresponding to oscillationless transformation, remains unchanged [2]. Again measuring the properties of bath (9 - 11) one can find Δm^2 and $\sin^2 2\theta$. In the case of 3ν - mixing the distortion factor will be the supperposition of two baths [3].

The density profiles of real matter filters are shown on Fig.2. Possible sources of mixed neutrinos are central region of the Sun, cores of collapsing stars, atmosphere of the Earth and, in principle, the accelerators. Consider the restrictions on neutrino parameters from the existing data.

3. Average ^{37}Ar production rate in Cl - Ar experiment

Suppresion factor for ^{37}Ar production rate (Q_{Ar}) is

$$R_{Ar} = Q_{Ar}/Q_{Ar}^{SSM} = (Q_{Ar}^{SSM})^{-1}\cdot \int dE\, P(E)\cdot \sigma(E)\cdot F(E), \quad (12)$$

where Q_{Ar}^{SSM} and Q_{Ar} are the production rates in standard solar model without and with oscillations correspondently (for Q_{Ar}^{SSM}: P = 1), σ is the cross section, F(E) is solar neutrino flux. In 2ν - case distortion P(E) depends on neutrino parameters Δm^2 and $\sin^2 2\theta$. Equation

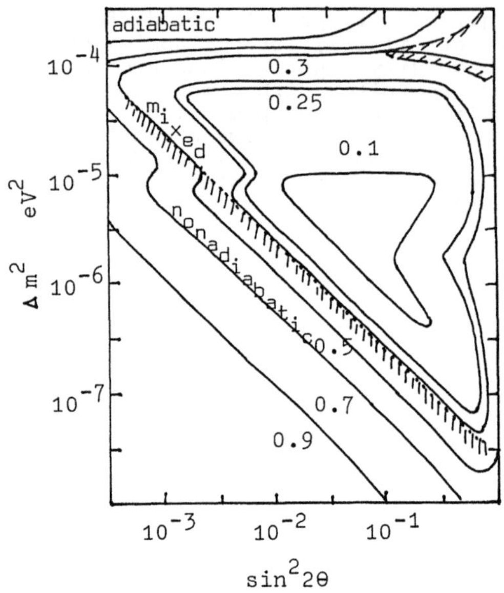

Fig.3 Lines of equal suppresion of ^{37}Ar production rate due to $\nu_e - \nu_\mu$ (ν_τ) oscillations (2ν - ISOSNU lines); figers on the curves are values of suppresion; dotted line is KAMIOKANDE - II restriction; dashed line is IMB - restriction

$$R_{Ar}(\Delta m^2, \sin^2 2\theta) = a = const$$

determines the line of equal suppression [2,10] (ISOSNU line [9]) on the $\Delta m^2 - \sin^2 2\theta$ plot (Fig.3). Values Δm^2, $\sin^2 2\theta$ for which $R_{Ar} = R_{Ar}^{exp} = 1/4 - 1/2$ are called solutions of solar neutrino problem. The corresponding points are placed between ISOSNU lines $a = 1/4$ and $a = 1/2$.

The system of mixed $\nu_e - \nu_\mu - \nu_\tau$ neutrinos has three resonances, but for solar neutrinos only two of them are essential; naimly resonances in $\nu_e - \nu_\mu$ and $\nu_e - \nu_\tau$ channels [11-13]. They are called l- and h-resonances correspondently. $\nu_e - \nu_\mu - \nu_\tau$ system is determined with respect to $\nu_e - \nu_e$ transition by two sets of parameters $(\Delta m^2, \sin^2 2\theta)$: $(\Delta m^2, \sin^2 2\theta)_l \simeq (m_2^2, \sin^2 2\omega)$ and $(\Delta m^2, \sin^2 2\theta)_h \simeq (m_3^2, \sin^2 2\varphi)$ [10,11]. Here ω (φ) is vacuum angle which describes mixing between ν_e and ν_μ (ν_e and ν_τ); m_2 (m_3) is the mass of neutrino, composing ν_μ (ν_τ) predominatingly. We will suppose the mass hierarchy $m_1 \ll m_2 \ll m_3$ here. The sets of parameters $(\Delta m^2, \sin^2 2\theta)_{l,h}$ are characteristics of l- and h-resonances (Δm^2 fixes the resonant density and $\sin^2 2\theta$ determines the width of resonances). So 3ν - system is fixed by positions of l- and h-resonances on $(\Delta m^2, \sin^2 2\theta)$ - plot. Each 3ν - solution of solar neutrino problem is determined by two points on this plot [10].

3ν - solutions can be analized in terms of 2ν - ones: l- and h- resonances are placed on definite 2ν - ISOSNU lines (Fig.3). There are two types of 3ν - solutions [11, 13].

1) Mixed solution: h - resonance is on the adiabatic branch of 2ν - ISOSNU line, l - resonance is on nonadiabatic one.

2) Nonadiabatic 3ν - solution: both h- and l- resonances are on nonadiabatic branches (Fig.4b).

The rules have been formulated for positions of l- and h- resonances [14]. In particular $R^{(3)}$ can be written as

$$R^{(3)} = R_{Ar}^h \cdot R_{Ar}^l + K , \qquad (13)$$

where $R^{l,h} = R_{Ar}^{(2)}(\Delta m^2, \sin^2 2\theta_{\ l,h})$ are 2ν - suppression factors. Factorization of $R^{(3)}$ ($K = 0$) means that if l- and h- resonances are on 2ν - ISOSNU lines a_1 and a_2 correspondently, then the total suppression is $R^{(3)} = a_1 a_2$. Note that K is zero if

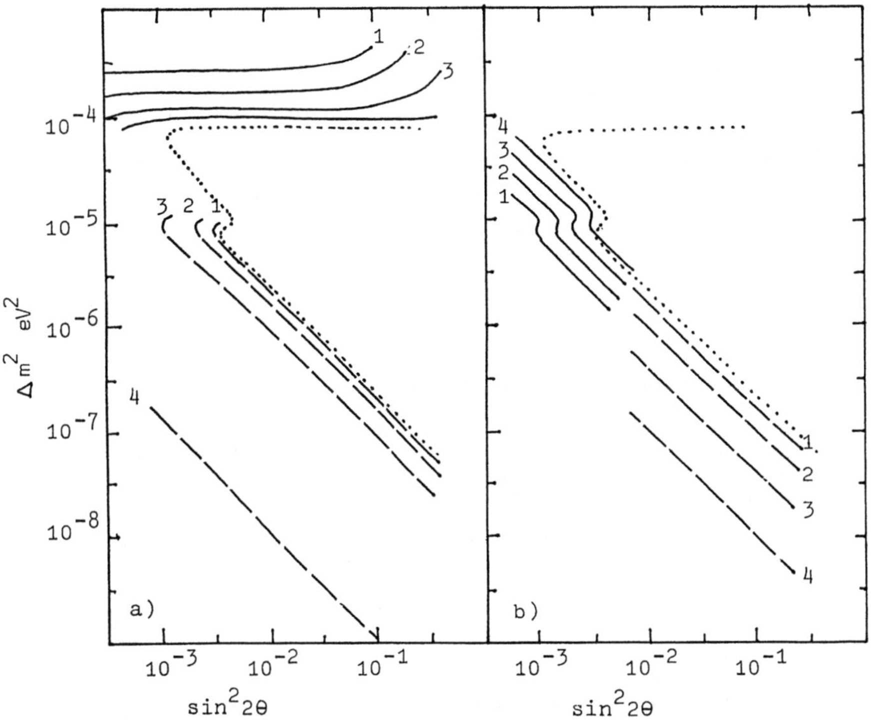

Fig.4 Solutions of solar neutrino problem in the case of $\nu_e - \nu_\mu - \nu_\tau$ oscillations; R_{Ar} =1/3,h- and l- resonances are correspondently on solid and dashed lines with the same numbers; a) mixied 3ν - solution, b) non-adiabatic 3ν - solution; dotted lines are 2ν - ISOSNU lines with a = 1/3

$$\sigma \cdot F(E) = A \cdot \delta(E - E_0)$$

(and $\sin^2\varphi \ll 1$). This approximation is good for experiments on detection of ^8B - neutrinos with high registration thresholds (KAMIOKANDE - II, SUDBERY). In the case of Cl - Ar experiment the factorization (13) is broken mainly by the existance of two comparable contributions to ^{37}Ar - production rate from ^8B - and ^7Be - neutrinos. It can be shown [14] that

$$R^{(3)} < R^h \cdot R^l \quad \text{(mixed solution)}$$
$$R^{(3)} > R^h \cdot R^l \quad \text{(nonadiabatic solution)}$$

and moreover in nonadiabatic case the violation of $R^{(3)}$ - factorization is smaller than that in mixed one.

If both resonances lie on the same 2ν - ISOSNU line a, then 3ν - suppression is approximately $R^{(3)} \simeq a^2$ (under factorization this result is exact). Consequently, to obtain given suppression R^{exp}_{Ar} one should place at least one resonance between 2ν - ISOSNU lines

$$a_1 = \sqrt{R^{exp}_{Ar}} \quad \text{and,} \quad a_2 = R^{exp}_{Ar}. \tag{14}$$

Conditions (14) determine the ISOSNU stripes on (Δm^2 - $\sin^2 2\theta$) - plot (Fig.3). When one of resonances moves away from 2ν - resonance region, another one approaches 2ν - ISOSNU line a = R^{exp}_{Ar}. If the suppression R_{Ar} = 1/4 - 1/2 in Davis's experiment is due to resonant oscillations only, then at least one resonance lies between 2ν - ISOSNU lines 0.25 and 0.7 (Fig.3).

4. Annual Modulations of ^{37}Ar - Production Rate

For solar neutrinos resonant conversion in the Earth results in regeneration of initial ν_e - flux (regeneration solution, see Fig.3). Due to Earth rotation day - night and seasonal modulation of ^{37}Ar - production rate are predicted [15, 5]. For $\sin^2 2\theta < 0.3$ strong annual modulations are in contradiction with Davis's data [5]. This anables to exclude the region of ν - parameters $\Delta m^2 = (1 - 3)\ 10^{-6}$eV, $\sin^2 2\theta = (0.3 - 3)\cdot 10^{-1}$. Note that in this region $R_{Ar} < 1/4$ and for standard solar model corresponding Δm^2, $\sin^2 2\theta$ are excluded by $\overline{R}_{Ar}$ - data too. The restrictions on annual modulations become important if the central temperature of the Sun T_c is larger, than T_c^{SSM} and we need more strong, than 1/4, suppression of R_{Ar}.

Annual waves (averaged over the day) for large values $\sin^2 2\theta$ are shown on Fig.5. For these $\sin^2 2\theta$ $R_{Ar} > 1/4$. Calculation have been done for latitude 43°. The shape of wave is sensitive to neutrino parameters, but typically it has the feature: maximal regeneration takes place in March and October [10]. So it is shifted with respect to the December, when the Earth shieldes the Sun maximally. The effect is explained by the following. For latitude 43° ν_e - regeneration is due to resonance in mantle. The last can be considered as the layer with a constant density (Fig.6) and size d. Maximal regeneration effect, integrated over some energy interval is achieved when ^8B - neutrinos spectrum is in resonance peak and moreover the probability P(E) filles the resonance peak $\sin^2 2\theta_m(E)$ maximally (Fig.6). The last condition is realized at

$$d/l_m^R \approx d\ \mathrm{tg}2\theta/l_o = 1/2\ . \qquad (15)$$

That is when the size of layer equals to the half of resonant oscillation length. For $\mathrm{tg}2\theta > 0.5$ the condition (15) is fulfilled in mantle, when zenith angle is $\cos\Psi < 0.6$ (rather then $\cos\Psi > 0.85$). The maximal time the Sun spends at angle $\cos\Psi \approx 0.6$ in March and in September. Note that when θ decreases, $\cos\Psi$ rises and maxima of regeneration approach to the December.

Using Davis's data on season variation [5] we conclude that

1) the predicted annual waves for large θ can not be excluded,
2) moreover there are indications on maxima of ^{37}Ar - production rate in March - April and September - October.

Note that annual waves due to resonant oscillations are in opposite phase to that, expected from neutrino spin precession in magnetic field of the Sun [16].

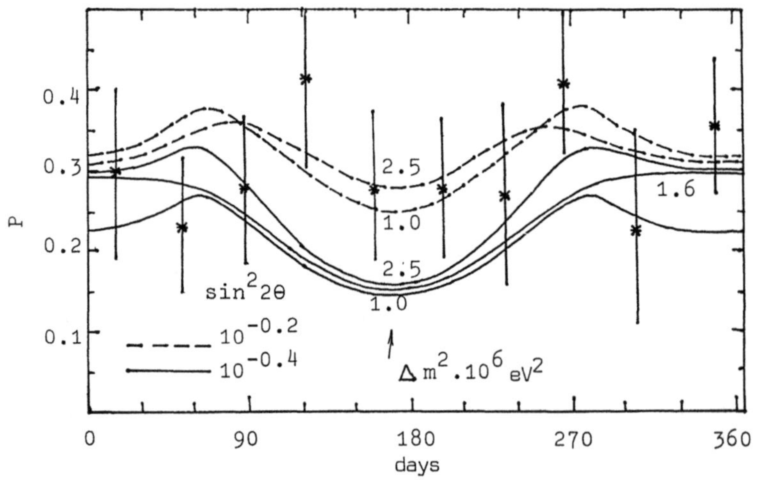

Fig.5 Annual waves; modulations of ^{37}Ar - production rate by the Earth matter averaged over the day (latitude is 43°)

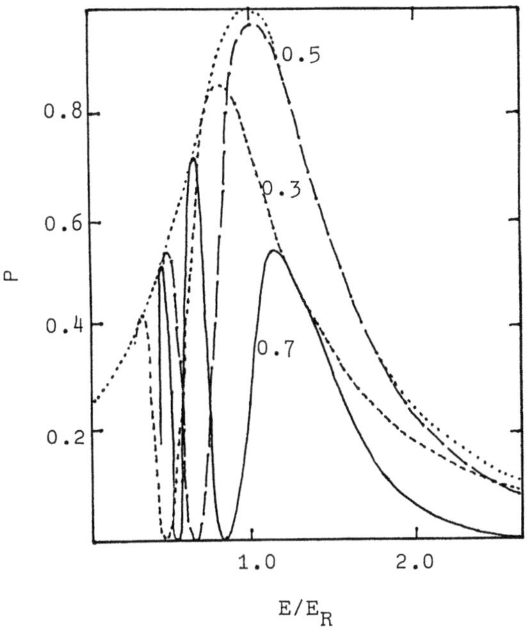

Fig.6 The probability of $\nu_\mu - \nu_e$ transition in the layer with constant density and the size d for different d/l^R_m (figers on the curves); $\sin^2 2\theta = 0.2$; the dotted line is the dependence of $\sin^2 2\theta$ on E/E_R

5. KAMIOKANDE - II Limit on ^8B - Solar Neutrino Flux

The suppression of number of νe νe events ($R_{\nu e}$) due to 2ν - oscillations for fixed values of R_{Ar} have been calculated (Fig.7). We used the threshold for electron registration $E_e = 9.5$ MeV and took into account the scattering of neutrinos ν_μ (ν_τ) which are produced in $\nu_e - \nu_\mu$ (ν_τ) - oscillations. For 3ν - case one can use 2ν - results (2ν - lines of equal suppression $R_{\nu e}(\Delta m^2, \sin^2 2\theta) = \text{const}$ [5, 17]) and factorization of $R^{(3)}$: $R^{(3)}_e = R^l_e \cdot R^h_e$.

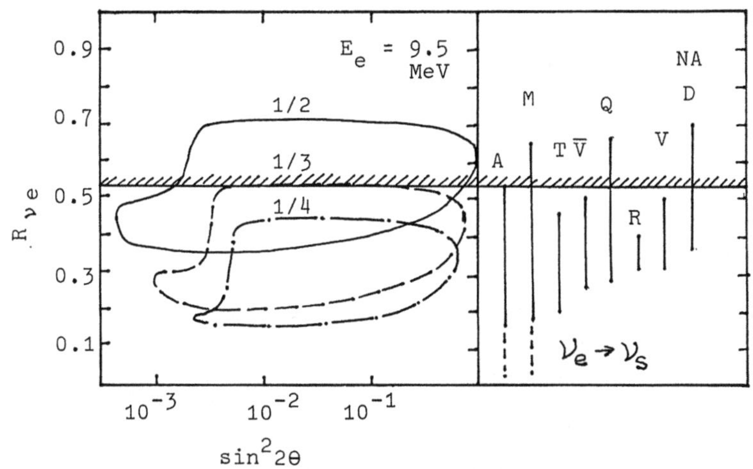

Fig.7 a) The dependence of suppression factor for number of νe - νe events (E_e > 9.5 MeV) due to ^8B neutrinos ($R_{\nu e}$) on $\sin^2 2\theta$ for different suppression R_{Ar} (figers on the curves); b) predictions of $R_{\nu e}$ in different solutions: A - adiabatic, $NA_{2,3}$ - nonadiabatic in 2ν - and 3ν - cases, $M_{2,3}$ - mixed in 2ν - and 3ν - cases, T_c - low central temperature of the Sun, V -, $\bar{V}$ - nonaveraged and averaged vacuum oscillations, R - regeneration, D - ν - decay (see [10])

KAMIOKANDE - II result: $R_{\nu e}(E_e > 9.5$ MeV$) < 0.53$ (90% c.l.) [6] is not yet crucial for choice of definite solution of solar neutrino problem (Fig.7). But it supports Davis's data and moreover the obtained limit permits to exclude some region of Δm^2 vs. $\sin^2 2\theta$ parameters in nonadiabatic solution with $R_{Ar} > 1/3$ (Fig.7a).

Futher improvement of KAMIOKANDE - II result will give the following discrimination of solutions (Fig.7b).

1) If $R_{\nu e} = 0.5 - 0.7$, then 2ν - and 3ν - nonadiabatic, 2ν - and 3ν - mixed, quasivacuum and ν- decay solutions survive.

2) If $R_{\nu e} = 0.25 - 0.5$ then nothing can be excluded.

3) If $R_{\nu e} = 0.15 - 0.25$ then the following solutions survive: 2ν - and 3ν - mixed, adiabatic and low T_c

4) If $R_{\nu e} < 0.15$, then the flavour oscillations are excluded, and 2ν - as well as adiabatic oscillations in sterile neutrinos remain only.

6. Possible Observation of ν- Burst Related to SN1987A

Resonant oscillations in matter of star as well as of the Earth may change the properties of ν- fluxes from gravitational collapse [2, 15, 18, 19]. The effects depend on channel of oscillations.

1) Suppression of $\bar{\nu}_e$ (or ν_e) signal due to oscillations in sterile states. The suppression factor for the energy in ν_α- neutrinos is

$$R_\alpha(\Delta m^2, \sin^2 2\theta) = (\mathcal{E}^o)^{-1} \int dt \int dE \sum_\beta F^o_\beta \cdot E \cdot P_{\beta-\alpha} , \quad (16)$$

where $\mathcal{E}^o_\alpha = \int dt \int dE\, F^o_\alpha \cdot E$, $F^o(E)$ is the generation flux of ν_β, $P_{\beta-\alpha}$ is the probability of $\nu_\beta - \nu_\alpha$ transition. For $\bar{\nu}_e - \bar{\nu}_s$ the sum in (16) equals to $F^o \cdot E \cdot P_{e-e}$. Using profiles (Fig.2) and Planck distribution with high energy cut off for $F^o(E)$ we have calculated lines of equal suppression: $R_e(\Delta m^2, \sin^2 2\theta) = a =$ const (Fig.8a). The appearence of two regions is related to that the effective density for $\nu_e - \nu_s$ - oscillations [2, 20]

$$\rho^{eff} = \rho(Y_p - Y_n/2)$$

changes sign in the region of strong neutronization. Here Y_p and Y_n are the numbers of protons and neutrons per nucleon. (It is effective density enters resonant condition). In the central part of star ($Y_n > 2Y_p$) resonant condition is fulfilled for $\Delta m^2 < 0$, in external layers ($Y_n < 2Y_p$) resonance appears at $\Delta m^2 > 0$. For $\Delta m^2 = (10^4 - 10^8)$ eV^2 (resonance in the core of star) the averaging due to inelastic collisions is essential. This results in copious sterile states generation and $R \to 1/2$.

The observable signature of $\mathcal{E}_{\bar{\nu}}$- suppression is the appreciable deviation of R^{exp}

$$R^{exp} = \begin{cases} 6\, \mathcal{E}^{exp}_{\bar{\nu}} / \mathcal{E}_{GR} \\ \mathcal{E}^{exp}_{\bar{\nu}} / \mathcal{E}_T = \mathcal{E}^{exp}_{\nu} / (7/16 \cdot \sigma \cdot T^4_\nu \cdot 4\pi R^2_\nu) \\ \min(\mathcal{E}^{exp}_{\bar{\nu}} / \mathcal{E}^{exp}_{\nu};\ \mathcal{E}^{exp}_{\nu} / \mathcal{E}^{exp}_{\bar{\nu}}) \end{cases} \quad (17)$$

from the unit. Here $\mathcal{E}^{exp}_{\bar{\nu}}$ ($\mathcal{E}^{exp}_{\nu}$) is experimentally observed energy of $\bar{\nu}$ (ν), $\mathcal{E}_{GR}$ is the maximal gravitational energy released in neutrinos (apparently $\mathcal{E}_{GR} \simeq (2-5)\, 10^{54}$erg), T_ν is the temperature of $\bar{\nu}_e$ signal, R_ν is the effective radius of neutrinosphere.

If underground events registrated in February 23 are caused by $\bar{\nu}_e$ from gravitational collapse, then according to criteria (17) there is no strong suppression of $\bar{\nu}_e$ signal. In any case one has $R^{exp} > 1/10$. Consequently Δm^2 vs. $\sin^2 2\theta$ region restricted by the line $a = 0.1$ (Fig.8a) should be excluded. Due to uncertainty in distribution $\rho(r)$ the edges of excluded region may be shifted by order of magnitude on Δm^2 vs. $\sin^2 2\theta$ plot.

Another method of searching for $\bar{\nu}_e - \bar{\nu}_s$ - oscillations is the comparision of the energy of neutronization peak ($\mathcal{E}^N$) and isotropical $\bar{\nu}_e$ signal. Signature is the deviation of ratio

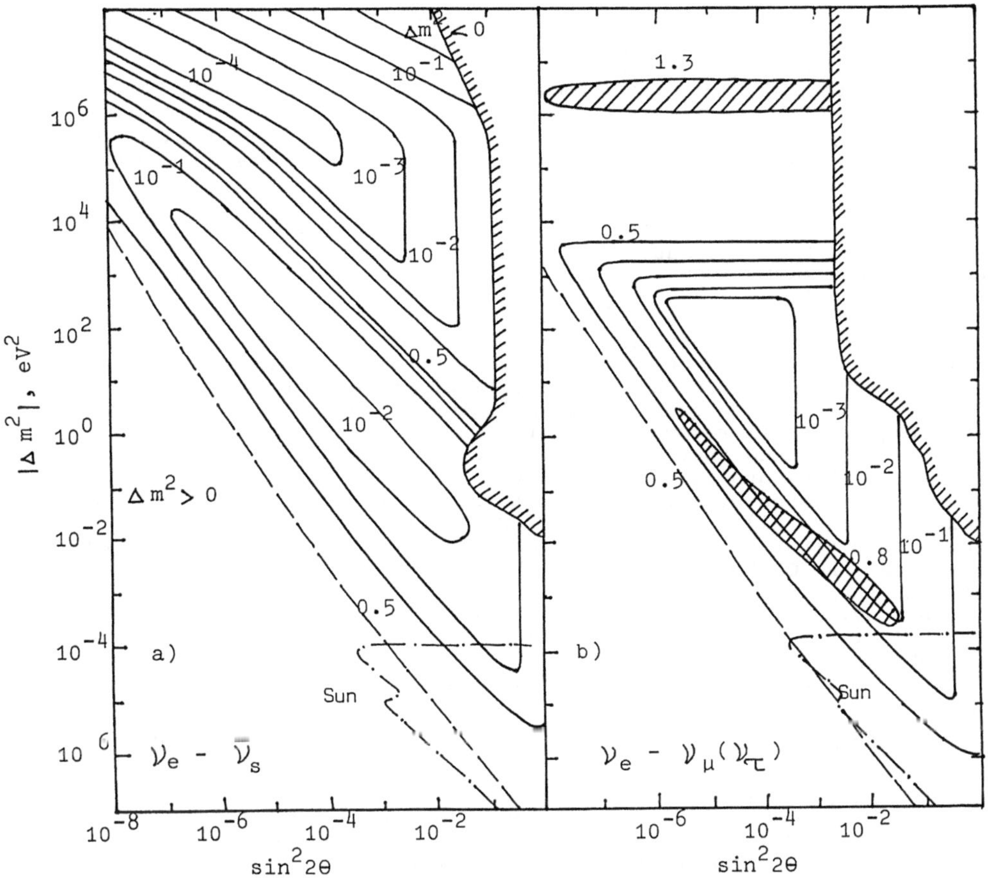

Fig.8 a) Lines of equal suppression (figers on the curves) of $\bar{\nu}_e$ - burst energy due to $\bar{\nu}_e - \bar{\nu}_s$ oscillations (stage of ν - opacity); b) lines of equal suppression of neutronization peak energy due to $\nu_e - \nu_\mu$ (ν_τ) oscillations: solid line corresponds to the profile 1 and dashed line does for profile 2 on Fig.2; the regions of nonsymmetric $\bar{\nu}_e \leftrightarrow \bar{\nu}_\mu$ spectra exchange on ν - opacity stage are shaded

$$\mathcal{E}^N/6\mathcal{E} \qquad (18)$$

from 0.1, predicted in standard picture. If two first KAMIOKANDE - II events are due to neutronization neutrinos, then

$$\mathcal{E}^N/6\mathcal{E}_{\bar{\nu}} > 0.3 - 1$$

and this can be considered as an indication on $\bar{\nu}_e - \bar{\nu}_s$ - oscillations with $R_{\bar{\nu}} = 1/3 - 1/10$.

Note that at Δm^2 vs. $\sin^2 2\theta$ near the excluded region $\bar{\nu}_e - \bar{\nu}_s$ - oscillations may result in distortion of $\bar{\nu}_e$ - spectrum.

2) Interchange of $\bar{\nu}_e$ and $\bar{\nu}_\mu$ ($\bar{\nu}_\tau$) spectra due to flavour oscillations. On the stage of ν - opacity comparable fluxes of $\bar{\nu}_e$, $\bar{\nu}_\mu$, $\bar{\nu}_\tau$ are emitted: $F^o(\bar{\nu}_\mu) = 1/2\ F^o(\bar{\nu}_e)$; $E(\bar{\nu}_\mu) = 2\ E(\bar{\nu}_e)$. If both $\bar{\nu}_e$ and $\bar{\nu}_\mu$ (ν_τ) spectra are on the bottom of the "bath" (see sect.2), then symmetric exchange of $\bar{\nu}_e - \bar{\nu}_\mu$ takes place: $R_{\bar{\nu}} = 1$. But due to $\sigma \sim E^2$ the number of $\bar{\nu}_e$ events in installations rises. On the edge of the bathes spectrum exchange is nonsymmetric and $R_{\bar{\nu}} \neq 1$ (Fig.8b) $\bar{\nu}_e \to \bar{\nu}_\mu$ dominates over $\bar{\nu}_\mu \to \bar{\nu}_e$ (or vice versa).

The signatures are i) appreciable difference in temperatures of ν_e and $\bar{\nu}_e$ signals, ii) the distortion of ν_e (or $\bar{\nu}_e$) spectra. SN1987A data are poor to give any conclusion.

3) Suppression of ν_e - peak from neutronization [18, 21,20]. The sum in (16) is $F_e^o \cdot E \cdot P_{e-e}$. Using E_ν = 20MeV and density profile for progenitor (see Fig.2) we have calculated the line of equal suppression of energy of ν_e - peak (Fig.8). (To calculate the suppression of number of νe - νe events in installations one should take into account ν_μ (ν_τ) - e - scattering for the case of flavour oscillations).

The signature here is the strong deviation of experimental $\mathcal{E}^N$ from that, evaluated from the mass of star. Another criterion is given in (18).

If two first KAMIOKANDE - II events are due to neutronization neutrinos, then there is no $\mathcal{E}^N$ - suppression. At least $R^N > 1/2$ and Δm^2 vs. $\sin^2 2\theta$ region restricted by line a = 1/2 on Fig.8b should be excluded.

The excluded region contains: positive BUGEY, PS191, BNL results, moreover it contains region of sensitivety of all planning oscillations experiments. Depending on density profile of progenitor it contains the region of adiabatic and quasivacuum solutions of solar neutrino problem or the region of all resonant oscillations solutions.

It should be noted that the considered interpretation of two first KAMIOKANDE - II events is faced with the following difficulties. i) The energy $\mathcal{E}^N$ evaluated by these events turns out to be much larger than the expected one. ii) Main argument for the interpretation is the correlation of two first events with the direction on LMC. But it was stressed that another KAMIOKANDE - II events show similar correlation (in contrast with the expected isotropy). IMB events have anisotropy too.

4) Distinctions of ν - signal on different installations due to Earth matter effect [15, 22, 21]. Distinctions in energy spectra and in total fluxes are expected.

If both KAMIOKANDE - II and IMB events are due to $\bar{\nu}_e$ from collapse then

$$T^{IMB} > T^{KAM}.$$

this may be considered as indication on $\bar{\nu}_\mu$ ($\bar{\nu}_\tau$) - $\bar{\nu}_e$ transition in the Earth. At 7h 35m UT SN1987A was at the following zenith angles 110° - KAMIOKANDE - II, 146° - Baksan and 132° - IMB. If the spectrum of $\bar{\nu}_\mu$ ($\bar{\nu}_\tau$) is placed on resonant peak (Fig.6) then $\bar{\nu}_\mu$ ($\bar{\nu}_\tau$) will convert into $\bar{\nu}_e$ and for IMB this effect is more stronger than for KAMIOKANDE - II.

7. Conclusion

Now resonant neutrino oscillations experiments give the following results.

1) If the suppression 1/4 - 1/2 of ^{37}Ar production rate in Cl - Ar experiment is due to resonant oscillations, then at least one resonance lies between 2ν - ISOSNU lines a = 0.25 and a = 0.7.

2) The consideration of season variations of ^{37}Ar production rate permits to exclude the region of Δm^2 vs. $\sin^2 2\theta$ parameters, where $R_{Ar} < 0.25$.

3) KAMIOKANDE - II limit on the flux of ^{8}B - neutrinos is not yet crucial for choice of definite solution of solar neutrino problem, but it exclude some region of Δm^2, $\sin^2 2\theta$ - parameters in nonadiabatic solution.

4) IMB analysis of atmospheric neutrino data permits to exclude the region of Δm^2 vs. $\sin^2 2\theta$ - parameters in adiabatic solution [7].

5) If the underground events in February 23 are due to $\bar{\nu}_e$ from gravitational collapse in LMC, then large region of Δm^2 vs. $\sin^2 2\theta$ parameters for $\bar{\nu}_e$ - $\bar{\nu}_s$ channel (ν_s is sterile state) is excluded.

6) If two first KAMIOKANDE - II events are due to ν_e from neutronization which is extremely problematic then very large region of Δm^2 vs. $\sin^2 2\theta$ for flavour oscillations can be excluded. In particular, depending on density profile of progenitor one can exclude

- all Δm^2 vs. $\sin^2 2\theta$ region of solutions of solar neutrino problem,
- part of solutions region,
- nothing in solar resonant region.

8. References

1. L. Wolfenstein: Phys. Rev. D17, 2369 (1978); ibidem D20, 2634 (1979)
2. S.P. Mikheyev and A.Yu. Smirnov: Yadernaya Fizika, 42, 1441 (1985) and ZhETF 91, 7 (1986)
3. S.P. Mikheyev and A.Yu. Smirnov: Nuovo Cim. 9C, 17 (1986)
4. R. Davis Jr.: Proc. of the 7th Workshop on Grand Unification/ICOBAN'86, ed. by J. Arafune, (Toyama, Japan 1986), p. 237
5. M. Cribier et al.: Phys. Lett. 182B, 89 (1986); M.L. Cherry, R. Davis Jr., and K. Lande: Proc. of 20th ICRC, ed. by A.E. Chudakov, (Moscow, USSR 1987) v. 4, p. 332
6. K. Hirata et al.: Preprint UT-ICEPP-87-04, UPR-0144E (1987)
7. J. Lo Secco et al.: Phys. Lett. 184B, 305 (1987)
8. K. Hirata et al. (KAMIOKANDE - II): Phys. Rev. Lett. 58, 1490 (1987); R.M. Bionta et al. (IMB): Phys. Rev. Lett. 58, 1494 (1987); V.L. Dadykin et al. (LSD): Pis'ma ZhETF 45, 464 (1987); E.N. Alexeyev et al. (BAKSAN): Pis'ma ZhETF 45, 461 (1987)
9. S.J. Parke and T.P. Walker: Phys. Rev. Lett. 57, 2322 (1986)
10. S.P. Mikheyev and A.Yu. Smirnov: Usp. Fiz. Nauk 153, 3 (1987) and references therein
11. T.K. Kuo and J. Pantaleone: Phys. Rev. Lett. 57, 1805 (1986); S. Toshev: Phys. Lett. 185B, 177, (1987)
12. T.K. Kuo and J. Pantaleone: Phys. Rev. D35, 3432 (1987)
13. A.Yu. Smirnov: Yadernaya Fizika 46, 1152 (1987)
14. S.P. Mikheyev and A.Yu. Smirnov: Submitted to Phys. Lett. B
15. S.P. Mikheyev and A.Yu. Smirnov: Proc. of 6th Moriond Workshop, ed. by O. Fackler and J. Tran Thanh Van (Tignes, France 1986), p. 355
16. M.B. Voloshin, M.I.Vysotsky and L.B. Okun: ZhETF, 91, 754 (1986) and Yadernaya Fizika 44, 677 (1986)
17. J.N. Bahcall, J.M. Gelb and S.P. Rosen: Phys. Rev. D35, 2976 (1987)
18. S.P. Mikheyev and A.Yu. Smirnov: Proc. of the 12th Int. Conf. NEUTRINO'86, ed. by T.Kitagaki and H.Yuta (Sendai, Japan 1986), p. 177
19. T.P. Walker and D.N. Schramm: Phys. Lett., 195B, 331 (1987)
20. S.P. Mikheyev and A.Yu. Smirnov: Pis'ma ZhETF, 46, 11 (1987)
21. J. Arafune et al.: Phys. Rev. Lett., 59, 1867 (1987); H.Minakata et al.: Tokyo Univ. Preprint TMUP-HEL-87-03 (1987); T.K. Kuo and J.Pantaleone: Preprint PURD-TH-87-08 (1987); S.P. Rosen: Los Alamos preprint April 1987; D. Nötzold: Phys. Lett. 196B, 315(1987)
22. P.O. Lagage et al.: Phys. Lett. 193B, 127 (1987)

Weak Interactions and Solar Neutrinos: Testing the Oscillation Hypothesis

J.W.F. Valle

Dept. de Física Teòrica, Universitat de València,
E-46100 Burjassot, València, Spain

Various proposals for understanding the reduced solar neutrino flux in terms of non-standard neutrino propagation properties are briefly reviewed in the context of weak interaction theory. Emphasis is given to neutrino oscillation models which may be tested experimentally, despite the small mass parameter involved. Spontaneous lepton number (and R parity) breaking in supergravity - the supersymmetric doublet Majoron model - solves the solar neutrino problem through matter-enhanced neutrino oscillations. The spectrum of supersymmetric particles is restricted in a way that will be probed by high energy collider experiments. In addition, low energy processes associated with the existence of the Majoron may signal the existence of an otherwise undetectably small neutrino mass.

1. Introduction

The discrepancy between the theoretical expectations of how the sun "shines" neutrinos and attempts to measure (some of) these neutrinos via capture in a chlorine detector by Davis and collaborators, has for long been a puzzle [1]. This solar neutrino problem may be resolved either by postulating new physics that modifies the solar parameters [2] or by assuming that non-standard weak interactions exist which modify neutrino propagation properties. The physics leading to the second type of solution includes neutrino decay [3] neutrino magnetic moment [4] and neutrino oscillations [5, 6, 7]. While it is possible to build consistent models for fast neutrino decay [8] the required value of the neutrino magnetic moment needed to solve the solar neutrino problem is inconsistent, in most models, with existing limits on neutrino mass. Neutrino decay has also been rendered unlikely, however, by the recent observation of neutrinos from the supernova [9] although it has been argued that the model may still survive if large mixings are allowed [10]. This leaves in any event neutrino oscillations as the most likely explanation of the solar neutrino problem in terms of non-standard neutrino properties.

Can the oscillation hypothesis by tested experimentally? Here one seems to be in a difficult position. If oscillations occur "in vacuo" then adequate suppression in the solar neutrino flux indicates values for the relevant mass squared differences that are too small to be probed [6].

A lot of interest has been recently devoted to the possibility of matter-enhanced oscillations [5]. These are possible due to the effect of coherent neutrino scattering in the solar medium even for values of the vacuum mixing angles which are very small. This resonant enhancement is possible, however, only if the neutrino mass difference lies in the range

$$10^{-7} \lesssim \left(\frac{\delta m}{ev}\right)^2 \lesssim 10^{-4}. \tag{1}$$

Detailed analysis of the range of parameters required for the resonant amplifiction of neutrino oscillations to take place has been given by many authors, see eg. ref.[11].

Again the oscillation hypothesis indicates mass diferences that are too small to be tested in existing experimental setups. Although effort will be pushed in this direction [12], it will be a number of years before one will be able to start probing the range specified in eq. (1). From this point of view it is of interest to ask what are the physics prospects for having an experimentally testable solution to the solar neutrino puzzle in terms of neutrino oscillations. Moreover one would like to know the origin for the relevant neutrino mass scale.

Introducing a small mass for neutrinos may be done in variety of ways eg. by adding heavy singlet fermions to the theory such as right handed neutrinos, at a mass scale **above** that of electroweak symmetry breaking. This could, for example, be the grand unification scale in SO(10) models [13] or some intermediate scale in superstring inspired E_6 models [14,15,16]. Alternatively in a theory without right handed neutrinos there may be new physics **below** the weak interaction scale. In such a theory the basic neutrino mass term is of Majorana type [17] and thus violates total lepton number by two units, $\Delta L=2$. If the fundamental source of L violation is spontaneous and $\Delta L=1$, the neutrino mass generated this way must have the form $m \propto v^2/M$ where v is the lepton number breaking vacuum expectation value (vev) and M denotes some mass scale associated with a lepton number blind sector. This situation occurs in supergravity, in which case the heavy states necessary in the see-saw mechanism are just the supersymmetric partners of gauge and Higgs fields (which should be at the Fermi scale) and the tiny neutrino mass is then related to new dynamics below the Fermi scale (i.e. to the Majoron, see below).

There are two generic types of models where the solar neutrino oscillation hypothesis may be experimentally checked. First it can be checked "kinematically" in models where a typical neutrino mass m is much larger than a typical neutrino mass difference, δm,

$$m \gg \delta m \tag{2}$$

One way to model this hierarchical difference in the values of m and δm is to attribute it to the presence of a large intermediate scale in superstring inspired models [15,16]. The electron neutrino is just one Weyl component of a four-component Quasi-Dirac neutrino [18], the other component may be another neutrino flavour or, as in ref.[7] can be a **sterile** neutrino. In this case the neutrino can be massive enough as to affect the endpoint beta spectrum in tritium decay experiments and also to play a role in providing the dark matter of the universe, without conflicting any experimental limit. The solar neutrino problem can then be solved by large vacuum oscillations from the active to the sterile component as neutrinos propagate from sun to earth.

Here I focus on the alternative possibility where one has

$$m = O(\delta m) \tag{3}$$

so that although neutrino mass effects will not be detectable directly, the theory contains some new "dynamical" degree of freedom whose "large" effects can be probed and used, in a sense, to "track" an otherwise undectably small mass. The prototype of this situation is when B-L is a true symmetry of the Lagrangian, broken only spontaneously. If ungauged [19], as in the standard model, spontaneous lepton number breaking generates a Goldstone boson - a Majoron - which we denote J [20,21]. The Majoron is a true dynamical degree of freedom so it has interactions which are related in a well defined way with the neutrino mass. Majoron emision generates new mechanisms of stellar energy loss [22]. Due to its flavour non-diagonal couplings it will also produce small changes in the decay parameters of the μ and τ lepton which could be seen in precision measurements [23]. The smallness of the neutrino mass, required by the resonance condition, eq. (1), is then dictated by constraints on the couplings of the Majoron (which involve v) that follow from a variety of considerations thus giving a dynamical basis for the see-saw mechanism [24].

2. The Supersymmetric Doublet Majoron

The Supersymmetric Doublet Majoron Model is described by the minimal supergravity superpotential

$$h_{ij}^u u_i^c Q_j H_u + h_{ij}^d d_i^c Q_j H_d + h_{ij}^e e_i^c \ell_j H_d + \mu H_u H_d \tag{4}$$

where the parameter μ is related with electroweak breaking driven by radiative corrections associated with the top quark. The first three terms give rise to masses for up and down-type quarks, and charged leptons, respectively, once the two Higgs fields H_u and H_d acquire their vevs v_u and v_d. Supersymmetry and lepton flavour are broken explicitly in the scalar potential via soft scalar mass terms and possibly cubic scalar self couplings, and also via SU(3)xSU(2)xU(1) invariant gaugino mass terms M_i , i=1,2,3. Gaugino masses break the continuous R invariance of the theory down to a discrete symetry, called R parity. R parity is **even** for all particles of the standard model (including the Higgs scalars) and **odd** for their supersymmetric partners. R parity too may be broken, either explicitly [25],or spontaneously by nonzero vevs for the scalar neutrinos [26] ,

$$v_i = \left\langle \ell_i^0 \right\rangle; \quad i = e, \mu, \tau. \tag{5}$$

The conditions for spontaneous R parity breaking are very restrictive. Recent analysis [27] indicates that sleptons lighter than about 65 GeV and a top quark heavier than about 70 GeV are required. This breaking may be far easier to achieve if one adds terms that break total lepton number explicitly as well. Spontaneously breaking of B-L, an ungauged continuous symmetry, generates a Nambu-Goldstone boson - a Majoron - given in ref.[24]. This also gives rise to a neutrino mass

$$m = \frac{\mu M \sum_i v_i^2}{2 v_u v_d M - M_1 M_2 \mu} \tag{6}$$

where

$$2M = g_1^2 M_2 + g_2^2 M_1$$

and g's are gauge coupling constants. Note that since B-L is broken by *one* unit via the scalar neutrino vev, eq.(5), it takes *two* such breakings to generate a (Majorana) mass for the (left handed) neutrino: hence the square in the see-saw formula, eq.(6). This contrasts with the non-supersymmetric Majoron model where a scalar Higgs triplet is added to generate neutrino masses directly and therefore linear in the lepton number breaking vev. In supersymmetry no triplet needs to be added and then the direct mass term vanishes (therefore the standard model ρ paramenter is identically one, at tree level). It is amusing to note that triplets are absent in superstring inspired models [14,15,16] . The full tree level neutral lepton mass matrix was given in ref.[24] and radiative corrections [28] are negligible for our purposes. Another striking difference makes the supersymmetric doublet Majoron model much more restrictive than the triplet Majoron: the fact that, to a good approximation, **one and only one** neutrino acquires mass, namely, that neutrino which is related by supersymmetry to the Majoron. Its mass is given by eq.(6). This simplifies considerably the structure of the CC weak interaction reducing the parameters relevant for describing the resonant neutrino oscillations in the sun to just **three** parameters: two mixing angles (the third angle is not a physical parameter, due to the mass degeneracy between the two massless neutrinos) and one neutrino mass parameter, m. Moreover, CP is conserved in the leptonic charged current. The matrix K describing the CC weak interaction, can be put in the canonical form [23,24]

$$K = \omega_{23}(\theta_{23})\omega_{13}(\theta_{13}) \tag{7}$$

where, for example, ω_{13} is a rotation by an angle θ_{13} in the 13 plane, etc

The mixing angles in eq.(7) are given in terms of the lepton number breaking vevs as

$$s_{13}=v_e/[v_e^2+v_\mu^2+v_\tau^2]^{1/2}$$

$$s_{23}=v_\mu/[v_\mu^2+v_\tau^2]^{1/2}$$

If electron neutrinos propagate through matter, then a small value of s_{13} can be resonantly amplified, due to the effect of the charged current. The value of s_{23} determines the channel where resonant conversion takes place. For $v_\tau \gg v_\mu \gg v_e$ both angles are all small so that the massive state is mostly ν_τ and resonant solar neutrino conversions occur from $\boldsymbol{\nu_e}$ into $\boldsymbol{\nu_\tau}$ [23,24]. For definiteness I assume that both angles are small. Because of the drastic simplification discussed above there are interesting experimental signatures that make the supersymmetric doublet Majoron model testable. Figure 1 shows how the gaugino-Higgsino mass spectrum is restricted by Davis´ experimental results.

The region where resonant solar neutrino oscillations will reduce the solar neutrino flux below 2.6 SNU is the region between the curves A and B which correspond to $\sin^2 2\theta_{13}= 0.2$ and v= 10KeV. Region C is

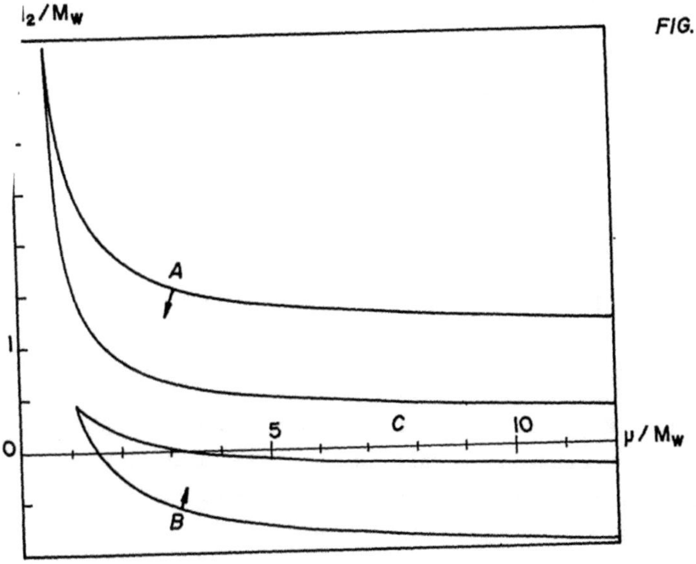

FIG. I

excluded by the nonobservation of *charginos* (supersymmetric partners of charged gauge and Higgs bosons) at PETRA. The supersymmetry breaking gaugino mass parameter M_2 and the Higgs mixing parameter μ are measured in units of the W mass. Also allowed are points where both M_2 and μ change sign. The allowed range of parameters where resonant amplification of solar neutrino oscillations can occur may be directly explored by searching for charged supersymmetric partners of gauge and Higgs particles, say, in electron-positron machines [24]. I now concentrate on additional tests of the model based on the dynamics of the Majoron and which involve low energy precision measurements [23].

3. Testing the Oscillation Hypothesis through Majoron Dynamics

The Majoron couples to quarks through its Higgs doublet admixture via the first two terms in eq.(4). These couplings are always flavour diagonal. It couples to charged leptons in the same way, via the third term in eq.(4). Thus in this model stellar energy loss will proceed via **single** Majoron emission in Compton-like process $\gamma+e\to e+J$. From the cross section given in ref.[22] we have

$$\frac{4}{\pi}G_F^2 m_e^2 v^2 a^2 \lesssim 3\times 10^{-26} \qquad (8)$$

where

$$a=\frac{1}{2}+fg_{ee}$$

which gives

$$va \lesssim 30 \text{ KeV}. \tag{9}$$

In addition, energy loss also proceeds via **double** Majoron emission in $\gamma+e \to e+J+J$. This process occurs with normal gauge strength couplings but is mediated by the heavy *charginos*. We have calculated the total cross section for this process,

$$\sigma = \frac{\alpha^3}{60 \sin^4\theta_W} \frac{E_\gamma^2}{M_W^4} f^2 g_{ee}^2 \tag{10}$$

in the approximation of large photon energies $E_\gamma >> m_e$. From this we obtain the additional limit

$$f^2 g_{ee}^2 \lesssim 10. \tag{11}$$

In the supersymmetric doublet Majoron model the Majoron also couples to charged leptons of *different* flavour. This occurs via mass mixing between the *R-even* leptons (e, μ, τ) and the *R-odd* fermion states . This is described by the mass matrix

	e_j^+	$\widetilde{H}_u^+$	$\widetilde{W}^+$
e_i^-	$h_{ij}v_d$	0	$g_2 v_i$
$\widetilde{H}_d^-$	$h_{ji}v_i$	μ	$g_2 v_d$
$\widetilde{W}^-$	0	$g_2 v_u$	M_2

(12)

Since fermions of different weak isospin are mixed in eq. (12) it follows that in this model the Majoron couplings to charged leptons is not flavour diagonal. *This is another important difference between the supersymmetric doublet Majoron model and the triplet Majoron model and which as we will see, may be tested experimentally.* The coupling of the Majoron to physical charged leptons is described by the effective interaction Lagrangian,

$$i \frac{g_2^2}{\sqrt{2}} \frac{v m_j}{m_W^2} \left(\frac{1}{2}\delta_{ij} + f g_{ij} \right) \overline{e}_{Li} e_{Rj} J + \text{H.C.} \tag{13}$$

where the function f is given by

$$f = \frac{1+x^2}{(1-xy)^2}$$

and, for simplicity, we have assumed $v_u \simeq v_d$. In eq. (13) x denotes the Higgs mixing parameter μ and y denotes the supersymmetry breaking gaugino mass parameter M_2, in units of the W mass. The coupling matrix g is a projection matrix, and may be written as

$$g_{ij} = (KPK^\dagger)_{ij} \tag{14}$$

where P is the matrix P=diag(0,0,1) and K is given in eq. (7).

How about limits on these **flavour non-diagonal** Majoron couplings? This leads us to study the decays $e_j \to e_i + J$, whose existence is predicted in the supersymmetric doublet Majoron model as a result of mixing among charged fermions of different weak isospin. The branching ratio for $e_j \to e_i + J$ relative to $e_j \to e_i + \nu + \nu$ is given by

$$B_{ij} = 96\pi^2 \left(\frac{v}{m_j}\right)^2 f^2 g_{ij}^2. \tag{15}$$

For the case of μ decay the branching ratio may be as large as

$$B_{e\mu} \approx 3 \times 10^{-7} \times (\frac{v}{30Kev})^2 \tag{16}$$

thus suggestively close to the present experimental limit of TRIUMF [29]. For τ decays, however, the corresponding allowed branching ratios are far below what can be experimentally probed.

In addition, **double** Majoron emission in μ and τ decays is possible in the present model thus implying a change in the charged leptons decay parameters [30]. The amplitude for this process is [31]

$$T_{ij} \approx -2\sqrt{2} G_F f g_{ij} \overline{u}(p_i)(\not{p}_i + \not{p}_j) L u(p_j) \tag{17}$$

where $P_{i,j}$ is the $e_{i,j}$ momentum. Requiring the fractional change in the Michel parameter for each case to be within the accuracy of present experimental determination gives, for the case of μ decay, the limit

$$f^2 g_{e\mu}^2 \lesssim 10^{-2} \tag{18}$$

while for the corresponding τ decays the limits are

$$f^2 g_{e\tau}^2 \lesssim .35 \tag{19}$$

and a similar bound on $f g_{\mu\tau}$. These bounds pose non-trivial constraints on the parameters of our model and from this point of view precision measurements of μ and τ decays are theoretically interesting.

In summary we find most encouraging that effects of Majoron emission in μ and τ decays can be at the cutting edge of experimental test! The tiny neutrino mass needed for the MSW effect is accompanied by "large" dynamical effects associated with the existence of the Majoron thus providing a dynamical testing ground for the model. Interestingly, there can be large observable effects for parameter values which solve the solar neutrino problem by the enhanced oscillation effect. Under reasonable assumptions, mixing angles θ_{23} and θ_{13} are small and in this case resonant solar neutrino conversions occur, in the adiabatic regime, from ν_e to ν_τ. A reduction

in solar neutrino flux below 2.6 SNU implies severe restrictions on the supersymmetric spectrum (Fig.1) and potentially large effects in μ and τ decays [23]. It is possible to obtain values for the parameters in our model which change the τ decay Michel parameter by an amount about a few times larger than present experimental limit allows and a BR for $\mu \rightarrow e+J$ of 10^{-7} or so [23]. In addition, for these parameter values, the lightest *chargino* weighs less than half of the Z mass and thus might be seen at accelerator experiments. A more complete analysis will be presented elsewhere [32]. Finally we note that slightly stronger constraints on the lepton number breaking vev have been recently suggested from the analysis of helium ignition in stars and from the recent neutrinos from SN1987 [33].

This talk is largely based on work done in collaboration with A Santamaria with finantial support from CAYCIT (Spain) under Plan Movilizador de Fisica de Altas Energias and CSIC-IFIC. I also acknowledge support from United States Department of Energy, under contract DE-AC03-76SF00098 at the Lawrence Berkeley Laboratory and from National Science Foundation at the Aspen Center for Physics.

References

1 J N Bahcall, Rev Mod Phys 50, 881 (1978)

2 G Gelmini, L Hall, M Lin, Nucl Phys B281 (1987) 726.

3 J N Bahcall, S T Petcov, S Toshev, J W F Valle, Phys Lett B 181 (1986) 369

4 L B Okun, M Voloshin, M Vysotsky, ITEP preprint 86-82, 1986

5 P Mikheyev, A Yu Smirnov, Nuovo Cimento Soc. Ital. Fis. 9C, 17 (1986), H Bethe, Phys Rev Lett 56, 1305 (1986), L Wolfenstein, Phys Rev D17, 2369 (1978)

6 S Glashow, L Krauss, Phys Lett B190 (1987) 199; L Krauss, F Wilczek, Phys Rev Lett 55 (1985) 122

7 R N Mohapatra, J W F Valle, Phys Lett B177, 47 (1986).

8 J W F Valle, Phys Lett B131 (1983) 87; G Gelmini, J W F Valle, Phys Lett B 142 (1984) 181; G Gelmini, D Schramm, J W F Valle, Phys Lett B146 (1984) 311; J W F Valle, Phys Lett B159 (1985) 49; S Nussinov, Phys Lett B185 (1987) 171.

9 K Hirata et al Phys Rev Lett 58 (1987) 1490; R Bionta et al, Phys Rev Lett 58 (1987) 1494

10 J Freeman, H Haber, K Freeze, SLAC PUB 4261 (1987); see also J N Bahcall, A Dar, A Piran, Nature 326, 135 (1987); S Pakvasa, R Raghavan, Hawaii preprint, 1987.

11 M. Cribier et al Phys Lett B182 (1987) 89; Z Phys C32 (1986) 499; W Haxton, Phys Rev Lett 57 (1986) 1271; S P Rosen, J Gelb, Phys Rev D34 (1986) 969; V. Barger et al Phys Rev D34 (1986) 980

12 F. Boehm, private communication

13 M Gell-Mann, P Ramond, R Slansky in Supergravity ed D Freedman (North Holland, 1979), T Yanagida in Workshop on Unified Theory and Baryon Number of the Universe ed O Sawada et al, KEK, 1979, R Mohapatra, G Senjanovic, Phys Rev Lett 44, 912 (1980). For recent discussions in the context of the MSW effect, see P Langacker, S T Petcov, G Steigman, S Toshev, Nucl Phys B282 (1987) 589; R. Johnson, S. Ranfone, J Schechter, Phys Lett B179 (1986) 355.

14 J W F Valle, Phys Lett B186 (1987) 73

15 R Mohapatra, J W F Valle, Phys Rev D34 (1986) 1642; J P Derendinger et al, Nucl Phys B267 (1986) 365; S Nandi, U Sarkar, Phys Rev Lett 55 (1986) 566; F del Aguila et al, Nucl Phys B272 (1986) 413

16 J W F Valle in International Symposium on Weak and Electromagnetic Interactions in Nuclei ed H Klapdor, Heidelberg, 1986, Springer-Verlag, p 927

17 J Schechter, J W F Valle, Phys Rev D22, 2227 (1980)

18 J W F Valle, Phys Rev D27 (1983) 1672; J W F Valle, M Singer, Phys Rev D28 (1983) 540; S Petcov, Phys Lett B110 (1982) 245; L Wolfeinstein, Nucl Phys B186 (1981) 147

19 If the gauge group contains B-L there is a new gauge boson, coupled to neutrinos. See J W F Valle, Phys Lett B196 (1987) 157.

20 Y Chikashige, R Mohapatra, R Peccei, Phys. Lett. 98B, 265 (1980)

21 G Gelmini, M Rocandelli, Phys Lett B99 (1981) 411

22 H Georgi, S Glashow, S Nussinov, Nucl Phys B193 (1981) 297; J Ellis, K Olive, Nucl Phys B223 (1983) 252 and references therein

23 A Santamaria, J W F Valle, preprint, LBL 23896, FTUV-7/87, 1987; to appear in Phys Rev Lett.

24 A Santamaria, J W F Valle, Phys Lett B195 (1987) 423

25 L Hall, M Suzuki, Nucl Phys B231 (1984) 419; I H Lee, Nucl Phys B246 (1984) 120.

26 R Mohapatra, C Aulakh, Phys Lett B119 (1983) 136; G G Ross, J W F Valle, Phys Lett B151 (1985) 375; J Ellis, G Gelmini, C Jarlskog, G G Ross, J W F Valle, Phys Lett B150 (1985) 142; S Dawson, Nucl Phys B261 (1985) 297.

27 A Bouquet, P Salati, Nucl Phys B284 (1987) 557, B Gato et al, Nucl Phys B260 (1985) 203

28 R N Mohapatra, C Aulakh, Phys Lett B 121 (1983) 147; J F Nieves, Phys Lett B 137 (1984) 67

29 A Jodidio et al, Phys Rev D34 (1986) 1967

30 G Kane, W Rolnick Nucl Phys B232 (1984) 21; W Buchmuller, F Scheck, Phys Lett B 145 (1984) 421

31 A Santamaria, J Bernabéu, A Pich, Phys Rev D36 (1987) 1408.

32 A Santamaria, J W F Valle, in preparation.

33 D Dearborn, D Schramm, G Steigman, Phys Rev Lett 56 (1986) 26; J Ellis, K Olive, Phys Lett B193 (1987) 525

Is the Detection of Cosmic Background Neutrinos Possible?

B. Müller

Institut für Theoretische Physik, Johann-Wolfgang-Goethe Universität, P.O.Box 111932, D-6000 Frankfurt, Fed. Rep. of Germany

Three different schemes for the possible detection of the neutrino component in cosmic background radiation are reviewed: coherent elastic scattering, coherent inelastic scattering, and scattering of ultrahigh energy cosmic rays from the background neutrinos. The limitations of the various schemes are discussed.

1. Introduction

Neutrinos of all known species, ν_e, ν_μ and ν_τ were copiously produced in the hot environment of the early universe. Due to weak interactions they remained in thermal equilibrium with other matter until the universe cooled to a temperature of about 2 MeV. At this time, about a quarter second after the big bang, the neutrinos decoupled, but continued to be adiabatically cooled due to the expansion of the universe. Because the photon background was later reheated when electrons and positrons annihilated, the neutrino background temperature should be somewhat lower today than the temperature of the electromagnetic background radiation [1]:

$$T_\nu \sim 0.7\, T_\gamma \simeq 1.9\ \text{K}. \tag{1}$$

Light neutrinos, i.e. those with mass much below 1 MeV, were still relativistic at the moment of decoupling. Since the cooling was adiabatic, according to $p(t)R(t) = p_o R_o$, where p is the neutrino momentum and R the radius of the universe, their energy spectrum today is still of relativistic form:

$$dN_\nu = d^3pF_\nu(p) \approx (e^{-p/T_\nu}+1)^{-1}, \tag{2}$$

so that the average momentum of a neutrino is

$$\langle p\rangle \int d^3p\, F_\nu(p)p = 3T_\nu \sim 5 \cdot 10^{-4}\text{eV/c}\ . \tag{3}$$

For massless neutrinos this is also the average kinetic energy, but for neutrinos of 10 eV mass it would imply a very low kinetic energy of only 10^{-8} eV. By integration over momentum space one finds that the density of these background neutrinos is

$$n_\nu = n_{\bar\nu} \sim 50\ \text{cm}^{-3} \tag{4}$$

for each neutrino species. For massless neutrinos, which move with the speed of light, this yields an enormous particle flux of $j_\nu = 3$ x

$10^{12}/cm^2\cdot s$ of electron neutrinos and antineutrinos, which continuously penetrates every surface, including our own bodies.

The difficulty of observing them rests with the energy dependence of weak interactions cross sections, which give

$$\sigma_\nu \sim G_F^2\, E_\nu^2 \sim \begin{cases} 10^{-62}cm^2 & (m_\nu = 0) \\ 10^{-54}cm^2 & (m_\nu = 10\ eV) \end{cases} \tag{5}$$

For massless neutrinos the scattering rate per target atom hence is

$$R_\nu/\mathrm{atom} \sim \sigma_\nu j_\nu \sim 10^{-42}/\mathrm{year}, \tag{6}$$

which yields absurdly small event rates even for kiloton targets (10^{33} atoms). The detection of the cosmic neutrino background by conventional scattering techniques thus appears completely hopeless.

There are two conceivable ways out of this situation:

(1) The first one starts with the observation that the average momentum, eq. (3) corresponds to a de Broglie wavelength

$$\langle\lambda_\nu\rangle \sim \langle p\rangle^{-1} \sim 2.5\ \mathrm{mm}\ , \tag{7}$$

which is of macroscopic dimension. One may therefore consider scattering processes in which many atoms act coherently, so that the reaction rate becomes proportional to the square number of target atoms N_T:

$$R_\nu \sim \sigma_\nu\, N_T^2 j_\nu \sim (G_F E_\nu N_T)^2\ j_\nu\ . \tag{8}$$

For a large number of atoms this might yield observable event rates. Furthermore, in the case of coherent scattering it may be possible to observe the scattering amplitude itself, which is linear in the Fermi coupling constant, i.e. of order $G_F E_\nu N_T$.

(2) In the second case one increases the cross-section by using the neutrino backgrond as a target for very high energy particles. Unfortunately the target density n_ν, eq. (4), is very thin. The mean interaction length L of a particle in the neutrino background is

$$L = (\sigma n_\nu)^{-1} \sim (G_F^2 E_{cm}^2 n_\nu)^{-1} \sim (G_F^2 E_\nu E_p n_\nu)^{-1}. \tag{9}$$

where E_{cm} is the total c.m. energy, and E_p is the projectile energy. For $E_{cm} \sim 100$ GeV one obtains $L \sim 10^{32}$ cm, which is more that the radius of the universe ($\sim 10^{28}$cm). Thus any technique using the neutrinos as target has to rely on cosmic rays. This is also true because the required projectile energy is enormous: $E_p \sim 10^{10}$ GeV for $m_\nu \simeq 10^{14}$ GeV for $m_\nu = 0$. This may give a feasible detection method, if the background neutrinos have a mass in the eV region.

2. Coherent Elastic Neutrino Scattering

The propagation of light through materials is conveniently described by an index of refraction, as long as the wavelength is larger than the size of atoms. The refractive index can be related to the forward scattering amplitude f(O) through

$$\nu = 1 + 2\pi\rho f(0)/k^2 \tag{10}$$

where ρ is the atomic density of the material and k the photon momentum or wavevector. If the scattering is elastic, n is real. This concept can be taken over to describe the propagation of low-energy neutrinos through material; in this case f(O) is determined by elastic charged and neutral current scattering. For massless neutrinos one has

$$\nu = 1 - G_F\rho g/E_\nu \ , \tag{11}$$

where g is a number that depends on the atomic constitution of the material (for antineutrinos $g < 0$). For cosmic background neutrinos in gold one finds $(1 - \nu) \sim 5 \times 10^{-11}$. R. Opher [2] noticed in 1974, that this result implies that neutrinos could be totally reflected from the surface of a target, if the angle of incidence θ is less than the critical angle

$$\theta < \theta_c = (2(1 - \nu))^{1/2} \sim 10\ \mu\text{rad} \ . \tag{12}$$

Opher suggested that the recoil force excerted on the target could be measured and thus allow for detection of the neutrino background. Although the range of angles is small, $\Delta\theta \sim G_F^{1/2}$, the total force would be of order $G_F^{3/2}$ constituting a decisive improvement over the G_F^2 dependence of incoherent scattering processes. Opher's proposal was elaborated by Lewis [3], who designed a detector that made ample use of neutrino "optics" in order to enhance the effect of the macroscopic recoil force.

Unfortunately, the ideas of Opher and Lewis contained a critical flaw, as was pointed out by Cabibbo and Maiani [4] and by Langacker et al [5]. The reason is that the concepts of geometrical optics are not applicable to laboratory-sized targets due to the smallness of $(1 - \nu)$. The interaction between the neutrinos and the target can also be understood as due to a repulsive potential $V_\nu = \pi\rho f(0)/E_\nu$. We then know that the reflection coefficient does not jump abruptly from zero to one at the critical angle, but there is a smooth transition between these values. It is still of order one for neutrinos reaching the sur-

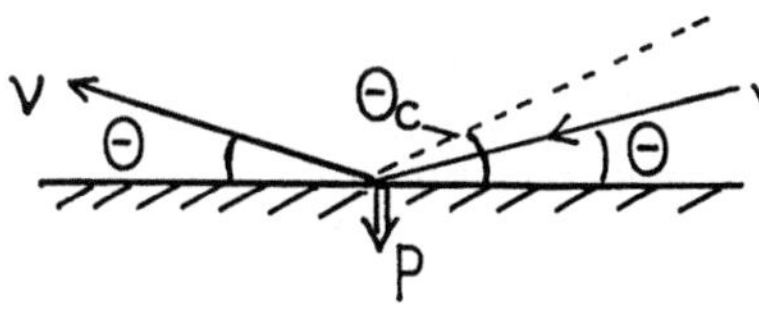

Fig. 1: Pressure exerted on a surface due to total reflection of neutrinos

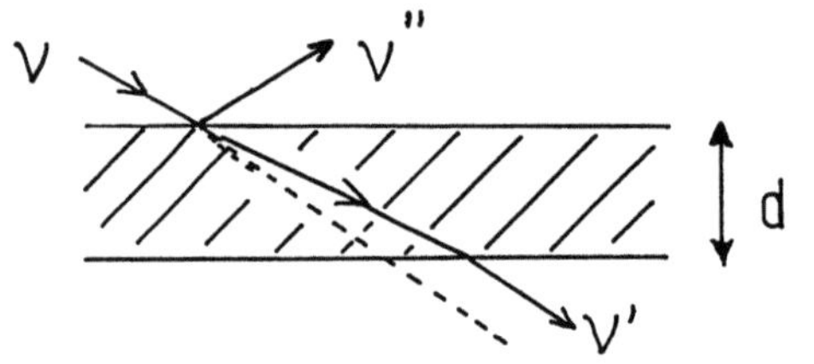

Fig. 2: Refracted wave (ν') and reflected wave (ν'') of neutrinos on a target of finite thickness d.

face from the opposite sign. Cabibbo and Maiani showed that the $G_F^{3/2}$ contribution cancels for a plane surface, unless the neutrino flux is collimated better than θ_c, which is not likely to be the case.

Langacker et al. went one step further in their analysis. Even for an ideal unidirectional flow of neutrinos, the reflection coefficient is not one, if the target has finite thickness d. The penetration depth of the refracted wave is $D = \lambda_\nu(1 - \nu)^{-1/2}$, which is of order 10 m. If $d \ll D$, most of the wave penetrates and there is no net pressure from reflection in order $G_F^{3/2}$. (Note that the actual proposals of refs. [2] and [3] involved large stacks of thin reflectors plates in order to achieve partial collimation of the neutrino flux.)

In general, for targets of size much less than $\lambda_\nu(1 - \nu)$ the intensity of the refracted wave is much less than that of the incident wave and the Born approximation should be valid. Under this condition, Langacker et al. were able to show that all effects of first order in $(1-\nu)$ cancel for any target of arbitrary shape, if the neutrino flux is homogeneous (not isotropic!). The pressure can then be related to the total net current through the surface which vanishes: This may be visualized by considering that the various contributions to the pressure from the neutrinos entering or leaving the surface elements cancel exactly. It is worthwhile to point out that, even if this cancellation would not occur, any remaining pressure would be cancelled, if the flux of antineutrinos is equal to the flux of neutrinos. This does not hold for effects of second order, G_F^2, because there the effects of neutrinos and antineutrinos have the same sign.

A similar analysis holds for the angular torque imparted on an arbitrary target, unless the target has a nonzero magnetization [5]. This effect was already noted by Stodolsky [6]. It relies on the parity-violating nature of the weak interaction. The interaction of a polarized electron with a neutrino is proportional to $(1 - \vec{\sigma}_e \cdot \vec{v})$ where v is the

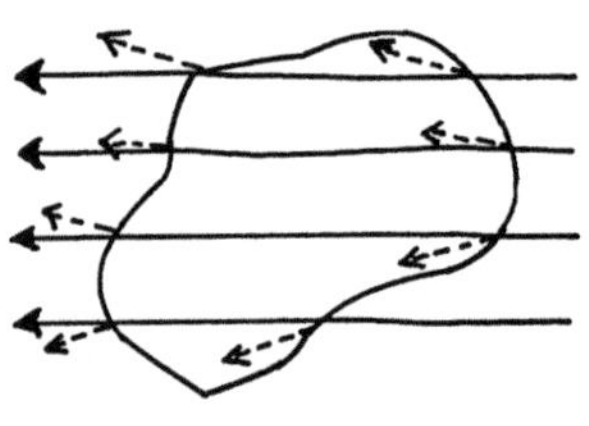

Fig. 3: Refraction of homogeneous neutrino wave at target surface. The refracted waves are indicated by dashed lines. Their contributions to the total pressure cancel exactly in first order Born approximation

relative velocity and $\vec{\sigma}_e$ the electron spin. Hence the two helicity states of an electron moving relative to the neutrino background are split in energy. The energy splitting between helicity states would lead to the precession of the spin of a moving electron, and to a torque on a ferromagnetic target, because the earth moves relative to the cosmic background. Stodolsky estimated that the two opposite orientations of 1 ton of ferromagnetic material would involve an energy difference of about 1 eV. (Note, however, that the effect of the antineutrino background is exactly opposite; hence no net effect is expected in the most likely case of lepton number symmetry in the cosmic background radiation. If there is lepton number violation, one would expect it on the same level as baryon number asymmetry, i.e. on the level of 10^{-9}!)

3. Coherent Inelastic Neutrino Scattering

Since all surviving effects of coherent scattering are of order G_F^2, one must rely on a possible N^2 dependence of the scattering probability, where N is the number of scattering centers. If the positions of the scatterers are denoted by $\vec{x}_i$, and f_i are the individual scattering amplitudes, the total scattering cross section will be proportional to

$$d\sigma/d\,\Omega \sim \left|\sum_i f_i \exp(i\vec{q}\cdot\vec{x}_i)\right|^2 \tag{13}$$

where $\vec{q}$ is the momentum transfer. If all f_i have equal phase, an N^2--dependence arises over spatial regions for which $q\cdot\Delta x<1$. For larger regions, an N^2 dependence can only occur for special scattering angles (cf. Bragg scattering), but the angle-integrated cross section will only grow like N (see the analysis of H. Lipkin [7]).

For neutrinos of MeV energy range q is of order $(100\ \mathrm{fm})^{-1}$, so that all nucleons of a large nucleus may contribute coherently [8-10]. For cosmic background one may naively think that mm-size regions could scatter coherently in view of the long wavelength, eq. (7). However, this is not so in inelastic processes because the final state involves lattice excitations, i.e. phonons. Their wavelength $\lambda_{ph} = 2\pi c_s/\omega$, where c_s is the speed of sound, is smaller by a factor $c_s/c \sim 10^{-5}$, thus reducing the possible number of coherent scatters tremendously.

To recover coherence on a macroscopic scale one has to consider scattering with very well-defined momentum transfer, corresponding to precisely defined scattering angle. Fortunately, there is a natural way of achieving this with phonons, namely by making use of stimulated emission. Full coherence is recovered by taking macroscopically occupied coherent quantum states as initial and final states of the target in the neutrino scattering process. The (inelastic) scattering cross section is then proportional to the product of phonon occupation number $n_{ph}^{(i)} n_{ph}^{(f)}$, which can be very large. Note that any standing sound wave

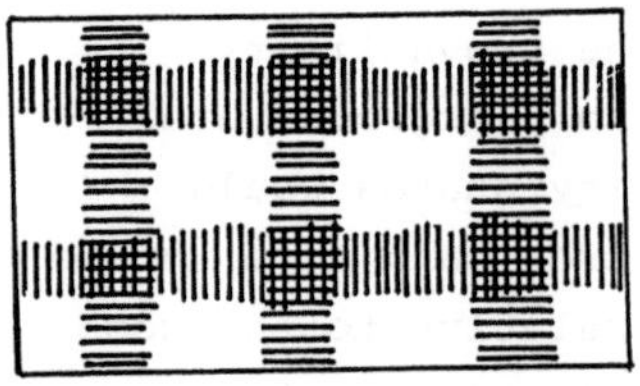

Fig. 4: Two orthogonal standing acoustic waves in a solid target

corresponds to a coherent state of acoustic phonons. (Optical coherent scattering on standing sound waves has been widely studied and has found many practical applications [11].) At frequency $\omega_{ph} \sim 1$ MHz the phonon wavelength is in the mm range, nicely matching that of the cosmic background neutrinos.

Coherent neutrino scattering on a detector containing two orthogonal sound waves was recently calculated by Tupper et al. [12]. The calculation proceeds in analogy to the theory of coherent neutron scattering [13]; the differential cross section becomes

$$\frac{d\sigma}{d\Omega_\nu dE'_\nu} = \frac{\sigma_o}{4\pi}\left(\frac{E'_\nu}{E_\nu}\right)^2 \delta(E'_\nu - E_\nu + E_f - E_i) \times$$

$$\times \left| \sum_{x_i} e^{-i\vec{q}\cdot\vec{x}_i} \langle f | e^{-i\vec{q}\cdot\vec{U}(\vec{x}_i)} | i\rangle \right|^2 , \qquad (14)$$

where E_ν (E'_ν) is the neutrino energy before (after) the scattering, i and f denote the initial and final phonon state and $\vec{U}(x_i)$ is the elongation of the i-th nucleus out of its equilibrium position. The $\vec{U}(x_i)$ are related to the phonon coordinates. σ_o is the elementary neutrino-nucleus cross section.

Now consider that the neutrino scatters Δn_1 phonons out of the first standing wave and Δn_2 into the second one. Then energy and momentum conservation yield

$$E_f - E_i = \Delta n_2 \omega_{ph}^{(2)} - \Delta n_1 \omega_{ph}^{(1)} \qquad (15a)$$

$$-\vec{q} = \Delta n_2 k_{ph}^{(2)} - \Delta n_1 \vec{k}_{ph}^{(1)} . \qquad (15b)$$

Note that because $\omega_{ph} = k_{ph} c_s$, but $E_\nu = k_\nu c$, neutrino scattering requires a Δn of order $c/c_s \sim 10^5$! Tupper et al. then find that the scattering rate is

$$R \approx j_\nu \frac{d\sigma}{d\Omega_\nu dE'_\nu} \sim N^2 j_\nu \sigma_o \left(\frac{\lambda_o}{2\lambda_{ph}^{(2)}} \frac{n_{ph}^{(2)}}{N}\right)^{\Delta n_2} \left(\frac{\lambda_o}{2\lambda_{ph}^{(1)}} \frac{n_{ph}^{(1)}}{N}\right)^{\Delta n_1} \qquad (16)$$

with

$$\lambda_o = \frac{2\pi}{Mc_s} \approx 10^{-8} \text{ mm} \qquad (17)$$

where M denotes the mass of a target nucleus. If we could make the last two factors in eq. (16) comparable to unity, the event rate $N^2 j_\nu \sigma_o$ would be of order one per minute with j_o determined by eq. (4) and $N \sim 10^{23}$. This would really become a practical experiment! Unfortunately, the trouble lies with these factors. To see why, let us require that the expressions in parenthesis in eq. (16) be at least unity. Considering that $\omega_{ph} = k_{ph} c_s$ we find that the total energy stored in each of the acoustic waves must be at least:

$$E_{ph} = n_{ph}\omega_{ph} \gtrsim N \frac{\lambda_{ph}\omega_{ph}}{\lambda_o} = \frac{2\pi c_s}{\lambda_o} N = Mc_s^2 \, N. \tag{18}$$

According to the virial theorem for harmonic oscillators $\frac{1}{2}Mc_s^2$ is just equal to the binding energy ε_o of a single atom in the solid lattice [14]. We thus must have

$$E_{ph} \gtrsim 2N\varepsilon_o \, , \tag{19}$$

i.e. the acoustic energy must exceed the total binding energy of the target. This is clearly impossible to achieve. On the other hand, if the factors are small compared to one, only $\Delta n_i = 1$ transitions are of interest. Then the N^2 factor in eq. (16) cancels and one obtains

$$R \sim n_{ph}^{(1)} \; n_{ph}^{(2)} \; j_\nu \sigma_o \, \frac{\lambda_o^2}{4\lambda_{ph}^{(1)}\lambda_{ph}^{(2)}} \; . \tag{20}$$

Thus, coherence over the total number of phonons is retained, indeed. Due to the small energy of a single phonon, their number could be very large. E.g. 10^{20} phonons of 1 MHz correspond to a total acoustic energy of about 10^{-8} Joule. Still, the rate according to eq. (20) would be only 10^{-18} per year!

The failure, due to the condition (19), to achieve full coherence over the numbers of atoms can also be understood in another way. The acoustic wave corresponds to a periodic density fluctuation $\Delta\rho(x)$. Stability of the target requires that $\Delta\rho \ll \rho_o$, the equilibrium density; otherwise nonlinear and dissipative effects set in. This observation also indicates a possible way to satisfy the requirements. The layered structure of a sandwiched target, say of alternating layers of gold and a material of light atomic weight, would simulate a standing density wave with $\Delta\rho \approx \rho_o$. (Such structures have been successfully applied to X-ray optics [15]). In order to retain coherence in the final state one would have to set up an acoustic phonon wave perpendicular to the layer structure. This will be difficult in view of the presence of the layer boundaries, but maybe not impossible. One might also consider setting the layered target in some sort of periodic motion, such as rotation.

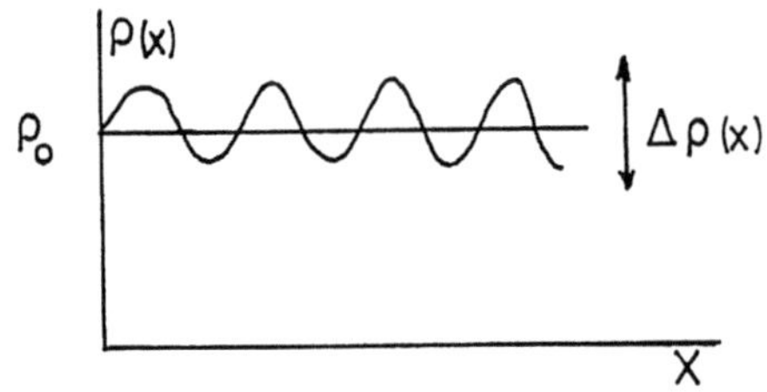

Fig. 5: A standing sound wave corresponds to periodic fluctuations in the density of the target

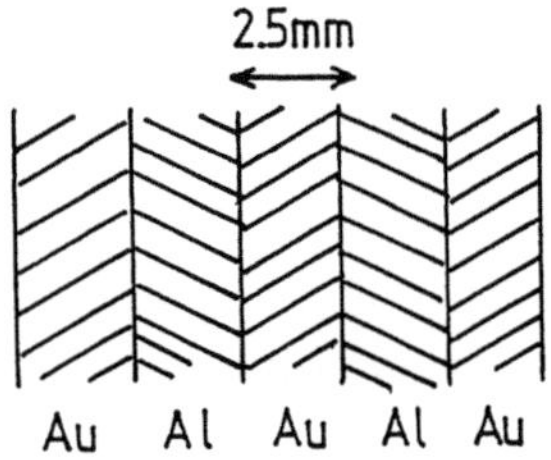

Fig. 6: A sandwich structure of alternating layers of materials of light and heavy atomic weight, eg. Au and Al, could simulate a large-amplitude density wave

These and other possibilities have not yet been sufficiently explored to allow final verdict on coherent acoustic phonon detectors for neutrinos.

4. Cosmic Ray Absorption by the Neutrino Background

The last scheme for detection of the cosmic neutrino background neutrinos that I will discuss here has been suggested by Tom Weiler [16, 17]. It is based on the fact that weak interaction cross sections grow with energy and become equal to electromagnetic cross sections at c.m. energies around 100 GeV. This corresponds to projectile energies of 5×10^{12}GeV/$(m_\nu+3T_\nu)$ where m_ν is the neutrino mass and T_ν is given by eq. (1). Such energies are available, although rarely, in cosmic ray particles. As mentioned in the introduction, even at these energies the scattering length is of the size of the universe. Charged or strongly interacting particles are not appropriate candidates because they also scatter on the photon background, thus only ultra-high energy neutrinos could show a characteristic signal of depletion due to scattering. The specific mechanism that most strongly contributes to depletion is ν-$\bar{\nu}$ scattering at the Z^o resonance. The energy-integrated cross section for this process is linear in G_F:

$$\bar{\sigma} = \frac{1}{M_Z^2} \int ds\sigma_{res}(s) = \frac{16\pi^2}{M_Z^3} \Gamma_{Z\to\nu\bar{\nu}} \sim G_F \ , \qquad (21)$$

and 80% of the final states are not neutrinos. Non-resonant scattering is proportional to G_F^2 and can be neglected.

Observationally, the presence of a neutrino background will show up by the reduction it causes in the cosmic ray flux at the appropriate

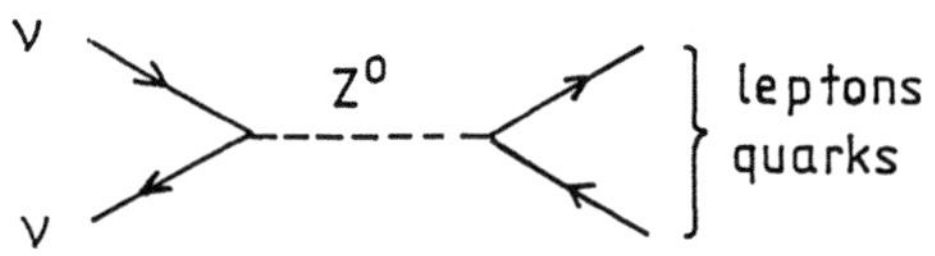

Fig. 7: Scattering at the Z^o-resonance leads to depletion of the ultrahigh energy component of cosmic ray neutrinos

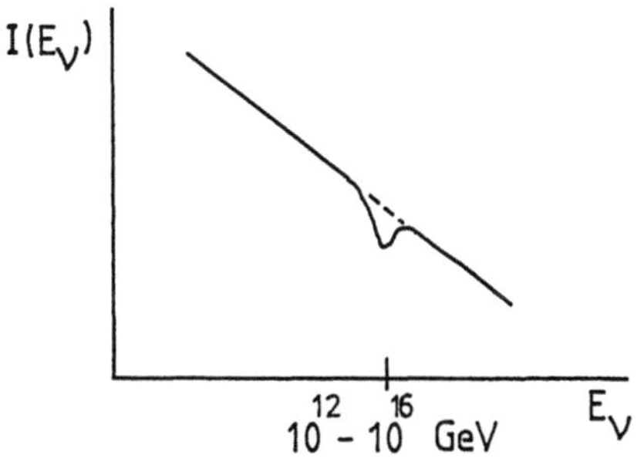

Fig. 8: Inelastic scattering on the cosmic neutrino background leads to an absorption dip in the spectrum of cosmic ray neutrinos of ultra-high energy. The position of the dip corresponds to c.m. energy in the Z^o resonance

energy (see Fig. 8). Where the dip occurs depends on the neutrino mass and on the age of the cosmic ray neutrinos. If they are of cosmological age, the absorption edge is shifted to lower energies, because of the increased background temperature at earlier ages, by a factor $(1+Z)^{-1}$ where Z is the Hubble red shift of the high energy neutrino source. If the high-energy part of the cosmic ray spectrum was created by quasars, the relevant red-shift may be $2<Z<4$. Weiler has estimated that an incident neutrino flux of about $10^{-13}/cm^2s$ is required for detection of the signal, which would be detectable by the Fly's Eye or DUMAND experiments. In addition, a clear absorption signal requires production of the cosmic rays within a relatively narrow age range of the universe, e.g. in an epoch where quasars were much more abundant than today. The combination of these requirements renders the detectability of the neutrino background by this method rather unlikely, but by no means impossible.

5. Conclusions

Detection of the cosmic background of relic neutrinos from the big bang remains a great challenge. If the cosmic background neutrinos could be detected they would serve as very valuable "eye-witnesses" of events that took place in the era of nucleosynthesis, which is the cornerstone of our present views of the early universe. Although no really practical detection scheme has yet been suggested, the situation does not appear to be entirely hopeless. It is worthwhile to set one's mind on this problem again from time to time.

6. Acknowledgements

I would like to thank M. Danos, J. Rafelski and G. Tupper for their collaboration and many valuable discussions.

References

1. G. Gelmini, Lecture at this Workshop
2. R. Opher, Astron. & Astrophys. 37, 135 (1974); 108, 1 (1982)
3. R.R. Lewis, Phys. Rev. D21, 663 (1980)
4. N. Cabibbo and L. Maiani, Phys. Lett. 114B, 115 (1982)
5. P. Langacker, J.P. Leveille and J. Sheiman, Phys. Rev. D27, 1228 (1983)
6. L. Stodolsky, Phys. Rev. Lett. 34, 110 (1975)
7. H. Lipkin, Phys. Rev. Lett. 58, 1176 (1987)
8. S. Adler, Phys. Rev. D11, 1155 (1975)
9. A. Drukier and L. Stodolsky, Phys. Rev. D30, 2295 (1984)
10. B. Cabrera, L.M. Krauss and F. Wilczek, Phys. Rev. Lett. 55, 25 (1985)
11. see e.g. J.G. Dil, Rep. Prog. Phys. 45, 285 (1982)
12. G. Tupper, B. Müller, J. Rafelski and M. Danos, Phys. Rev. D35, 394 (1987)
13. see e.g.: G.Z. Squires, Introduction to the Theory of Thermal Neutron Scattering, Cambridge Univ. Press, 1978
14. Kittel, The Quantum Theory of Solids, Wiley, New York 1987
15. see e.g. E. Spiller, AIP Conf. Proceedings 119, Laser Techniques in the Extreme Ultraviolet, ed. S.E. Harris, T.B. Lucatorto, (1984) p. 312; A.V. Vinogradov and Ya.B. Zel'dovich, Opt. Spectr. 42, 404 (1977)
16. T. Weiler, Phys. Rev. Lett. 49, 234 (1982); Astrophys. J. 285, 495 (1984)
17. First considerations were due to: J. Bernstein, M. Ruderman and G. Feinberg, Phys. Rev. 132, 1227 (1963); B.P. Konstantinov and G.E. Kocharov, Sov. Phys. JETP 19, 992 (1964); R. Cowsik, Y. Pal and S.N. Tandon, Phys. Lett. 13, 265 (1964).

Part V

Neutrinos from Collapsing Stars

Neutrinos from Collapsing Stars and SN1987A

W. Hillebrandt

Max-Planck-Institut für Physik und Astrophysik,
Institut für Astrophysik, Karl-Schwarzschild-Str. 1,
D-8046 Garching, Fed. Rep. of Germany

Weak interaction processes leading to the production and emission of neutrinos from collapsing stars are reviewed. Based on recent numerical simulations it is shown that: 1. Only electron neutrinos are emitted during the core collapse phase and the short deleptonization burst with a typical luminosity of about 10^{53}erg s^{-1} and 10^{54}erg s^{-1}, respectively; 2. during a first cooling phase lasting for a few seconds neutrinos and anti-neutrinos of all flavours are emitted with typical energies between 10 and 20 MeV and a total luminosity per species of about a few times 10^{52}erg s^{-1}. These theoretical predictions are compared with the neutrino energies and luminosities inferred from supernova 1987A.

1. Introduction

Models of type II supernovae have received a lot of attention over the last 30 years, because, besides being the most energetic local astronomical events, they are thought to be the birthplace of neutron stars as well as of the bulk of the heavy elements in the galaxy [1].

Spectra and light curves of average type II supernovae can be explained in a quite natural way if one assumes that the exploding star is rather massive ($M \gtrsim 8M_{\odot}$) and has an extended red giant envelope [2]. The surprising fact that the progenitor of SN1987A was a blue rather than a red supergiant may be attributed to lower than solar metallicities in the Large Magellanic Cloud and, therefore, is not in contradiction to this general picture [3]. However, from the observation of electromagnetic radiation we cannot learn much about the explosion mechanism because photons are emitted from the stellar photosphere at a radius of at least 10^{12} cm whereas all theoretical models predict that the physical processes leading to the explosion originate in the deep interior of the star [1].

The standard model of type II supernova explosions is based on the assumption that the core of a highly evolved massive star collapses to nuclear matter density [1]. The gain in gravitational binding energy during collapse is of the order of 10^{53}erg but the energy seen in kinetic motion of the ejecta and in electromagnetic radiation is only of the order of 10^{51}erg. Almost all the energy available will therefore be emitted in form of neutrinos, which because of their small cross-sections may provide a good tool to test the dynamics of the stellar core. In fact, as we will demonstrate later, neutrino luminosities and energies are strongly dependent upon the hydrodynamical evolution of the interior of the star and, therefore, if the temporal change of both quantities could be measured this would impose strong constraints on theoretical models. However, the small number of neutrino events detected from SN1987A [4], though being consistent with theoretical expectations, does not allow for such a detailed analysis.

2. The Physics of Stellar Collapse and Supernova Explosions

Supernovae are rare and recent estimates predict a rate of roughly one event every (40 ± 20) years in giant spiral galaxy like our own one [5]. Among all supernovae roughly one half are of type II, and only those are thought to emit large amounts of neutrinos. Since the birthrate of pulsars may be somewhat larger [6] it is possible that a significant fraction of stellar collapses escape optical detection, either because of interstellar absorption or because the progenitors are massive white dwarfs which may not produce an optical outburst. It is interesting to note that if SN1987A had exploded near the galactic center it would probably have been seen as a neutrino source only. On the other hand, the number of pulsars found in young supernova remnants is surprisingly low indicating that type II supernovae may not always leave neutron stars behind [6].

Despite of these uncertainties, we will adopt the viewpoint that type II supernovae gain their energy from the gravitational binding energy of a newly born neutron star. This assumption restricts the main sequence masses of the progenitor stars to values above $8M_\odot$, because otherwise either they would end as white dwarfs or would be disrupted by explosive carbon burning prior to collapse [7]. Stars more massive than the limit given here form central iron-cores of about (1.4 ± 0.2) $M_\odot$ at the end of hydrostatic evolution, the exact value depending on the main sequence mass and many details of the stellar evolution codes [8]. These cores become dynamically unstable because they have exhausted their nuclear fuel and electron captures on heavy nuclei and on free protons reduce the pressure.

The further evolution of those dense ($\rho \simeq 10^{10} gcm^{-3}$) and hot ($T \simeq 10^{10} K$) stellar cores is rather well understood [1]. Due to the low initial entropy nucleons are mainly bound in heavy nuclei. During the early phase of collapse neutrinos produced by e^--captures leave the star without interacting with the stellar gas, because their mean free path, essentially determined by coherent scattering off nuclei, is larger than a typical dimension of the core. Only at densities above a few times $10^{11} gcm^{-3}$ becomes the neutrino mean free path short compared to typical length scales, and since at the same time neutrino diffusion times become longer than the collapse timescale to a first approximation neutrinos can be treated as trapped in the star. Both, before and after neutrino trapping, the entropy is essentially conserved. Prior to trapping neutrinos carry away most of the entropy generated by non-equilibrium electron captures, and after trapping β-equilibrium is quickly obtained. Consequently, nucleons remain bound in heavy nuclei and leptons provide the main contribution to the total pressure until the nuclei dissolve into a homogenous fluid of nucleons near nuclear matter density. It is only then that the non-relativistic nucleons dominate the pressure and the collapse is stopped and reversed.

Pre-supernova stellar evolution models predict that at the onset of collapse the electron concentration is about 0.42 to 0.44 [8]. Until neutrino trapping these numbers are reduced by approximately 0.04 to 0.06, due to e^--captures and neutrino losses. At nuclear matter density the lepton concentration in the stellar core is still almost the same, but now about 20 to 30% of the leptons are electron neutrinos. Up to that phase of the dynamical evolution no significant amounts of electron anti-neutrinos or μ- and τ-neutrinos have been produced, because the electron chemical potential is high and the temperatures are relatively low [1]. The rather high final lepton fraction is crucial for the further evolution. One can show quite generally [1] that only an inner part of the stellar core collapses homologously with subsonic velocities, while the outer parts fall in with supersonic velocities. The position of the sonic point is approximately given by the Chandrasekhar mass which is proportional to the square of the lepton fraction. The mass of the homologous core, therefore, is roughly $(0.7 \pm 0.1)\ M_\odot$, the actual value depending on details of the equation of state, neutrino-transport, etc.. This inner part of the original iron core comes to rest about 1ms after bounce, a shock wave forms at the sonic point which propagates outwards both in mass and in radius. The matter passing through the shock front is heated to entropies of about 6 to 10 k_B/nucleon and, consequently, the preexisting heavy nuclei are dissociated into free protons and neutrons. Since initially the energy of the shock was about $(6 \pm 2) \times 10^{51}$erg one can estimate that at most 0.5 to 0.6 $M_\odot$ of heavy elements can be dissociated because the binding energy per nucleon is typically 8 MeV. This in turn implies that the shock cannot successfully propagate unless the mass of the stellar core is smaller than about $(1.2 \pm 0.2)\ M_\odot$, even if neutrino losses are neglected [1]. It is not surprising, therefore, that the only models, which so far give prompt explosions by the core bounce mechanism had iron cores below 1.3 $M_\odot$ [1,9]. The corresponding main sequence masses of the progenitors will be below 12 $M_\odot$ and thus smaller than the progenitor of SN1987A, which had approximately 20 $M_\odot$ on the main sequence. It is therefore rather unlikely that the prompt explosion mechanism gave raise to SN1987A [10].

An alternative possibility is that a few hundred milliseconds after core-bounce energy transport by neutrinos may revive a stalling shock [11]. This mechanism is commonly called a delayed explosion. The basic idea is again simple. The shocked matter, composed of free neutrons and protons, will be irradiated by a strong neutrino flux originating from the neutrino sphere. Typical neutrino energies will range from 9 to 15 MeV over timescales of the order of a second and the neutrino luminosity will be of the order of $(2-4) \times 10^{53}$erg s^{-1} (see also section 3). Some of these neutrinos and anti-neutrinos will be captured by free neutrons and protons, respectively, and will be reemitted with lower (thermal!) energies, thus heating the matter behind the accretion front. Numerical simulations have shown [11], however, that even in cases when the heating was sufficient to cause an explosion the explosion energy was always too small ($(2-3) \times 10^{50}$ erg) to explain typical type II supernova light curves and in particular the very energetic outburst of SN1987A [10].

So in conclusion it seems that the explosion mechanism of type II supernovae is still not well understood. One has to keep these uncertainties in mind if attempts are made to predict neutrino spectra and luminosities from theoretical models. They will also strongly influence interpretations of the neutrinos seen from SN1987A.

3. Neutrinos from Collapsing Stars and Newly Born Neutron Stars

As we have discussed in the previous section neutrinos are essentially trapped in the collapsing stellar core and, therefore, no measurable signal is expected from this early phase of a typical galactic supernova. In fact, we can estimate the expected neutrino flux in a more or less model independent way. During the core collapse phase the neutrino sphere is typically at a radius of 90 km and the neutrino temperature will be around 4MeV [1]. In the black body limit we, therefore, obtain an electron neutrino luminosity of approximately 10^{53} s^{-1} and a duration of $\Delta t \simeq 10^{-2}$S. Because in water Cerenkov detectors like the KAMIOKANDE-experiment these neutrinos can only be detected via neutrino-electron scattering, for which the cross-section is of the order of 10^{-44} (Ev/MeV) cm^2, a supernova at a distance of 3kpc would give just 1 count in 2000t of water. Note that the distance to the galactic center is about 7.5 kpc.

The situation changes dramatically when the hydrodynamical shock front reaches densities below $10^{11} gcm^{-3}$ where the neutrino mean free path becomes large compared to the density scale height of the star. During collapse the central core of the star has gained at least 10^{53}erg of gravitational binding energy which ultimately will be radiated away mostly on timescales of the order of a few seconds in form of thermal neutrinos [12].

a) Neutrinos from the deleptonization burst

As has been mentioned earlier, the nucleonic component of the matter that has gone through the shock front is composed of free neutrons and protons, because of the high entropy there [1]. This means that the neutrino scattering cross sections are significantly reduced and neutrinos can diffuse on timescales of a few ms to the shock front, where they pile-up until the shock passes the neutrino sphere. But only those neutrinos can diffuse which are created in the hot shocked gas, because the diffusion times in the unshocked material are significantly longer (of the order of several tenths of a second) than the shock propagation time. Once the shock has passed the neutrino sphere most of these neutrinos are emitted on a hydrodnamical timescale (~ms) [13].

Their total as well as their average energy can be estimated from the processes that lead to their production [13]. In the shocked gas the temperature will be high and, therefore, neutrino degeneracy is removed. This leads to an increase of e^--captures on free protons and, thus, the matter becomes more neutron (and neutrino) rich. About half of the electrons will be converted into electron neutrinos. The average energy per neutrino produced by e^--captures on free protons is roughly 2/3 of the electron Fermi-energy, giving neutrino energies of the order of 10 to 20 MeV. Neutrino-electron scattering tends to thermalize the neutrino energy distribuion and, thus, the average neutrino energies may be somewhat lower. From the fact that by the time the shock has reached the neutrino sphere it has crossed about 0.5 $M_\odot$, we can estimate that the total energy in neutrinos from e^--captures will be of the order of 2×10^{51}erg. The actual numbers may be a factor of two higher or lower, depending on the strength of the shock, with the general tendency that the energy emitted in neutrinos increases with increasing shock strength. But typically we expect an electron neutrino luminosity of about 10^{54}erg s^{-1} and a duration of the burst of less than 5ms. It is important to note that the neutrino spectrum during this burst will be non-thermal.

Our general considerations are in good agreement with the results of numerical simulations [1]. Again we can estimate the number of neutrinos which could be seen in existing neutrino detectors and find for the KAMIOKANDE experiment 60 counts $(\langle E\nu\rangle/10\ \mathrm{MeV})\ (D/\mathrm{kpc})^{-2}$ and a factor of 20 less for the Mont Blanc detector. Consequently, there is a certain chance that neutrinos from the deleptonization burst can be detected if the distance to a supernova is less than or of the order of a few kpc. On the other hand, it is very unlikely that any of those neutrinos have been seen in the KAMIOKANDE detector from SN1987A ($D \simeq 50$kpc). This result is in contradiction to interpretations which attribute the first two events detected by KAMIOKANDE to ν_e's emitted during this burst [14]. There remains the problem, however, that in both cases the positron (electron?) was clearly directed away from the Large Magellanic Cloud, indicating ν_ee-scattering rather than $\bar{\nu}_e$p-reactions [4].

b) The cooling phase

After the deleptonization burst the electron concentration in certain mass zones behind the shock front has dropped to values of about 0.1 and consequently the electron chemical potential is close to kT [1]. Since the temperature is still high ($T \gtrsim 1$ MeV) electron-positron pairs form in equilibrium and anti-neutrino producing reactions such as $e^+ + n \rightarrow \bar{\nu}_e + p$ and $e^+ + e \rightarrow \bar{\nu}_e + \nu_e$ become important. Moreover, electron-positron pairs will also decay into μ and τ neutrinos. Therefore, we expect that most of the thermal energy of the proto-neutron star will be carried away by neutrinos of all flavours [12,15]. Finally, once the temperature has dropped to values below 1 MeV URCA neutrinos and neutrinos from nucleon pair bremsstrahlung will take over.

Again we may ask whether or not neutrinos from the cooling phase can be detected. Unfortunately only a few numerical studies have been performed which, however, follow the evolution over about 1s after core bounce only [15,16], and those computations which deal with the cooling of a newly born neutron star start from ad hoc initial models and are inconsistent with the results obtained from hydrodynamical simulations [12]. We can, therefore, give estimates of the expected neutrino flux only, and it is somewhat surprising that these estimates are in reasonable agreement with the detections from SN1987A.

We expect from the numerical simulations that for a time interval of about 0.5s the electron neutrino luminosity will be around 10^{52} to 5×10^{52}erg s^{-1} and that the electron antineutrino luminosity is about the same [15]. The luminosity in μ-and τ-neutrinos may be roughly a factor of two to four lower. Average energies will range from 10 or 12 MeV for ν_e's to 15 MeV for $\bar{\nu}_e$'s and 20 MeV for ν_μ's and ν_τ's. These numbers seem to be rather insensitive to the details of the numerical models and may, therefore, be typical. The total number of ν_e's and $\bar{\nu}_e$'s emitted from the newly born neutron star during this time will thus be in the range from a few times 10^{56} to 10^{57}. The expected number of counts in the KAMIOKANDE experiment is therefore $(6\pm4)(\langle E^{\nu}e\rangle/10\text{MeV})(D/5\text{kpc})^{-2}$ ν_e's and $(60\pm40)(\langle E^{\bar{\nu}}e\rangle/10\text{MeV})^2$ $(D/5\text{kpc})^{-2}$ $\bar{\nu}_e$'s. For SN1987A ($D\simeq50$kpc) this translates into less than 0.1 counts of ν_e's and less than 2 counts of $\bar{\nu}_e$'s, whereas the actual countrate for the first 0.4s was 4 [4]. So, within the limited statistics of a few events, the agreement is quite good.

At even later times the predictions become extremely model dependent. They will, for example, be different for prompt and delayed explosions, the equation of state at high densities will play an important role, and the final mass of the neutron star will influence the neutrino fluxes and spectra. It is conceivable, however, that a neutrino luminosity of about 10^{52}erg s^{-1} or somewhat higher can be sustained for a couple of seconds [12], and most of the neutrinos seen in the KAMIOKANDE experiment should have originated from this early cooling phase, provided the neutrinos have no restmass or their restmass is less than a few eV/c^2 [17]. It is difficult to understand, however, why after about 10s high energy neutrinos still have been seen from SN1987A, since a typical neutrino diffusion time will be of the order of 2 or 3s only. Moreover, the large gap of about 7s between the arrival times of events number 9 and 10 in the KAMIOKANDE detector is difficult to understand. One is, therefore, tempted to assume that the last three events were noise and did not originate from SN1987A.

A few tens of seconds after core bounce the average energy of neutrinos emitted from the newly born neutron star will have dropped to values below 5 MeV and the neutrino luminosity will be less than 10^{51}erg s^{-1} [12]. It is obvious that these neutrinos cannot be detected by existing experiments, even if the supernova exploded at an average place in our galaxy.

4. Summary and Conclusions

We have demonstrated that the collapse of a massive star to neutron star densities will manifest itself in a fast burst of ν_e's of energies around 10 to 15 MeV, followed by significant emission of neutrinos and anti-neutrinos of all flavours over a timescale of several seconds. However, because the cross section for $\bar{\nu}_e$ p-reactions is about a factor of 100 higher than for ν_e e-scattering, only $\bar{\nu}_e$'s will give a clear signal in existing neutrino detectors unless the collapsing star is at distance of less than 5kpc.

The expected neutrino luminosities and spectra are very sensitive to many details of the astrophysical models and the physical input data, but the observed neutrino flux from SN1987A indicates that the basic ideas are essentially correct. The observations have shown, in particular, that massive stars indeed do collapse, thereby emitting a large fraction of their gravitational binding energy in form of thermal neutrinos.

However, certain aspects of the neutrino events seen from SN1987A are difficult to explain within the standard scenario. If indeed two neutrino bursts were emitted from SN1987A, separated by more than 4 hours, this would call for significant modifications of the simple core-collapse picture. The apparent anisotropy of the positrons (electrons?) in both the KAMIOKANDE and IMB detectors cannot easily be understood, the effect is too large to be of purely statistical origin. It may either indicate that the neutrino luminosity of the supernova was much higher than predicted by the standard theory or may show that even the $\bar{\nu}_e$ p-reaction is not fully understood at intermediate neutrino energies. Finally, the time distribution of the neutrino events needs to be explained.

Some of these questions can be answered after the supernova envelope has become transparent and we have learned more about the nature of the compact remnant left over by the explosions. Others should be answered by particle physics.

References

1) Hillebrandt, W. 1982, in "Supernovae: A Survey of Current Research", M.J. Rees and R.J. Stoneham (eds.), NATO-ASI C90, Reidel, p.123

Hillebrandt, W. 1986, in "High Energy Phenomena Around Collapsed Stars", F. Pacini (ed.), NATO-ASI C195, Reidel, p.73

Wilson, J.R., Mayle, R.W., Woosley, S.E., and Weaver, T.A. 1986, Ann. N.Y. Acad. Sci. 470, 267

Bruenn, S.W. 1985, Ap.J. Suppl. 58, 771

2) Arnett, W.D. 1982, in "Supernovae: A Survey of Current Research", l.c., p.221

3) Hillebrandt, W., Höflich, P., Truran, J.W., and Weiss, A. 1987, Nature 327, 597

Truran, J.W. and Weiss, A. 1987, Lecture Notes in Physics 287, Springer, p.293

Arnett, W.D. 1987, Ap.J. 319, 136

4) Aglietta, M. et al. 1987, Europhysics Lett. 3, 1315
Bionta, R.M. et al. 1987, Phys. Rev. Lett. 58, 1494
Hirata, K. et al. 1987, Phys. Rev. Lett. 58, 1490
5) Tamann, G.A. 1982, in "Supernovae: A Survey of Current Research", l.c., p.371
van den Bergh, S. 1987, in "ESO Workshop on the SN1987A", J. Danziger (ed.), ESO, Garching, p.557
6) Taylor, J.H., and Manchester, R.N. 1977, Ap.J. 215, 885
Lyne, A.G. 1986, in "High Energy Phenomena Around Collapsed Stars", l.c., p.121
7) Arnett, W.D. 1977, Ap.J. Suppl. 35, 145
Weidemann, V., and Koester, D. 1983, Astron. Astrophys. 121, 77
Iben, I. 1985, Q.J. Royal Astron. Soc. 26, 1
8) Weaver, T.A., Woosley, S.E., and Fuller, G.M. 1985, in "Numerical Astrophysics", J. Centrella et al. (eds.), Jones and Bartlett, p.374
Woosley, S.E., and Weaver, T.A. 1986, Ann. Rev. Astr. Ap. 24, 205
Hashimoto, M. and Nomato, K. 1987, Univ. of Tokyo, preprint
9) Baron, E., Cooperstein, J. and Kahana, S. 1985, Nucl. Phys. A 440, 744
10) Hillebrandt, W. 1987, in "ESO Workshop on SN1987A", l.c., p.301
11) Wilson, J.R. 1985, in "Numerical Astrophysics", l.c., p.422
Bethe, H.A., and Wilson, J.R. 1985, Ap.J. 295, 14
12) Burrows, A., and Lattimer, J.M. 1986, Ap.J. 307, 178
13) Hillebrandt, W. and Müller, E. 1984, in "Neutrino Physics and Astrophysics", K. Kleinknecht and E.A. Paschos (eds.) World Sci. Publ., p.229
14) De Rujula, A. 1987, Phys. Lett B 193, 514
15) Mayle, R.W. 1985, Ph.D. Thesis, Lawrence Livermore National Laboratory
16) Hillebrandt, W. 1985, in "Proc. 3rd Workshop on Nuclear Astrophysics", W. Hillebrandt et al. (eds.), MPA 199, p.59
17) Sato, K., and Suzuki, H. 1987, Phys. Rev. Lett. 58, 2722

Detection of the Neutrino Signal from SN1987A Using the INR Baksan Underground Scintillation Telescope

E.N. Alexeyev, L.N. Alexeyeva, I.V. Krivoscheina, and V.I. Volchenko

Institute for Nuclear Research of the USSR Academy of Sciences, 60th October Anniversary Prospect 7a, SU-117312 Moscow, USSR

1. Introduction. Since June 1980 a continual Galaxy observation to search for electron antineutrino signal accompanying a massive star core collapse and neutron star formation is running at the underground scintillation telescope of the INR Baksan Neutrino Observatory. There were no events undoubtedly interpreted as a collapse neutrino burst. However, the LSD group /1/, the KAMIOKANDE-II /2/ and the IMB /3/ collaborations reported the detection of the neutrino burst from the supernova 1987A in the Large Magellanic Cloud on February 23 at the times 2:52:36 UT, 7:35:35 UT and 7:35:41 UT respectively. In the fiducial mass (200 tons) of the Baksan telescope we observed only 1 event at the LSD time and a burst of 5 events spread over 9.1 seconds at the KAMIOKANDE-II - IMB time.

In this paper we analyse and compare our data with the other detector data.

2. The Detector. The detailed description of the Baksan telescope have been published elsewhere /4/. Here we only remind of its main properties.

The telescope is located at a depth of 850 m.w.e. in the mountains of the North Caucasus. It consists of 3156 standard detectors arranged in eight planes forming a parallelepiped shape with two internal horizontal layers. Each of the 3156 elements (70x70x30 cm^3) contains oil-based liquid scintillator (C_nH_{2n+2}, $n \sim 9$) that is viewed by one 15cm photomultiplier tube. The total target mass is 330 tons.

The detection efficiencies are $\sim$8% for 8 Mev electrons, $\sim$50% for 10 Mev electrons and $\sim$80% for electrons energies $\gg$12 Mev. The energy threshold of the individual detectors is 10 Mev.

The telescope is triggered by any event above the threshold. The main sources of background are cosmic ray muons and discharges in the photomultiplier tubes. The counting rate due to the first of these sources is considerably eliminated by the use of special selection of events, so-called "single detector mode" /4,6/. The scintillators of three internal layers (1200 detectors, 130 tons of scintillator) were chosen as a trigger mass in analysis of the telescope information because of their low background in the "single detector mode" ($\sim$0.013 sec^{-1}).

The fiducial mass that includes the trigger mass and a part of the external scintillators is 200 tons. Counting rate of single background events is $\sim$ 0.033sec^{-1}. An additional condition is applied to an external detector event: its amplitude must be less than 50 Mev.

Amplitude, coordinates and times information is recorded for each telescope event in on-line computer.

The Baksan detector is the most sensitive to electron antineutrinos through the reaction $\tilde{\nu}_e + p \rightarrow n + e^+$ on free protons of the scintillator. Electron neutrinos of $E_\nu > 30$ Mev can also be detected through their captures by carbon $\nu_e + {}^{12}C \rightarrow {}^{12}N + e^-$.

3. <u>The Data and Analysis.</u> During the "live" observational time of 5.7 years no signals with number of events more than 7 spread over $\leq$ 20 seconds have been detected in the trigger mass. According to the stellar core collapse model /5/, we can expect about 35 neutrino events in the trigger mass or about 54 events in the fiducial mass, if a distance to a star is 10 Kpc.

In the period of February 1-23, 1987 previous to the SN 1987A we found no pulse clusters with the rate of background imitation < 1 week^{-1}. Then we recieved information that the other detectors recorded the neutrino signal from this supernova. We looked through our data again.

Figure 1 shows the time sequence of the events recorded by the LSD neutrino detector and the Baksan pulses at the same time.

The Baksan arrival time are defined with millisecond precision, but the analysis of the clock made for a long time shows that the clock could be shifted forward by (0-54) seconds at the moment of the SN. So, the Baksan overal time uncertainty is (-54 sec, +2 sec). It is seen that only one Baksan event ($E_e \sim 11$ Mev) is close to the LSD time. However, this is not surprising if the average energy of collapse-produced neutrinos $\tilde{\nu}_e$ is considerably less than 10 Mev.

The Baksan events recorded at the KAMIOKANDE-II - IMB time are given in Table 1. In fig.2(b) we show the time sequence of all Baksan events detected in the time interval from 7:35:30 UT to 7:36:30 UT.The event occurring at 7:36:06.571 (dashed line) is supposed to be of background origin /6/. Data from the KAMIOKANDE-II,IMB and LSD detectors are also shown in fig.2.

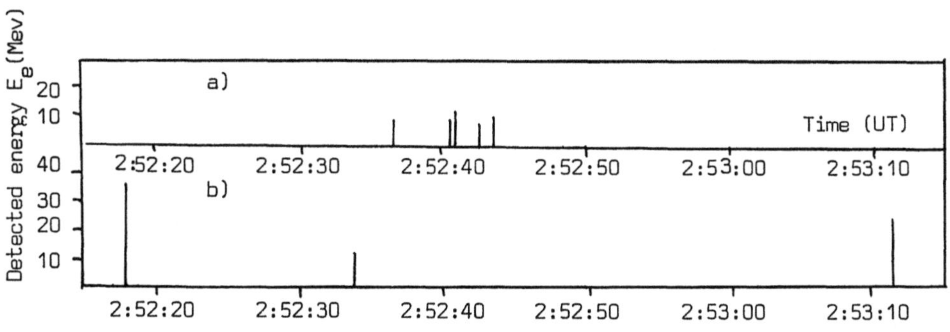

Figure 1. The time sequences of events detected by the LSD (a) and the Baksan telescope (b) at 2:52:36 UT on February 23, 1987.

Table 1. The Baksan events recorded at $7:36:11^{+2\,sec}_{-54sec}$ UT on February 23,1987.

	Time (UT)	Energy E_e (Mev)
1.	7:36:11.818	12 ± 2.4
2.	7:36:12.253	18 ± 3.6
3.	7:36:13.528	23.3 ± 4.7
4.	7:36:19.505	17 ± 3.4
5.	7:36:20.917	20.1 ± 4.0

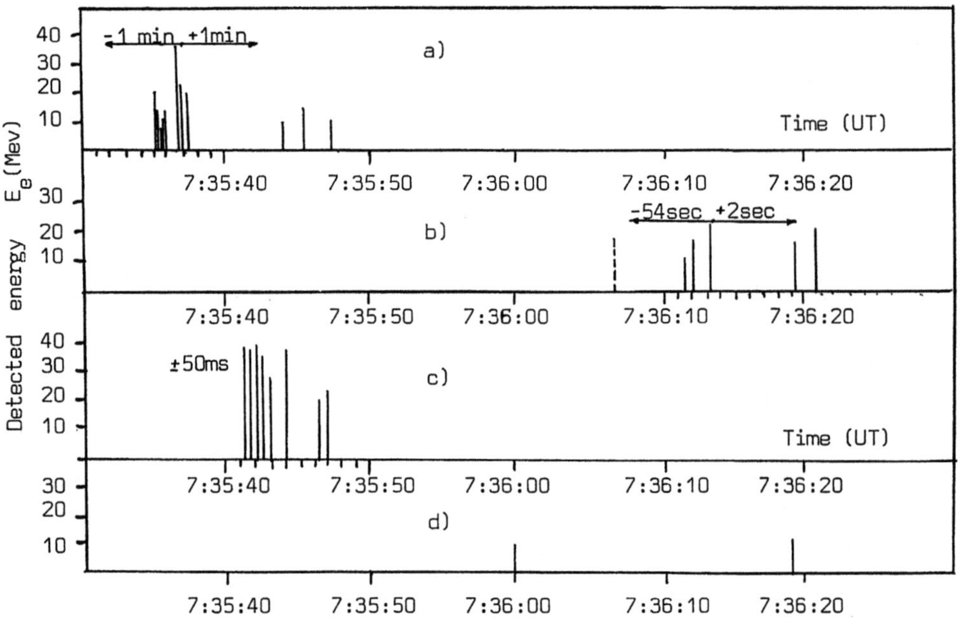

Figur 2. The time sequences of events detected by the KAMIOKANDE-II (a), the Baksan telescope (b), IMB (c) and LSD (d) detectors at 7:35 UT on February 23, 1987.

Within errors the signals of these three detectors are simultaneous as the KAMIOKANDE overall time uncertainty is ±1 minute and this value of IMB detector is ± 50 msec.

The rate of background imitations of 5 pulses per 9 seconds is ~ 0.7 day^{-1}. The random probability of such signal to get into any one-minute interval is $\sim 5 \cdot 10^{-4}$. The last value decreases considerably if we take into account the temporal shape of the signal.

So, setting t = 0.0 as the time of first event of each burst we plot the ensemble of data with their energies and reported energy errors as a function of time in Fig.3. The 11 KAMIOKANDE - II events are clustered in three bunches. The 5 Baksan events repeat the KAMIOKANDE - II temporal and energy-time shapes of the signal. The relation of KAMIOKANDE-II - Baksan events is almost conserved in the each of three bunches :~ (3 : 1).

The IMB events have another distribution.

Though the Baksan signal is only 5 events, nevertheless we extract some properties of the burst and compare them with the other detector data.

a) The Temporal Structure of the Signals

We test the temporal structure of each detected signal by the criterion of likelihood relation /7/, that was fitted to our goal /8/ :

$$\lambda = \frac{{}^{s}W}{{}^{b}W} \qquad [1]$$

where ${}^{b}W$ is a probability of signal to be of background origin, so-called "background function" ; ${}^{s}W$ is a probability of a signal to be of expected origin.

In the case of $\lambda > 1$ one can adopt the hypothesis that the observed signal is of his interest. This likelihood relation is, in fact, the statistical filter of a beforhand chosen shape of a signal.

Let us to consider arrival times of events, setting t= 0.0 for the first of them As a "background function" we take the Poisson probability of $r = k - 1$ events (k is the total number of events) to occur at the moments $t_1, t_2, \dots, t_r$ measured relatively the time of first event within time interval T_o, if the whole signal is r events within T_o. The "background function" is

$${}^{b}W(t_1, t_2, \dots, t_r / T_o, r) = \frac{r!}{T_o^{\,r}} \, dt_1 \cdots dt_r \qquad [2]$$

In the same way we obtained the "signal function":

$${}^{s}W(t_1, \dots, t_r / T_o, r) = \frac{r!}{r^r} \prod_{i=1}^{r} f_\nu(t_i) \, dt_i \qquad [3]$$

where $f_\nu(t)$ is a time dependence of neutrino emission.
Equations [2] and [3] are combined to obtain λ :

$$\lambda = \frac{T_o^{\,r}}{r^r} \prod_{i=1}^{r} f_\nu(t_i) \qquad [4]$$

With this criterion we fitted several different functional forms to the observed events including the three-bunch hypothesis. In the last case we assume that

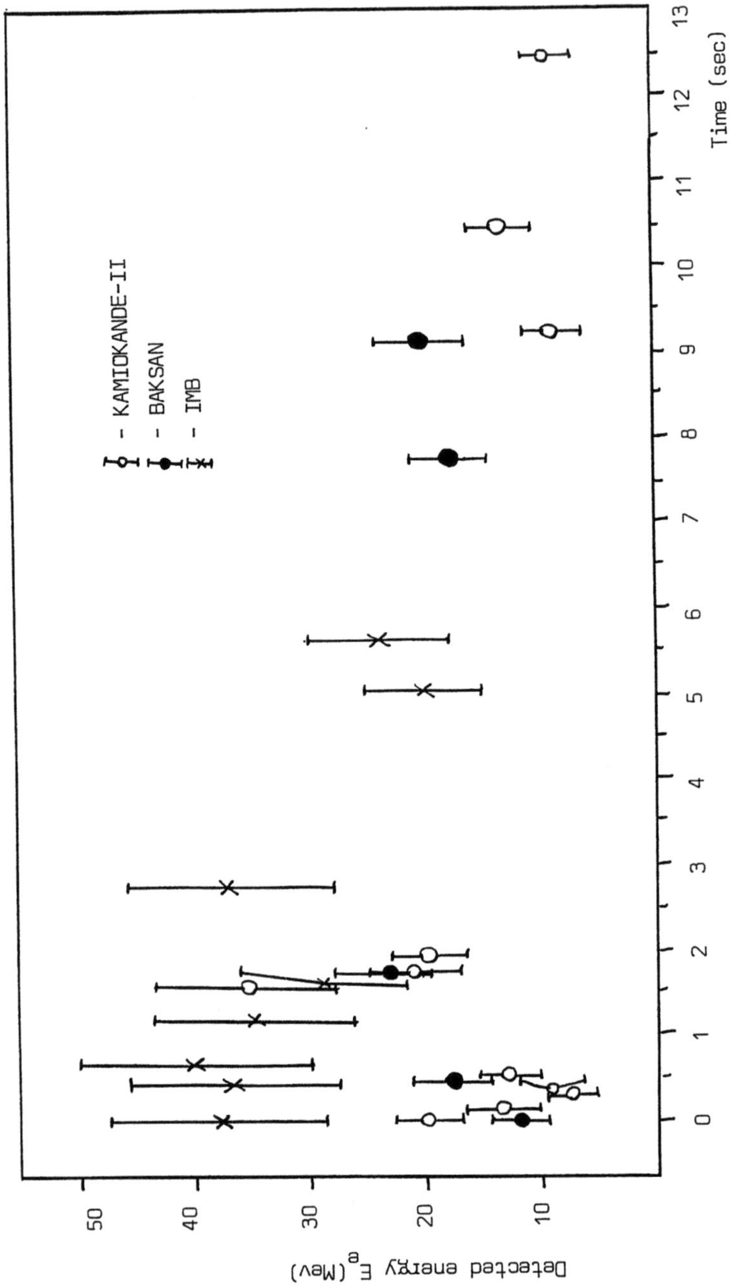

Figure 3. Energies of all events detected at 7:35 UT on February 23 versus time. t = 0.0 is set as the time of the first event of the each signal.

the neutrino emission is constant in time during the each bunch. In the gaps of (0.6 - 1.5) sec and (2 - 9) sec within which KAMIOKANDE-II has no events, we adopt the number of events to be equal to 1. Functions $f_\nu(t)$ are always normalized to the observed numbers of the events. The total duration of the neutrino signal, T_o, is supposed to be of 13 sec.

Table 2. Likelihood relation values (λ) of supposed functional shapes of the neutrino signal.

Functional shape	KAMIOKANDE-II $k = 11$	BAKSAN $k = 5$	IMB $k = 8$
$e^{-\frac{t}{\tau_0}}$ $T_0 = 13$sec	$\tau_0 \sim 5$ sec $\lambda = 36.0$	$\tau_0 \sim 6.5$sec $\lambda = 4.0$	$\tau_0 \sim 2.5$sec $\lambda = 292$
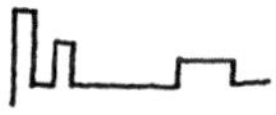 $T_0 = 13$sec	$\lambda = 2.4\cdot 10^6$	$\lambda = 16.0$	$\lambda = 0.5$

The test results with the largest values of the likelihood relation λ are summarized in Table 2. The Table 2 shows that an exponentially decaying signal is the most convenient to the IMB data while the bunching is the best fit to the KAMIOKANDE-II data. The temporal structure of the Baksan events supports apparently the bunching hypothesis. The huge value of λ of the KAMIOKANDE-II data ($2.4\cdot 10^6$) and its support with the Baksan data give proofs of the non-random shape of the neutrino signal.

b) Energy Estimates

Using the detected positron energies, we estimate the effective fit temperature of stellar neutrinosphere assuming the spectrum of Fermi-Dirac shape with different absorption coefficient α /5/. The average neutrino energy of the observed events is

$$\langle E_{\tilde\nu_e}\rangle \overset{det}{=} \frac{\int_0^\infty F(E_{\tilde\nu_e},\alpha)\cdot\sigma(E_{\tilde\nu_e})\cdot\eta(E_{\tilde\nu_e})\cdot E_{\tilde\nu_e}\cdot dE_{\tilde\nu_e}}{\int_0^\infty F(E_{\tilde\nu_e},\alpha)\cdot\sigma(E_{\tilde\nu_e})\cdot\eta(E_{\tilde\nu_e})\cdot dE_{\tilde\nu_e}} = \frac{\sum_i (E_{e,i}+\Delta m)}{K} \qquad [5]$$

where $F(E_{\tilde\nu_e},\alpha) = \frac{E^2_{\tilde\nu_e}\cdot \exp[-(E_{\tilde\nu_e}/kT)^2]}{1+\exp(E_{\tilde\nu_e}/kT)}$; kT is the effective neutrino temperature, (Mev) ; $E_{\tilde\nu_e}$ is neutrino energy, (Mev) ; $E_{e,i}$ is energy of i-th detected event, (Mev) ; $\Delta m = m_n - m_p = 1.293$ Mev ; k is the number of the observed events; $\sigma(E_{\tilde\nu_e})$ is the cross-section of $\tilde\nu_e$ absorption /9/ :

$\sigma(E_{\tilde\nu_e}) = 9.576\cdot(E_{\tilde\nu_e} - \Delta m)\cdot\sqrt{(E_{\tilde\nu_e}-\Delta m)^2 - m_e^2}\cdot 10^{-44}$ cm² ; $\eta(E_{\tilde\nu_e})$ is the detection efficiency,

$$\eta(E_{\tilde{\nu}_e}) = \begin{cases} 0.95 \cdot \left\{1 - \exp\left[-\left(\frac{E_{\tilde{\nu}_e} - \Delta m}{9.3}\right)^4\right]\right\} & \text{, KAMIOKANDE-II} \\ 0.8 \cdot \left\{1 - \exp\left[-\left(\frac{E_{\tilde{\nu}_e} - \Delta m}{10.17}\right)^9\right]\right\} & \text{, BAKSAN} \\ 1 - 3\exp\left(-\frac{E_{\tilde{\nu}_e} - \Delta m}{16}\right) & \text{, IMB /10/} \end{cases}$$

The temperatures, kT, relevant for each detector, were obtained by solutions of eq. [5] with α of 0.0; 0.02; 0.04. The average energy of the emitted neutrinos $^{em}\langle E_{\tilde{\nu}_e}\rangle$ and the total $\tilde{\nu}_e$ energy, $L_{\tilde{\nu}_e}$, corresponding to the observed number of the events are:

$$^{em}\langle E_{\tilde{\nu}_e}\rangle = \frac{\int_0^\infty F(E_{\tilde{\nu}_e},\alpha)\cdot E_{\tilde{\nu}_e}\cdot dE_{\tilde{\nu}_e}}{\int_0^\infty F(E_{\tilde{\nu}_e},\alpha)\cdot dE_{\tilde{\nu}_e}} = \begin{cases} 3.15\ kT, & \alpha = 0.0 \\ 2.78\ kT, & \alpha = 0.02 \\ 2.54\ kT, & \alpha = 0.04 \end{cases} \quad [6]$$

$$L_{\tilde{\nu}_e} = \frac{4\pi\cdot R^2}{6.24\cdot 10^5}\cdot\frac{k\ ^{em}\langle E_{\tilde{\nu}_e}\rangle}{N_p\cdot\langle\sigma(E_{\tilde{\nu}_e})\cdot\eta(E_{\tilde{\nu}_e})\rangle}\ \text{ergs} \quad [7]$$

where R is the distance to the LMC, R = 50 Kpc; N_p is the number of target free protons, $N_p = 1.42\cdot 10^{32}$ (KAMIOKANDE-II), $N_p = 1.87\cdot 10^{31}$ (BAKSAN), $N_p = 3.32\cdot 10^{32}$ (IMB) ; $\langle\sigma(E_{\tilde{\nu}_e})\cdot\eta(E_{\tilde{\nu}_e})\rangle$ is the product of $(\sigma(E_{\tilde{\nu}_e})\cdot\eta(E_{\tilde{\nu}_e}))$ averaged over the each type of the thermal neutrino spectrum.

The results of calculation are given in Table 3. The table shows that for the KAMIOKANDE-II data kT is 2.5 - 4.2 Mev, $^{em}\langle E_{\tilde{\nu}_e}\rangle$ is 8 - 11 Mev and $L_{\tilde{\nu}_e}$ is $(3 - 8)\cdot 10^{52}$ ergs, depending on the absorption coefficient α . The same Baksan parameters are : kT is 2.9 - 5.2 Mev, $^{em}\langle E_{\tilde{\nu}_e}\rangle$ is 9 - 13 Mev and $L_{\tilde{\nu}_e}$ is $(9\text{-}27)\cdot 10^{52}$ ergs. The temperatures and the appropriate $^{em}\langle E_{\tilde{\nu}_e}\rangle$ are much the same within errors for these two detectors. However, the Baksan $L_{\tilde{\nu}_e}$ is $\sim$3 times higher than the KAMIOKANDE-II one. It is quite possible that we were lucky and caught the antineutrinos in excess. But the difference is too large.

The kT and $^{em}\langle E_{\tilde{\nu}_e}\rangle$ derived from the IMB data are systematically higher, and the corresponding $L_{\tilde{\nu}_e}$ are systematically lower than the above values.

c) Periodicity of the Detected Neutrino Events

The harmonic structure of the neutrino burst was examined by a few authors. A period of 8.9 msec was found in the data of the KAMIOKANDE-II and the IMB detectors /11/. Another periods of $\sim$ 0.2 sec with evaluated statistical significance of $\sim 3\sigma$ was found for the KAMIOKANDE-II events only /12/. We also examine the periods of $\sim$ 0.2 sec. A comparable analysis of phase diagrams of events observed by

Table 3. Derived temperatures (kT), average energy of emitted antineutrinos ($^{em}\langle E_{\tilde{\nu}_e}\rangle$) and total $\tilde{\nu}_e$ energy ($L_{\tilde{\nu}_e}$).

Detector	k	$^{det}\langle E_{\tilde{\nu}_e}\rangle$ (Mev)	α = 0.0			α = 0.02			α = 0.04		
			kT (Mev)	$^{em}\langle E_{\tilde{\nu}_e}\rangle$ (Mev)	$L_{\tilde{\nu}_e}\cdot 10^{-52}$ (ergs)	kT (Mev)	$^{em}\langle E_{\tilde{\nu}_e}\rangle$ (Mev)	$L_{\tilde{\nu}_e}\cdot 10^{-52}$ (ergs)	kT (Mev)	$^{em}\langle E_{\tilde{\nu}_e}\rangle$ (Mev)	$L_{\tilde{\nu}_e}\cdot 10^{-52}$ (Ergs)
KAMIOKANDE	11	16.7±1.1	2.8±0.3	8.8±1.0	5.8±1.8	3.4±0.3	9.5±0.8	5.3±1.6	3.9±0.3	9.9±0.8	5.0±1.6
BAKSAN	5	19.4±1.7	3.3±0.4	10.4±1.3	18.6±8.5	4.0±0.5	11.1±1.4	17.2±7.9	4.6±0.6	11.7±1.5	16.1±7.4
IMB	8	33.8±2.9	4.5±0.7	14.2±2.2	2.9±1.1	5.7±0.8	15.8±2.2	2.3±0.8	6.7±0.9	17.0±2.3	2.0±0.7

α - absorption coefficient of the thermal neutrino spectrum,

k - number of the observed events,

$^{det}\langle E_{\tilde{\nu}_e}\rangle$- average detected neutrino energy.

the KAMIOKANDE-II, the IMB and the BAKSAN detectors shows evidence of this period.

Let us consider the phase diagrams of all these detector data using the phase

$$\varphi = \frac{t_o - t_i}{P_o} - N$$

where P_o is 0.202 sec and N is the number of whole periods; t_o is zero time point and t_i is arrival time of i-th event. Fig.4 shows the phase diagrams of all detectors. The ordinate axis is the event energies. As it is seen from this figure the events form two groups with the gap of $P_o/2$ between them. The second group consists of the first KAMIOKANDE-II event, of the third Baksan event and of the first, the second, the third and the sixth events of the IMB detector. The first event of the KAMIOKANDE-II is supposed to be electron neutrino /2/. Following this idea we can suggest that all events of the second group were produced by electron neutrinos. Respectinely, the events of the first group is supposed to be produced by electron antineutrinos. These events are given in Table 4.

Is this picture true? We have no answer at the present time, as the work is in progress. But as regards the third Baksan events it is quite possible it to be produced by ν_e as this event has the highest energy, 23.3 ± 4.7 Mev (Table 1) that is very close to the energy threshold of the $\nu_e\,{}^{12}C \rightarrow {}^{12}N\,e^-$ reaction.

It seems very attractively to assume the pulser of such period in the centre of the SN at least at the time of neutrino emission. And then we can imagine the neutrino and antineutrino beams due to some interactions with strong magnetic field of young pulser. But all these ideas require the detail considerations.

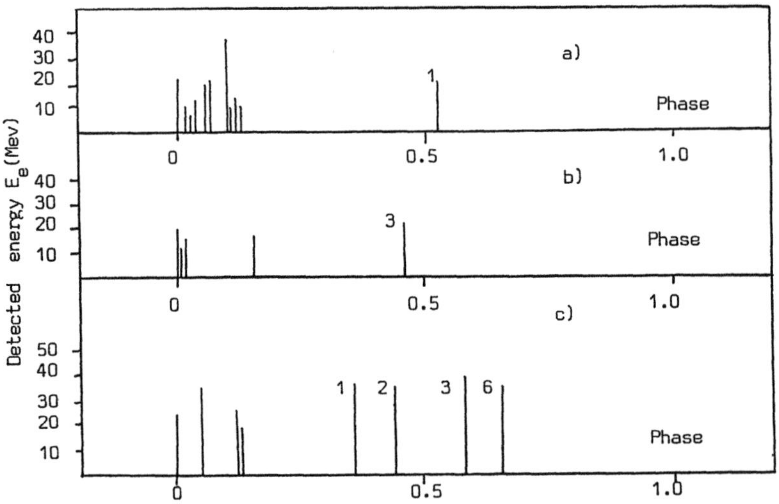

Figur 4. The phase diagrams of the events observed by the KAMIOKANDE-II (a), by the Baksan telescope (b) and the IMB detectors (c). Figures the number of the events within the time sequences.

Table 4.

Detector	first group (number of events)	Second group (number of events)	Angles (second group)
KAMIOKANDE-II	10	1	18 - 18 °
Baksan	4	1	-
IMB	4	4	74°,52°,56°,52°

4. Conclusion. We have compared the data of the KAMIOKANDE-II, the IMB and the Baksan detectors from the SN 1987A LMC and arrive at the next conclusions.

1. The effective neutrino temperatures derived from the KAMIOKANDE-II and the Baksan telescope data are much the same. The temperature range is (2.5 - 5.0) Mev depending on the shape of the thermal neutrino spectrum. In this sense it seems surprising that there is the apparent difference in the estimates of the total $\bar{\nu}_e$ energy obtained for these two detectors. Besides the effects of small-number statistics, this make us to suppose that the KAMIOKANDE-II fiducial mass is probably overestimated.

2. The temporal structure of the KAMIOKANDE-II - BAKSAN events is in favoir of three-bunch shape of the neutrino signal.

3. The properties of the IMB signal are somewhat distinct from the KAMIOKANDE -II one and the BAKSAN one.

4. Phase diagrams of these three detector data show evidence of the period $P_o \approx 0.202$ sec. The events form two groups with the gap of $\sim P_o/2$ between them. The question requires the detail consideration.

We are grateful to G.T.Zatsepin and R.Svoboda for useful discussions on the subject of this paper.

References

1. Aglietta M.et.al. : IAU Circular No.4323,(1987)
2. Hirata K. et.al.: Phys.Lett. Vol.58,1490(1987)
3. Bionata R.M. et.al.: Phys.Rev.Lett. 58 , 1494 (1987)
4. Alexeyev E.N. et.al. : Proc. 16th ICRC, Kyoto, Japan, 10 , 282 (1979) ; Proc. 12th Int.Conf."Neutrino-86",Sendai,Japan,270(1986)
5. Imshennik V.S. and Nadyozhin D.K.: Itogi Nauki i Tekhniki, ser.Astronomiya (Sov) 21 , 63 (1982)
6. Alexeyev E.N. et.al.: JETP Letters (Sov), 45 , 461 (1987)

7. Middleton D.: Trans. TRE; PGIT-3, 26 (1954) ; Davenport B., Root J. and W.L. "An Introduction to the theory of random signal and noise",McGraw-Hill Book Company, (1958)
8. Alexeyev E.N. et.al. : Preprint INR П- 0196, (1981)
9. Domogatsky G.N.,Nadyozhin D.K.,Eramzhan R.A.: Preprint IAM, No. 43, (1977)
10. Burrows A. and Lattimer J.M.: Preprint of the Steward Observatory No.725,(1987)
11. Harwit M. et.al. : Nature (1987) in press
12. Ogelman H. and Buccheri R.: Astron.Astrophys. (1987), in press

;he Neutrino Burst from SN1987A
:cted in the Mt. Blanc LSD Experiment

;lietta[1], *G. Badino*[1], *G. Bologna*[1], *C. Castagnoli*[1], *A. Castellina*[1],
)adykin[2], *W. Fulgione*[1], *P. Galeotti*[1], *F.F. Kalchukov*[2],
Kortchaguin[2], *P.V. Kortchaguin*[2], *A.S. Malguin*[2], *V.G. Ryassny*[2],
Ryazhkaya[2], *O. Saavedra*[1], *V.P. Talochkin*[2], *G. Trinchero*[1],
'netto[1], *G.T. Zatsepin*[2], *and V.F. Yakushev*[2]

uto di Cosmogeofisica del CNR, Torino, Italy and
uto di Fisica Generale dell'Università di Torino, Italy
.tute of Nuclear Research, Academy of Sciences of the
R, Moscow, USSR

.s paper we discuss the event, consisting of 5 interactions recorded during 7 ls, detected in the Mont Blanc Underground Neutrino Observatory (UNO) on ıry 23, 1987. The updated pulse amplitudes, and the background imitation ›ility of the event are reported. It is also shown that some interactions led at the same time in other underground experiments, with lower detection .ency, are consistent with the Mont Blanc event.

RODUCTION

ıerground Neutrino Observatory (UNO) has been built[1] by our two Insitutes with aiu aim to search for bursts of low energy neutrinos from stellar collapses. NO is a Liquid Scintillation Detector (LSD), running[2] since October 1984 in ›nt Blanc Laboratory, at a depth of 5200 hg/cm^2 of standard rock underground. 'ery large coverage of rock, and an additional shield, provide very good :ound conditions, and allow us to operate the UNO at a very low energy ıold.

ɜnt, considered as a candidate of a neutrino burst, was detected[3] on real time ; its occurence on February 23.12 ($2^h52^m36^s$ UT), 1987. On February 24.23 ›n in Las Campanas Observatory (Chile) reported observation[4] of an optical ıova (SN 1987A) in the Large Magellanic Cloud, 50 kpc faraway. Optical data ıte that no star brighter than magnitude 12 was present at February 23.08, and the supernova was of magnitude 6.1 at February 23.44.

ninary results of our detection have been published[5-7] previously; here we t the updated energy values of the 5 pulses in the burst, and discuss the :ound imitation probability of the event. We compare also our data with the ts obtained in other underground experiments, and discuss the connected ino outflow from the stellar collapse which originated supernova SN 1987A.

: MONT BLANC UNDERGROUND NEUTRINO OBSERVATORY

ont Blanc neutrino telescope, (5,200 hg/cm^2 of standard rock deep)underground Liquid Scintillation Detector (LSD) consisting of 72 counters in 3 layers, as ee fig.1) with 6x7 m^2 area and 4.5 m height. The total active mass is 90 tons e liquid scintillator C_nH_{2n+2} (with n = 10), containing 8.4 10^{30} free protons. ow energy local radioactivity background from the surrounding rock has been ed by shielding each scintillation counter and the whole detector with more 200 tons of Fe slabs.

Liquid scintillator is watched from the top of each counter by 3 photomulti- 's, in a 3-fold coincidence within 150 ns. Our calibrations[1], both from muons

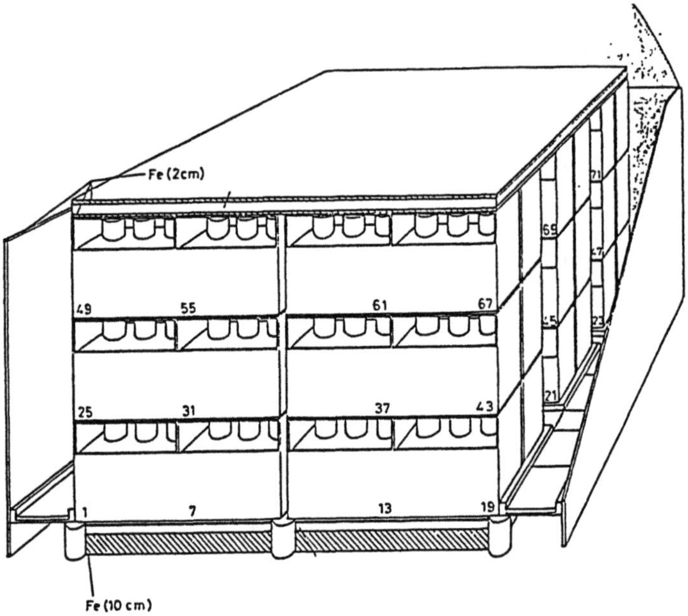

Fig. 1.- The 90 ton Liquid Scintillation Detector in the Mt.Blanc Laboratory

and with a ^{252}Cf source, show that a 1 MeV energy loss yields on the average 15 photoelectrons in 1 scintillation counter.

The low background of the LSD experiment and the high energy resolution of the scintillation counters allow us to operate the UNO at very low energy thresholds: fig. 2 shows the LSD detection efficiency as a function of the visible energy of the detected pulses. The electronic system consists of 2 levels of discriminators for each scintillation counter. A high-level (HEL) discriminator for pulses above the energy threshold 6-7 MeV for the 56 surface counters, and 5 MeV for the 16 internal ones, with a total trigger rate of 0.012 Hz. A low-level (LEL) discriminator for pulses above the energy threshold 0.8 MeV is active only during a 500 μs wide gate, opened for all the 72 counters by the main high-level trigger. Two ADCs per counter measure the energy deposition in the scintillator in 2 overlapping energy ranges. A TDC gives the relative time of the interactions with a resolution of 100 ns. Three memory buffers, 16 words deep, for the 2 ADCs and the TDC of each scintillation counter, allow us to record all pulses without dead time. On-line

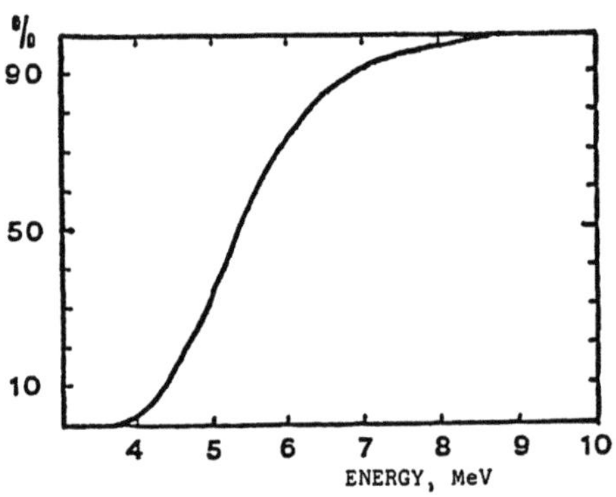

Fig.2.- The LSD detection efficiency

software prints any burst of pulses satisfying our operational definition of a neutrino burst, namely a burst of pulses above a given multiplicity in a given time. This recording system allows us to detect both products of the reaction:

$$\nu_e + p \rightarrow n + e^+$$
$$\qquad\qquad \llcorner\!\!\longrightarrow n + p \longrightarrow d + \gamma$$

with the free protons of the scintillator (positron and gammas in a delayed coincidence within 500 μs), which gives the main signal in detecting neutrinos from collapsing stars. For positron detection, the pulse amplitude is given by:

$$E_e = T_e + m_e c^2 \simeq E_{\bar{\nu}} - 1.3 \text{ MeV}$$

The gammas from the (np,dγ) capture reaction, with energy E_γ = 2.2 MeV, are emitted with an average delay of 200 us after the main interaction; the efficiency to detect a gamma, in the same counter where the neutron was produced, is about 40% on the average.
The absolute time in the LSD is recorded by using the signal broadcasted by the Italian Standard Time Service (IEN Galileo Ferraris). The accuracy of the absolute time is better than 2 msec.
The energy calibration of the LSD experiment is made by using both the 2.2 MeV peak of gammas from the (np,d γ) capture reaction and the distribution of energy losses of near vertical muons crossing the detector, peaked at 150 MeV. The accuracy of both calibration techniques is about $\pm$ 10%, and is limited at low energies by the ADCs resolution, and at high energies by the low muon statistics.

3. THE NEUTRINO BURST DETECTED IN THE LSD EXPERIMENT ON FEB. 23.12

Since January 1, 1986, the LSD experiment has been running with an average efficiency of 90% (and almost 99% since October 1986, see fig.3). Recently, the detector shielding has been partly increased for test purposes, with paraphin and lead in order to further decrease the low energy background from the surrounding rock. Trigger pulses are analysed in order to have a long term statistics and to

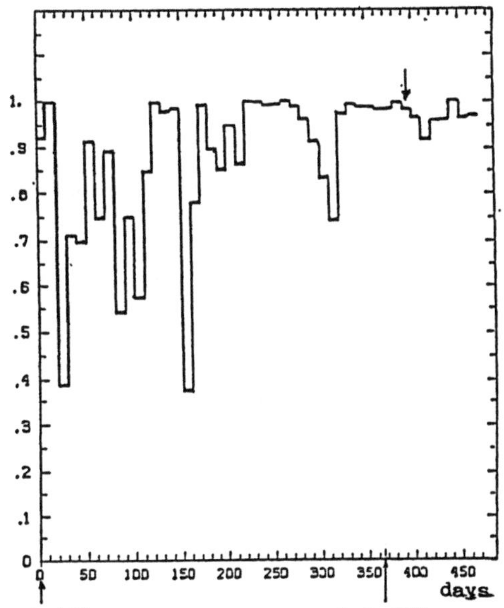

Fig.3.- The LSD running efficiency since 1.Jan.1986 for 480 days.

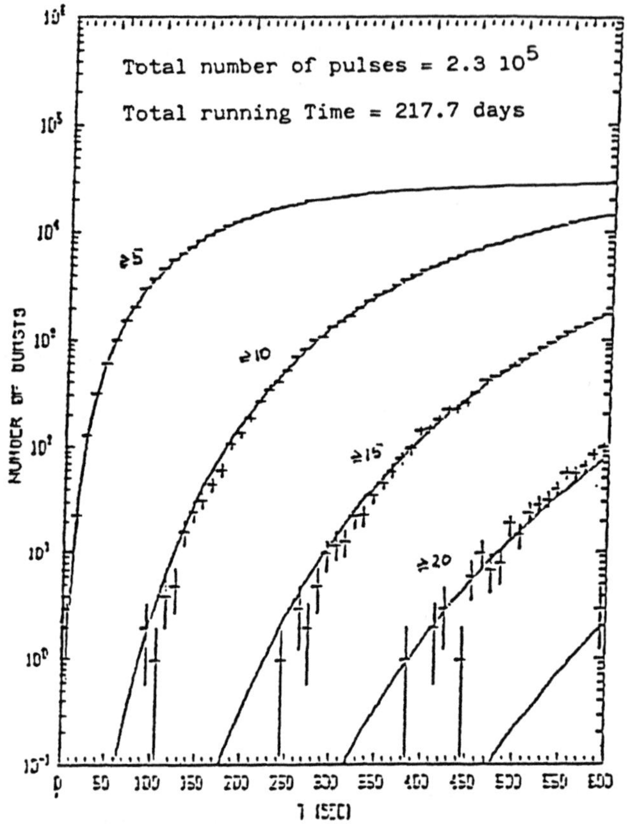

Fig.4.- Experimental distribution of bursts of pulses with multiplicity above 5, 10,15,and 20 as a function of their duration (sec),with a binning of 10 sec measured during 217.7 days.

search for bursts. The experimental time distributions of these pulses, grouped in bursts above a given multiplicity, are plotted in fig. 4 as a function of their duration and with a bin width of 10 sec. The distributions of fig. 4 refer to a data-taking period of 217.7 days (from September 28, 1986, to May 23, 1987). The smooth behaviour as well as the agreement with the predicted Poisson distributions (represented by the continuous curves) show that the trigger counting rate is stable and the detector is properly working during this time.

For monitoring purposes the detector counting rate is also checked on-line every 100 triggers. By analysing these data, and from the tape analysis several days before and after the event discussed here, we have been able to verify that the apparatus was running properly throughout the entire period.

On February 23.12, 1987, ($2^h 52^m 36^s$ UT), an event, consisting of a burst of 5 pulses and printed in real time at the occurence, was recorded in 5 different counters (3 of them internal) during 7 seconds. Table I gives the event number, the counter number, the absolute universal time (with an accuracy better than 2 msec), and the visible energy of the detected pulses (with an average accuracy of about 15%). A low energy pulse, with energy $E = 1$ MeV, accompanying the 3rd interaction, was recorded 278 μs after the main pulse in the same scintillation counter. From the measured efficiency to detect γ's from neutron capture, we expect on the average 2 such pulses in the 5 counters involved in the burst.

Table I - Characteristics of the pulses in the burst detected on February 23rd, 1987

Event no.	Counter no.	Time (UT)	E_{vis} (MeV)
994	31	$2^h52^m36^s.79$	6.2
995	14	40.65	5.8
996	25	41.01	7.8
997	35	42.70	7.0
998	33	43.80	6.8

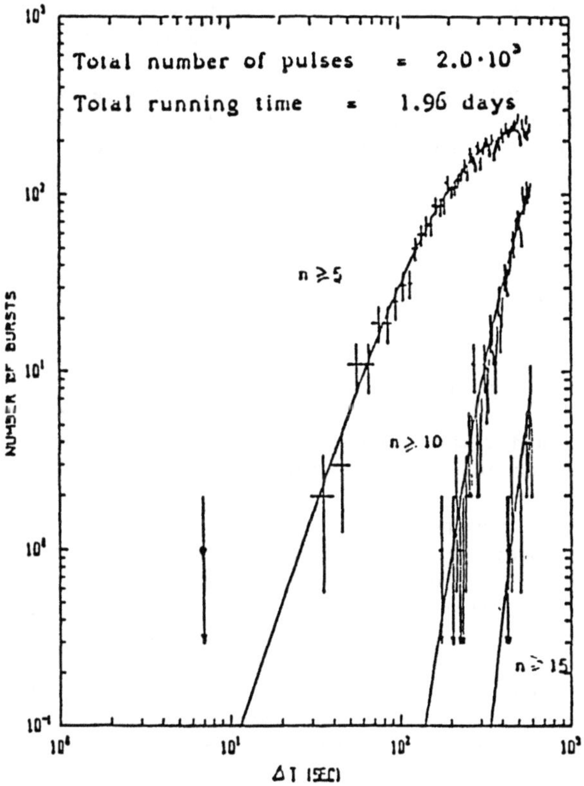

Fig.5.- The same as fig.4 but for the run including the event of Feb.23.1987.

Fig. 5 shows the distribution of the number of bursts with multiplicity n ≥ 5, n 10, and n ≥ 15, recorded during 1.96 days of data-taking encompassing the event discussed here. The full lines are computed according to a Poisson distribution of the trigger counting rate, with a binning of 10 seconds, and mean value given by our average row trigger rate, which has the stable value of 0.012 Hz during this run as for the previous ones. From fig. 5 excellent agreement between the expected and measured distributions is found, except for the point at t = 7 sec, which has been added just to show the event considered here. The imitation rate of this event, evaluated in the on-line analysis from the row trigger rate, is 0.7 per year, or $6.5\ 10^{-4}$ in the time interval corresponding to the uncertainty of the instant of collapse, i.e. the 8.2 hours interval from February 23.08 to 23.44, as suggested by the optical observations.

4. COMPARISON WITH OTHER EXPERIMENTS

Neutrino detection from supernova SN 1987 A were reported also by the Kamiokande II[9], IMB[10], and Baksan detectors[11]. Since the energy threshold in the IMB water Cerenkov detector is 20 MeV, there is no possibility to correlate the Mont Blanc event with pulses recorded in this experiment. The Baksan Scintillation Telescope, with an energy threshold of about 10 MeV, and a background counting rate of 0.033 Hz, detected one pulse with energy 10.8 MeV at February 23, UT $2^h52^m34^s \pm 2$ sec.
The comparison between the LSD and Kamiokande II data is hampered by the high degree of uncertainty in the Kamiokande linkage to universal time ($\pm$ 1 min). However, assuming that the IMB and Kamiokande events at UT 7^h35^m coincide, it is possible[8] to obtain the necessary time correlation. It follows that during a 10 seconds time interval encompassing the Mont Blanc event time, the Kamiokande II experiment recorded 4 pulses, 2 of them with visible energies 12 MeV and 8 MeV respectively, energies which are considered significative in the signal detected at 7^h35^m UT. From the Poisson distribution of the events given by the Kamiokande II collaboration[9], the random coincidence in time with the LSD event occurs about once every 17 years, corresponding to a rate of $5\ 10^{-5}$ random coincidences in the 8.2 hours uncertainty of the collapse.

5. DISCUSSION

The energy spectrum of neutrinos escaping from the collapsing core can be approximated by a Fermi Dirac distribution:

$$\varphi(\ \bar{\nu}_e/\text{secMeV})\ \propto \varepsilon^2 \exp(-\alpha\varepsilon^2)/(1+e^{\varepsilon}\ ); \text{ where } (\varepsilon = E_\nu/KT)$$

where E_ν is the $\bar{\nu}_e$ energy (in MeV), and T the temperature of the neutrinosphere. The correction factor $\exp(-\alpha\,\varepsilon^2)$ takes into account neutrino absorption in the stellar envelope above the neutrinosphere.
The central temperature KT of a type II presupernova is supposed to be less than 1 MeV, and the temperature of the neutrinosphere, after neutrino trapping, is of the order of a few MeV.Depending on the values of the absorption parameter α, the energy spectrum of the neutrinos emitted from the collapsing stellar core can be different from the spectrum produced during the collapse. Because of the energy dependence of the neutrino cross section, high energy neutrinos are more absorbed than the low energy ones, and bigger the mass of the star envelope stronger is the shift of the spectrum towards low energies. The neutrino interactions recorded in the LSD experiment agree with a temperature of the neutrinosphere of $\sim$2 MeV and absorption parameter $\alpha \approx 0.1$.
The total energy involved in the burst can be estimated assuming that the 5 pulses recorded in the LSD are due to $\bar{\nu}_e$ capture processes. This assumption seems natural as the $\bar{\nu}_e$ capture cross section is about 2 orders of magnitude higher than the scattering cross section of ν_e with the electrons of the scintillator. Adopting a distance of 52 kpc for SN 1987A, the total energy outflow in $\bar{\nu}_e$ recorded in the LSD experiment is $\sim 6\ 10^{53}$ erg. The total energy outflow, integrated over the energy spectrum (3), is: $1.2\ 10^{54} \leq E \leq 6.3\ 10^{54}$ erg (at 90% c.l.)

The limited statistics of the number of pulses in the burst may explain why this value is rather high in comparison with the predicted theoretical expectations.
The expected number of $\bar{\nu}_e$ interactions in the other underground experiments can be estimated by using their experimental detection efficiency and the energy spectrum (3). Table II gives the expected and recorded number of pulses in the other experiments which reported neutrino detection from SN 1987 A. In spite of the larger number of target protons in comparison to the LSD, the number of recorded

Table II - Expected and detected ν_e-interactions in other experiments.

Experiment	Expected	Detected
BAKSAN	1.1 $\pm$ 5	1
KAMIOKANDE	10 $\pm$ 8	2
IMB	0	0

interactions is rather small because of the lower detection efficiency of these experiments at low energies. Hence, from the experimental point of view, there is no contradiction between the event detected in the Mont Blanc neutrino Observatory and in the other detectors at the same time.

6. CONCLUSIONS

The neutrino burst recorded on real time in the Mont Blanc Underground Neutrino Observatory during the occurence of supernova SN 1987 A, indicate that the collapse had a duration of several seconds, during which a high luminosity burst of neutrinos was emitted by a low temperature neutrinosphere. Even if at large distance, this stellar collapse produced a significant signal in the LSD experiment because of its very low energy threshold and high efficiency to detect low energy pulses. A second burst, detected 4.7 hours after the Mont Blanc one, is not contradictory from the experimental point of view, and indicate that the stellar collapse in the Large Magellanic Cloud developed on at least two stages.

REFERENCES

1. G.Badino et al., Nuovo Cim.,7C,573,1984
2. M.Aglietta et al., Nuovo Cim.,9C,185,1986
3. M.Aglietta et al., IAU Circ.no.4323
4. I.Shelton, IAU Circ.no.4316
5. M.Aglietta et al., Europhys.Lett.3,1315,1987
6. V.L.Dadykin et al., JETP Lett., (in russian), 45,464,1987
7. M.Aglietta et al., Europhys.Lett., 3,1321,1987
8. A.De Rujula, CERN TH-4702,1987
9. K. Hirata et al., Phys.Rev.Lett., 58,1490,1987
10. R.M.Bionta et al., Phys.Rev.Lett., 5P, 1494, 1987
11. L.N.Alexeeva et al., JETP Lett. (in russian), 45, 461, 1987

Supernova 1987A and the Neutrino Mass

H. Meyer

University of Wuppertal, D-5600 Wuppertal 1, Fed. Rep. of Germany

1. INTRODUCTION

G.I.ZATSEPIN [1] in a short note pointed out the possibility to use the time spread of a Supernova (SN) neutrino pulse to set a limit on the ν_e-mass. His estimates showed, that limits could be achieved much superior to the then existing mass limit for electron neutrinos of ≅ 200 eV. He made an estimate for the size of a detector needed if a supernova occurred in our own galaxy. He recommended a mass of 1.000 tons and scintillator as the detector material. To detect supernovae from more distant galaxies a detector of at least 10^6 tons is needed, which for practical reasons he argued should be based on cerenkov light detection in water. Zatsepin assumed a neutrino pulse length of ≈ 3 x 10^{-2} sec, sufficiently short to get interesting limits on ν-masses for a galactic supernova. But more recent expectations for the neutrino pulse duration are of the order of (5-10) sec [2] and therefore a larger distance is needed to get the same mass limit. This requires detection of a SN outside our own galaxy. More detailed considerations on possible mass determinations before SN1987A were given by T. Piran [3], and later by Reinartz and Stodolsky [4]. However nobody would have anticipated the flood of papers which appeared very shortly after the neutrinos from SN1987A were actually observed [5].

The supernova SN1987A kindly chose to happen in the large magellanic cloud (LMC) the nearest neighbour to our galaxy, probably just too far away for the 90 tons scintillator detector (LSD) in the Mont Blanc tunnel [6] however just close enough for the two big water cerenkov detectors KamiokandeII [7] in the Kamioka leadmine (Japan, 3.000 tons) and IMB III [8] the Morton Salt mine (Ohio,USA, 7.000 tons).

Furthermore some miracles helped the water detectors to finally have their spectacular success. Both underwent considerable improvements only very recently just in time to receive the neutrino pulse from SN1987A and both detectors were taking data those few all-important seconds although IMB suffered from a power failure of about 1/4 of their PMT's. Kamiokande under normal circumstances would have been switched off for changing tapes at exactly the time of the neutrino pulse but this was not the case at that lucky day.

2. GENERAL CONSIDERATIONS

Although the basic idea to obtain ν-mass limits or even detect a finite neutrino mass is indeed very simple, the details of any particular SN event can be very complicated. First, all the physical processes generating SN-neutrinos just after the final collapse of a massive star to a compact object must be clear. Secondly the fate of the neutrinos on their travel path from the LMC to earth should be understood and finally the actual observations in the detectors should allow an unambiguous interpretation.

The progenitor of SN1987A is apparently identified as Sk-69°202, a blue giant of about 20 $M_{\odot}$ [9]. In the final development of such a star just prior to the SN-event a core of about 1.4 $M_{\odot}$ consisting mainly of iron is formed which collapses down to a very hot neutron star on a very short time scale. The density in the newly born neutron star is large enough ($\rho \approx 10^{14}$ g /cm^3 more than in ordinary nuclei) such that neutrinos will have a short interaction length inside the neutron star. Detailed models have been worked out on the process of neutrino radiation from the hot neutron star [10]. I will base some of the forthcoming arguments on neutrino spectra as obtained by Mayle, Wilson and Schramm [11]. An initial spike of ν_e from the conversion of protons to neutrons is most likely not intense enough for detection in the experiments although the first two of the Kamiokande events have frequently been attributed to this initial ν_e-pulse [12]. The main signal is assumed to originate from neutrino pairs produced in the cooling process of the neutron star and all possible neutrino flavours ν_e, ν_μ, and ν_τ are generated with the same probability (for massless neutrinos). Since limits on the number of ν-generations are available from cosmological arguments [13], from experiments at the e^+e^-- storage rings PETRA and PEP [14] and from the production cross sections of the vectorbosons (W,Z) at the CERN-Sp$\bar{p}$S [15], it seems at this time rather safe to ignore the possibility of 4 or more neutrino generations [16].

The energy spectra of different neutrino species given in [11] provide enough detail for a first estimate of the signals to be expected in water and scintillator detectors however more work on other models is clearly needed to get a better understanding of the uncertainties involved. For example the remnant could be something else than just an ordinary neutron star. A "strange matter" star or even a black hole are viable possibilities and no detailed calculations of neutrino spectra for those cases are available. Effects from a magnetic field and a rapid rotation in the very moment of creation of a neutron star are also not yet incorporated in the models although it is generally believed that they are not of great importance.

One important effect seems rather well understood and is incorporated in most models. Electron-neutrinos and -antineutrinos at early times of neutrino emission from the neutronstar are rather soft in their energy spectra since inverse β-decay on free neutrons and protons strongly damp the high energy end of the spectra as compared to ν_μ, ν_τ which suffer from neutral current scattering only. How large precisely this difference between different neutrino flavors is, is a matter of considerable importance.

Even more complications could arise if the ν-generations were mixed, not unlikely to occur if neutrinos have mass, I will however ignore this possibility in what follows (see Ref.[29]).

The influence of the intergalactic space on the behavior of the neutrinos on their way to earth could introduce further uncertainties if neutrinos were unstable or carry a residual charge or a magnetic moment. Of more direct importance is the absolute value of the distance to LMC which is known only with an uncertainty of about 30 % [17].

The detectors and the details of the observations involve still more uncertainties like the exact response of the detectors near threshold, the consideration of background events, the measurements of the absolute time and, as it became pretty clear for the water experiments, the adverse effects of power failures.

The electron neutrino mass is constraint by laboratory measurements to the range of a few tens of eV **[18]**. Therefore to a very good approximation the travel time t from LMC to earth is given in terms of a possible ν-mass m_ν and for neutrino energies E_ν of order 10 MeV by

$$t = t_{em} - t_{det} \cong \frac{d}{C}\left(1+\frac{1}{2}\frac{m_\nu^2}{E_\nu^2}\right) \qquad (1)$$

with t_{em} the time of emission at the SN, t_{det} the time of detection on earth, d the distance to LMC and c the velocity of light. We see that neutrinos with higher energy come earlier than neutrinos with lower energy and this is the basis of most of the attempts to determine a possible neutrino mass contribution to the time duration of the observed neutrino pulses. For a qualitative assessment we get from equ. (1) the time difference for two neutrinos of energies E_1 and E_2 as

$$\Delta t = t_1 - t_2 = \frac{d}{C}\frac{m_\nu^2}{2}\left(\frac{1}{E_1^2} - \frac{1}{E_2^2}\right) \qquad (2)$$

or approximately

$$\Delta t = \frac{1}{2}\,\frac{d}{C}\,\frac{m^2}{E^2} \quad (E_2 >> E_1 = E) \qquad (3)$$

using d= 5O kpc and convenient units we obtain

$$\Delta t = 2{,}575\left(\frac{d}{kpc}\right)\left(\frac{m_\nu}{10eV}\right)^2\left(\frac{10\,MeV}{E}\right)^2 [sec] \qquad (4)$$

For a neutrino mass of about 10 eV and neutrino energies of order 10 MeV a neutrino pulse of very short time duration at the SN would be about 2.6 sec long after traveling a distance of 50 kpc.

3. OBSERVATIONS

Four experiments report on neutrino signals related to the Supernova SN1987A, the liquid scintillator based experiments in the Mont Blanc tunnel **[19]** and in the Baksan underground laboratory **[20]** and the water based experiments Kamiokande II in the Kamioka mine in Japan [21] and IMB III in the Morton Salt mine[**22**]. Obviously there are problems with the absolute time of observation of the LSD experiment {2h52m36s UT} on Feb.23, almost 5h earlier than the three other experiments and - in addition- with the relative timing of KAII{7h35m35sUT}, IMB III{7h35m41sUT} and Baksan{7h36m11sUT}. If all four observations really consist of neutrino signals from SN1987A it will be very difficult to reconcile them with the standard picture of the birth of a neutron star in the supernova event. Although a collapse to a neutron star at the LSD time and later a formation of a black hole at the time of KAII and IMBIII (or an even more complicated sequence of events) cannot be ruled out **[23]** it is reasonable for the purpose of neutrino mass determinations to concentrate on the second neutrino event and just leave the LSD observation without further explanation as I will choose to do. Furthermore the time shift of about 30 sec between the Baksan and IMB observations is incomprehensible with the claimed absolute accuracy of both time measurements (± 50msec). This is shown in FIG.1 below. The significance of the Baksan data however is certainly lower than the IMB data and I will

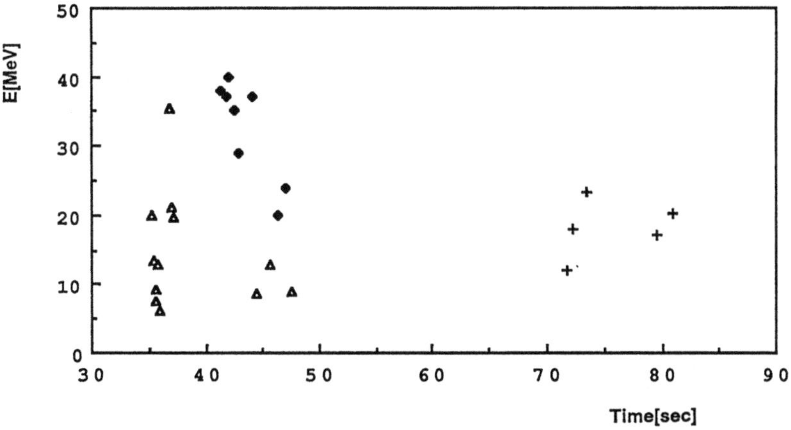

FIG1 The observation time vs. energy for Kamiokande (Δ), IMB (♦) and Baksan(+). after 7h35m30s UT on 23Feb. 1987

therefore consider in what follows only the data sets from Kamiokande and IMB. Here we have the further problem that the KAII time is uncertain by ±1 min and only the IMB observation has sufficiently precise timing well within the error box of KAII. However the KAII observation has very high significance in itself and the IMBIII data is furthermore consistent in detail with the KAII data set therefore it is generally agreed that both experiments detected the SN1987A neutrinos.

The IMBIII experiment has - for the detection SN neutrinos - a rather high threshold of about 20 MeV and with reference to eqn. (4) we see from the $[E_V]^2$ dependence that limits on neutrino masses will in general be higher than for the KAII experiment with its much lower threshold of about 7 MeV. If a more precise timing of the Kamiokande experiment would have been available, of course an analysis combining both experiments would have been of great value. An arbitrary fixing of the relative timing of the experiments as it has frequently been applied is a very strong assumption and cannot at present be justified.

The data is displayed in FIG.2 (KAII upper part and IMBIII lower part) showing the observed energy vs. the direction of the electron w.r.t. the position of the SN 1987A.

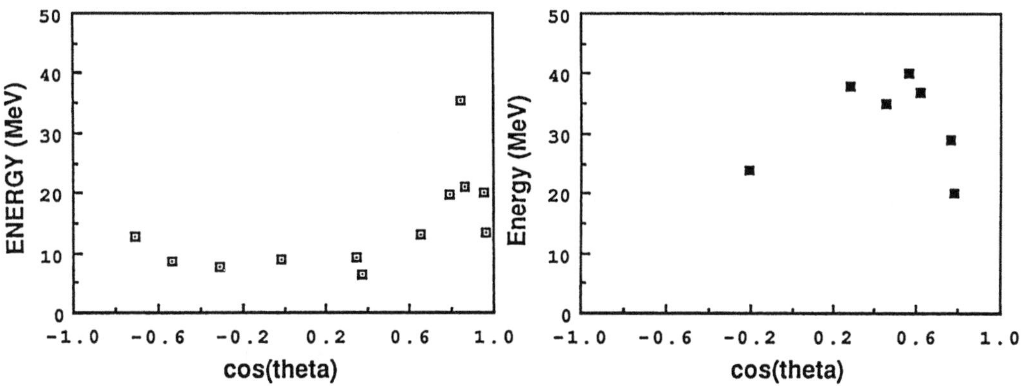

FIG. 2 The observed energy of the neutrino events from KAII (left part) and from IMBIII (right part) versus the direction of SN1987A .

4. COMPARISON WITH EXPECTATIONS

For a more quantitative assessment of the data I will refer to the calculations of Ref. [**11**] and select the model C with $M = 25\ M_\odot$. The ν-spectra show approximate Fermi-Dirac distributions and as mentioned earlier, ν_μ ($\bar{\nu}_\mu$) and ν_τ ($\bar{\nu}_\tau$) carry more energy than ν_e ($\bar{\nu}_e$) by about a factor of two.
The response of a water detector is based on observing positrons from

$\bar{\nu}_e + p \Rightarrow n + e^+$ with the cross section as given in Ref. [**24**] and electrons from elastic scattering (of all ν-types)

$\nu + e^- \Rightarrow \nu + e^-$ with cross sections as given in Ref. [**25**]

The angular distribution of positrons from the inverse ß-decay on free protons is nearly isotropic and most of the electron antineutrino energy is seen as the positron energy while in the elastic scattering processes of ν_e , ν_μ , $\bar{\nu}_\mu$, ν_τ , $\bar{\nu}_\tau$ the angular distribution is strongly peaked in the forward direction and on average only about 1/2 of the ν-energy is transferred to the electron. For $\bar{\nu}_e$ elastic scattering on average only about 1/3 of the neutrino energy is seen as the energy of the electron and with much less forward peaked angular distribution .

The relative number of positrons $N(e^+)$ to the number of electrons $N(e^-)$ is

$$\frac{N(e^+)}{N(e^-)} = \frac{\sigma_{\bar{\nu}_e p} * N(p) * \frac{dN}{dE_{\bar{\nu}_e}}}{\left(\sigma_{\nu_e} * \frac{dN}{dE_{\nu_e}} + \sigma_{\bar{\nu}_e} * \frac{dN}{dE_{\bar{\nu}_e}} + \sigma_{\nu_{\mu,\tau}} * \frac{dN}{dE_{\nu_{\mu,\tau}}}\right) N(e^-)}$$

with dN/dE_ν taken from Ref. [**11**] and the neutrino crossections σ_ν from Ref. [25], for $\nu_{\mu,\tau}$ the average of neutrino and antineutrino crossections are taken. N(P) is the number of free protons in the detector and $N(e^-)$ the number of electrons, for water we have $N(p)/N(e^-) = 1/5$. This ratio is shown in FIG. 3 and shows a dominance of the inverse ß decay signal from
$\bar{\nu}_e + p \Rightarrow n + e^+$. The contribution from elastic neutrino electron scattering is only about (5 -10)% of the total signal.

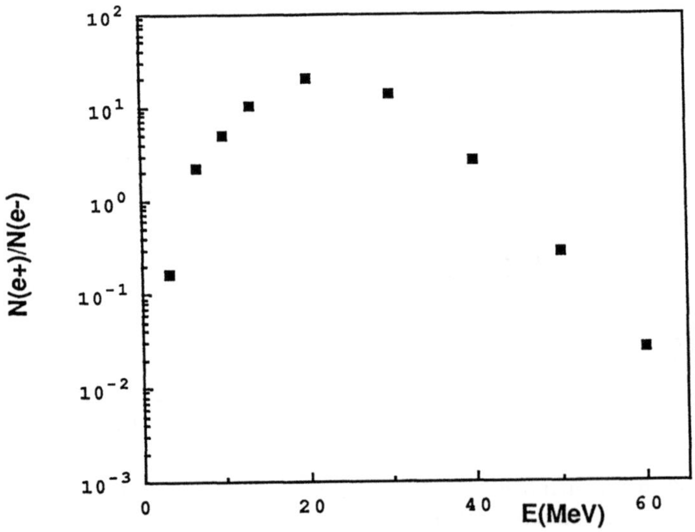

FIG. 3 The ratio of the positron to electron yield in a water detector versus the energy of the neutrinos

We see that above a neutrino energy of about 40 MeV the signal from neutral current scattering dominates, however the relative contribution to a signal in water is only a few percent. The data from KAII and IMBIII are shown in FIG. 2. The real question coming up here is the experimental finding that the angular distribution is not quite isotropic but favors the direction towards SN1987A and moreover the "directional" events have the highest energies. This does not really correspond to, what is predicted from the neutrino spectra of Ref.[**11**] it rather seems to indicate a much softer energy distribution of ν_e (and $\overline{\nu}_e$) and a dominance of events from neutrino electron scattering at the higher energies.

It is also clear that the data of FIG.2 is - although with low probability - still consistent with isotropy and therefore with dominantly absorption events especially in view of a possible bias in the angular distribution of the IMB data due to the missing PMT`s (see footnote (c) to table III of Ref. [**22**]). But a more detailed evaluation of this possible bias in the IMB data as well as giving the angular uncertainty for each individual event is highly desirable for a more quantitative assessment.

Using (in accordance with standard models) the assumption that all events observed by IMBIII and KAII are due to $\overline{\nu}_e + p \Rightarrow n + e^+$ the total energy emitted in ν s at the source can be evaluated. This involves taking account of the exact trigger threshold behavior of the two experiments and an extrapolation from the observed part of the neutrino spectrum into the region below threshold. The result is (1.5 - 4.0) x 10^{53} erg, (Ref.[**26**]) in excellent agreement with expectations for the formation of a neutron star.In case the high energy events from both IMBIII and KAII were due to elastic scattering $\nu + e \Rightarrow \nu + e$ the total energy in ν `s emitted from the SN-event came out to be higher by about a factor of 10. This seems incomprehensible with the total energy available from the gravitational field in the process of forming a standard neutron star. One would be forced to the conclusion that a neutron star was not created in the event of SN1987A, but presumably more mass was involved, in dramatic contrast to the standard view of the development of a large star like Sk-69°202. New light on this important issue can be expected from the discovery of the true nature of the remnant in SN1987A most likely to be expected from detailed observations in the X-ray region.

5. MASS LIMITS

Under the proposition that all events are due to $\overline{\nu}_e$ absorption the energy-time correlation seen in the data can be used to set limits on the $\overline{\nu}_e$-neutrino mass. The IMBIII data are of higher energy than KAII and because of the E_ν^2 - dependence in eqn (4) they are much less useful than the KAII data. If KAII had precise absolute timing both data sets together would have been extremely valuable. Given this situation using the KAII data alone sets the best limits.

Very many papers have appeared on this subject and finite ν-masses as well as limits ranging from < 6 eV to < 32 eV have been claimed[**5**]. Since no clear cut ordering of early detection times with higher neutrino energies can be seen in the data and furthermore since cooling of a neutron star does already produce a trend in the direction of lower neutrino energy at later times *only the highest value* - **m_ν < 32 eV** - *for the neutrino mass limit should be taken seriously* [**25**].

In many of the papers better limits on neutrino masses are achieved by ignoring some events from the data sample, with some preference to consider the last three events of KAII as background events. Such strong assumptions on the true nature of the published events can certainly be made to indicate the dependence of the results on particular events in the data sample for a conservative statement on neutrino mass limits however none of the published events should be discarded from the analysis although the observed grouping of the KAII data is indeed very suggestive for further speculations on the detailed nature of the SN event.

It is however amusing to realize that a sizeable contribution of neutrino scattering events (from ν_μ , ν_τ) to the data (due to small time delay with the electron neutrinos) would result in mass limits for ν_μ and ν_τ of about 30 eV, a very dramatic improvement over the present experimental results, by a factor of 10^3 for the ν_μ [27] and 10^6 for the ν_τ [28].

This not yet quite satisfactory situation with the KAII and IMBIII data would become extremely complicated for neutrinos with finite masses and mixing angles. Then oscillations and neutrino decays had to be taken into account. Furthermore the neutrinos could have a magnetic moment or even a residual charge. Those complications have already partly been covered in the literature [29] and will not be discussed here, I think just the experimental observations present enough open problems.

6. ARE IMPROVEMENTS POSSIBLE

Where do we expect future improvements regarding the foregoing discussion and in particular with respect to neutrino mass determinations?

Looking at eqn. (4) a better determination of the distance (entering only in the square root) will certainly help and can rather safely be expected to be obtained from future observations of SN1987A. A more extented and detailed publication of the experimental results from both the Kamiokande and the IMB experiments would also be invaluable, but dramatic developments may be expected in case the remnant is not a standard neutron star! Somewhat related is the problem of the number of neutrino generations which enters the evaluation of the total energy emitted in neutrinos, here a definite answer can be expected in the next few years from experiments at very high energy e^+e^- - (or $p\bar{p}$-) colliders. Also a true solution of the solar neutrino problem might be very useful to further constrain the possibilities.

Most dramatic and important however would certainly be the observation another SN neutrino pulse in the near future with more and better detectors operational and hopefully taking data for exactly those few seconds when sufficiently energetic SN neutrinos path through the detectors.

Acknowledgements:

I would like to thank V. Klapdor for the invitation to this meeting and for the kind hospitality. I also would like to acknowledge interesting conversations with W. Hillebrandt, P. Langacker, E. Alexejev and V. Lubimov.

References:

1) G. I. Zatsepin **JETP Lett. 8**, 205 (1968)

2) R.F. Sawyer et al. **Ap.J. 230,** 859 (1979) A. Burrows et al. **Ap.J. 251,** 325 (1981) A. Burrows et al. **Ap.J. 307,** 178 (1987)

3) T. Piran **Phys. Lett. B102**, 299 (1981)

4) P. Reinartz et al. **Z. f. Physik C27,** 507 (1985)

5) J. N. Bahcall et al. **NATURE 326**, 135 (1987) W. Hillebrandt et al. **Astron. Astrophys. 177,** L41 (1987) K. Sato et al. **Phys. Rev. Lett.58,** 2722 (1987) J. Arafune et al. **Phys. Rev. Lett. 59,** 367 (1987) J. N. Bahcall et al. **Nature 326,** 476 (1987) W. D. Arnett et al. **Phys. Rev. Lett. 58**, 1906 (1987) W. Hillebrandt et al. **Astron. Astrophys. 180,** L20 (1987) W. D. Arnett **Ap.J. 319,** 136 (1987) R. Schaeffer et al. **NATURE 300,** 142 (1987) A. Burrows et al. **Ap.J. 318,** L63 (1987) J. Arafune et al. **Phys. Rev. Lett. 59,** 1864 (1987) J. Ellis et al. **Phys. Lett. B193, 525** (1987) A. De Rujula **Phys. Lett. B193**, 514 (1987) E. W. Kolb et al. **Phys. Rev. D35**, 3598 (1987)

6) G. Baldino et al. **Nuovo Cimento C7,** 573 (1984)

7) A. Suzuki in **12th Int. Conf. on Neutrino Physics and Astrophysics** Sendai, June 1986

8) J. M. LoSecco **Science 224**, 56 (1984) T.J. Haines et al. **Phys. Rev. Lett. 57,** 1986 (1986)

9) R. Gilmozzi et al. **NATURE 328**, 318 (1987)

10) S. A. Colgate et al. **Ap.J. 143,** 626 (1966) W. D. Arnett **Canadian J.Phys. 45**,1621 (1967) J.R. Wilson **Ap.J. 163,** 290 (1971) T.J. Mazurek **NATURE 252,** 287 (1974) K. Sato **Prog.Theo.Phys. 54,** 1325 (1975) R.F. Sawyer et al. **Ap.J. 230,** 859 (1979) A. Burrows et al. **Ap.J. 251,** 325 (1981) A. Burrows et al. **Ap.J. 259,** 330 (1982) H. Bethe et al. **Ap.J. 295,** 14 (1985)

11) R. W. Mayle et al. **Ap.J. 318,** 288 (1987)

12) J. Arafune et al. **Phys. Rev. Lett. 59,** 1864 (1987)

13) G. Steigman et al. **Phys. Lett. B176,** 33 (1986)

14) H.J. Gerber in Proc. of the **Int. Europhysics Conf. Uppsala Vol II,** p.937(1987) S.L. Wu **CERN-EP/87-235,** talk at the 1987 International Symposium on Lepton and Photon Interactions at High Energies, Hamburg.

15) P. Jenni **CERN-EP/87-226** 10 December 1987 L. DiLella **CERN-EP/88-02** 6 January 1988

16) R. Schaeffer et al. **NATURE 300,** 142 (1987) J. Ellis et al. **Phys. Lett. B193,** 525 (1987) D. N. Schramm **FERMILAB-PUB-87/91-A,** June 1987

17) A. R. Sandage et al. **Ap.J. 167**, 293 (1971) G. De Vaucouleurs **Ap.J. 223**, 730 (1978) E. W. Kolb et al. **Phys. Rev. D35**, 3598 (1987)

18) M. Fritschi et al. **Phys. Lett. 173B,** 485 (1987)

19) E. Alexeyev, private communication and this volume

20) M. Aglietta et al. **Europhys. Lett. 3,** 1315 (1987) M. Aglietta et al. **Europhys. Lett. 3,** 1321 (1987)

21) K. Hirata et al. **Phys. Rev. Lett. 58,** 1490 (1987)

22) R. Bionta et al. **Phys. Rev. Lett. 58,** 1494 (1987)

23) W. Hillebrandt et al. **Astron. Astrophys. 180,** L20 (1987) S. Nussinov et al. **NATURE 329,** 134 (1987) A.De Rujula **Phys. Lett. B193**, 514 (1987)

24) P. Vogel **Phys. Rev. 29,** 1918 (1984)

25) E. W. Kolb et al. **Phys. Rev. D35**, 3598 (1987) L.F. Abbott et al. **CERN-TH.4799/87** July 1987

26) see References 5 ,16 and 25

27) R. A. Abela et al. **Phys. Lett. B146,** 431 (1984)

28) H. Albrecht et al. **Phys. Lett. B163,** 404 (1985) ARGUS-Coll. **DESY 87-148** November 1987

29) L. Wolfenstein **Phys. Lett. B194**, 197 (1987)
J. A. Friedman et al. **SLAC-PUB-4261**,September1987(Revised)
J. Arafune et al. **Phys. Rev. Lett. 59,** 1864 (1987) D. Noetzold **Phys. Lett. B196**, 315 (1987) M. Fukugita Preprint **RIFP-718** October 1987.

Part VI

Future Neutrino Projects

(A Few) Future Neutrino Experiments

E. Bellotti

Laboratori Nazionali del Gran Sasso, INFN, Università degli Studi, Via Celoria, 16, I-20133 Milano, Italy

Many experiments and projects on neutrino physics have been discussed at this Conference; I will describe a few other non accelerator experiments which are, according to the title of this paper, in the design or construction stage. I will concentrate on double beta decay and solar neutrinos future detectors.

1. Double beta decay

According to papers presented at this conference, the situation of double beta decay can be summarized as follows [1]:

- the new analysis by MOE et al. [2] of data on ^{82}Se leads to a half-life for the two neutrino mode which is in agreement with the radiogeochemical measurements;
- experiments on ^{76}Ge [3] are approaching a limit of 10^{24} years for the neutrinoless mode which is practically the highest value that can be reached with present detectors;
- theoretical problems in the computation of nuclear matrix elements seem now understood [see the theoretical papers presented at this same conference].

The question is: how to improve experimentally this situation? Answer is quite obvious:

- make use of large sample of isotopically enriched materials;
- find new methods which allow a wide choice of elements, the use of large samples and, at the same time, show good energy resolution and stability over long periods, important features for the neutrinoless decay mode detection.

Germanium, as also it has been stressed by AVIGNONE [4], and Xenon enriched samples can be efficently used. Regarding Xenon, at present there are three experiments in progress which use, or plan to use, enriched ^{136}Xe: the Soviet experiment at BAKSAN [5], the CALTECH TPC, now in operation with normal Xenon [3], and the Milano experiment at Gran Sasso, also at present in operation with normal Xenon. The present limit obtained by the Milano group, which operates a multicell proportional chamber [6], on the neutrinoless decay mode is $T_{1/2}>10^{22}$ years (Fig.1); after the substitutions of a few wires, the chamber has been filled with normal Xenon again to find the better operation conditions (pressure and H.V.); in a short time - a few weeks from now - it will be filled with Xenon enriched at ~65% in ^{136}Xe. If the background conditions and energy resolution will remain the same as with natural Xe (but they could improve) a limit 10 time higher than the quoted above will be reached in a few months of good operation.

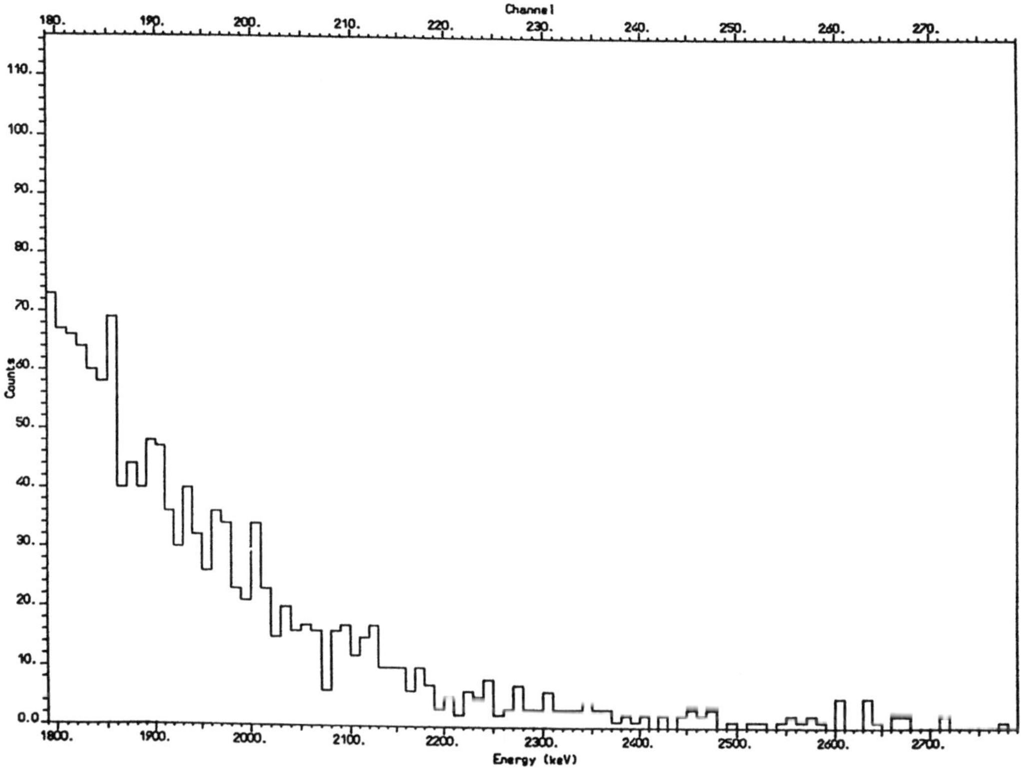

Fig. 1 - Part of the energy spectrum obtained with a Xe multicell proportional counter.

Regarding new techniques, it seems to me that cryogenic devices are promising detectors to search for double beta decay; the interest in cryogenic detectors - superheated superconducting granules, tunnel junctions, bolometers - is increasing as it has been demonstrated at the Workshop on Cryogenic Detectors in Munich a few months ago [7] where potentiality of these detectors has been also discussed. I will briefly discuss the case of bolometers in view of their use in double beta experiments [8].

1.1 Bolometers

The heat capacity of an insulating crystal is given by:

$$C_V = 1.2\cdot 10^{16} \times (m/A) \times (T/\Theta)^3 \quad \text{MeV/K}$$

where m is the mass of the sample (g), A the atomic weight, T the absolute temperature and Θ the Debye temperature (see Table for Debye temperatures of a few interesting materials).

Schematically the structure of a bolometric detector is shown in Fig. 2; the thermistor, which can be a part of the crystal, is biased with a low voltage. Deposition of energy in the sample causes an increase in temperature and therefore a

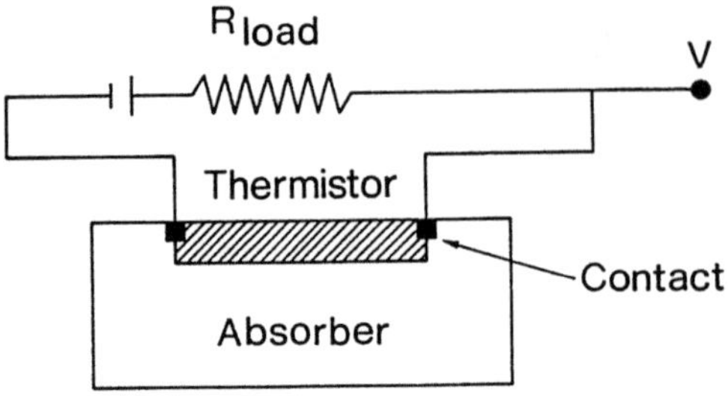

Fig. 2 - Scheme of a bolometer.

TABLE

material	A	Debye temperature (K)
Si	28.1	640
Ge	72.6	370
Zr	91.22	291
Mo	95.94	460
Te	127.	153
Nd	144.29	163

decrease in the resistance of the thermistor. The resulting voltage signal is detected and analyzed by an electronic chain, like in other radiation detectors.

To be definite, consider a Te sample of 127 g (m/A=1); the heat capacity at 0.01 K, a temperature which can be reached with dilution refrigerator, is 3.4 10^3 MeV/K; an energy deposition of 2 MeV causes an increase in temperature of 0.6 mK, a variation which can be measured with the technique above sketched.

According to a simple model of these detectors, the expected energy resolution is given by:

$$E= \xi\sqrt{k\ T^2 C_V}$$

where k =8.6 10^{-11} MeV K^{-1} is the Boltzmann constant and ξ is a numerical factor of the order of a few units. Putting numbers in such a formula, resolutions of a few eV are obtained. The validity of this formula requires that all the energy lost by the particle and absorbed in the bolometer is transformed in heat, without any loss. This hypothesis seems not to be valid, both theoretically and experimentally. The problem of the energy resolution of these detectors is therefore an open problem.

Many encouraging results have been obtained in past years; let me list some of them.

The University of Wisconsin - Goddard Space Flight Center Coll. [9] measured the ^{55}Fe X-ray lines with a TeCdHg adsorber and obtained an energy resolution of 35 eV.

N. CORON et al. [10] used a very thin (0.02 mm) diamond adsorber and measured alfa particles from various sources; they obtained an energy resolution of 36 keV (Fig. 3) at 1.2 K. They also demonstrated the good linearity on a very large energy

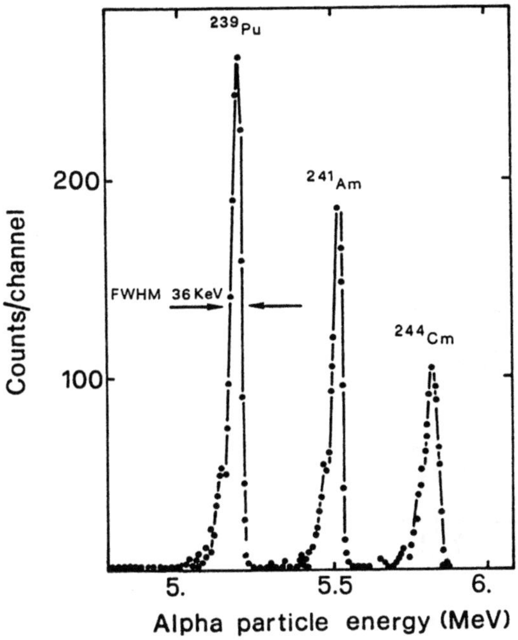

Fig. 3 - Energy spectrum obtained with a diamond detector.

interval (from 20 keV to 6 MeV) [11]. They measured also alfa particles from a ^{224}Ra source; due to the fact that some daughter nuclei entered in the absorber, some alfa- lines present a satellite peak at the high energy side, corresponding to the total transition energy (alfa and recoil energy) of the decay (Fig. 4) [12].

J.C. OVERLEY et al. [13] measured alfa particles and obtained an energy resolution of 60 keV FWHM.

E. FIORINI et al. [14] tested various absorber (silicon, germanium, diamond) and measured as well x-rays and alfa particles. Fig. 5 shows the energy spectrum obtained with an absorber significantly larger than the previous ones, namely 2x3x.5

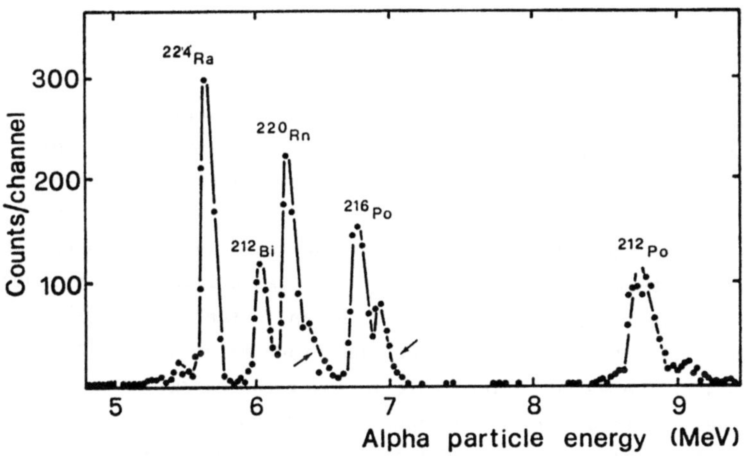

Fig. 4 - Energy spectrum of alpha particles; arrows indicate peaks corresponding to the total (alpha and recoil nucleus) energy.

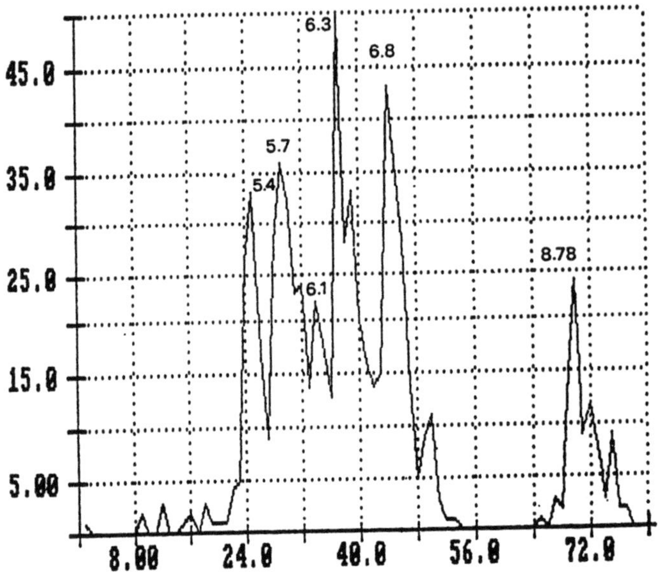

Fig. 5 - Energy spectrum of alpha particles.

mm^3; energy resolution is not as good as in the measurement by Coron et al., however also in this case the high energy satellites are visible.

Many points have still to be understood, namely: validity of the Debye law at very low temperatures, energy resolution and long term stability, which is very important for long lasting measurements as they are needed for most of the applications. However I think in a short time it will be possible to use cryogenic crystal of ≥ 1 cm^3 of Te or Ge and therefore to start a real measurement on double beta decay.

2. Solar neutrinos - direct experiments

At present, two results on ^8B neutrino flux from the Sun are available. The first one is from the famous Davis experiment; neutrino flux was measured by a radiochemical technique through the reaction:

$$^{37}Cl(\nu_e,e^-)^{37}Ar$$

The energy threshold being 840 keV, this measurements was essentially sensitive to ^8B neutrinos. In only one figure, the result of a many years lasting measurement is: 2.1 ± 0.3 SNU's, to be compared to $6.\pm2.$ SNU's as predicted from the Standard Solar Model.

Recently a limit on the same flux has been given by the KAMIOKANDE II Collaboration [15]. At it is known, the Kamiokande detector was originally designed to search for proton decay events (Fig. 6); however, thanks to the large number of phototubes covering the walls, the purity of water and other materials and to the very accurate data analysis, the authors were able to lower the background and the energy threshold at values which allow to distinguish solar neutrinos interactions from background events. Solar neutrinos are detected through the purely leptonic reaction:

$$\nu e^- \rightarrow \nu e^-$$

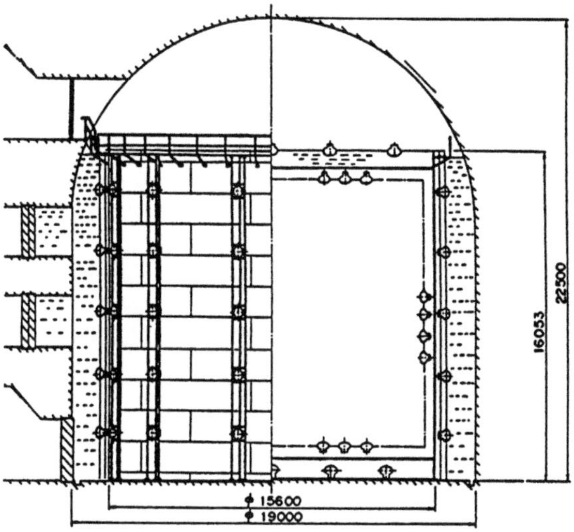

Fig. 6 - The KAMIOKANDE-II detector.

It is important to notice that also at low energy, the electron is emitted essentially in the same direction of the incoming neutrino; f.i. an electron neutrino of 10 MeV gives in 40% of the cases an electron of energy greater than 6 MeV which is emitted at an angle less than 20°, which means in a solid angle around the direction of the interacting neutrino of 3% of total solid angle.

One of the difficulties in devices like water Cerenkov detectors, is a reliable energy calibration; in this experiment, electrons emitted in high Q-values beta decays of nuclei like ^{15}C produced by cosmic rays as well electrons from μ-decay were used to calibrate in energy the detector (Fig. 7).

Background is quite high (Fig. 8); it was both by restricting the fiducial volume and by rejecting events which are time correlated to muons passing through the detector.

Figure 9 shows the angular distribution of electrons of energy >10.5 MeV and the expected signal, according to the SSM. No clear excess is visible and therefore a limit of 3.2 10^6 $\nu_e(cm^2/s)$ was set; this limit corresponds to about one half of the expected flux.

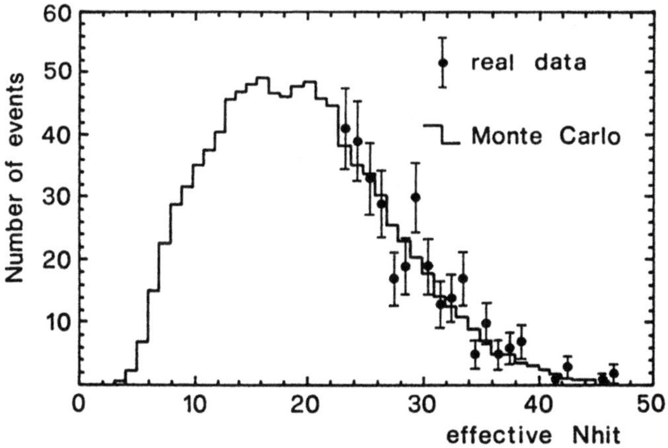

Fig. 7 - Energy calibration using spallation events.

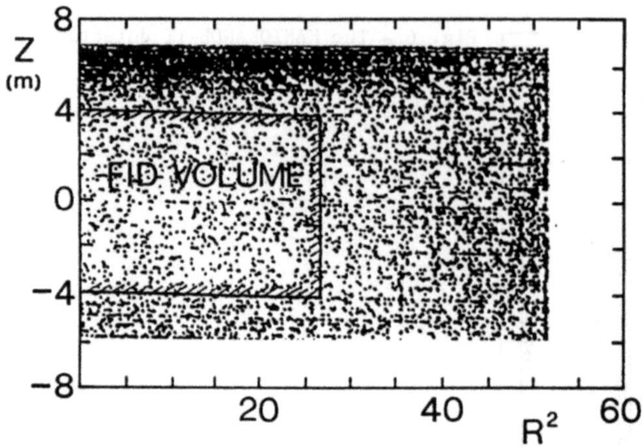

Fig. 8 - Spatial distribution of low energy events.

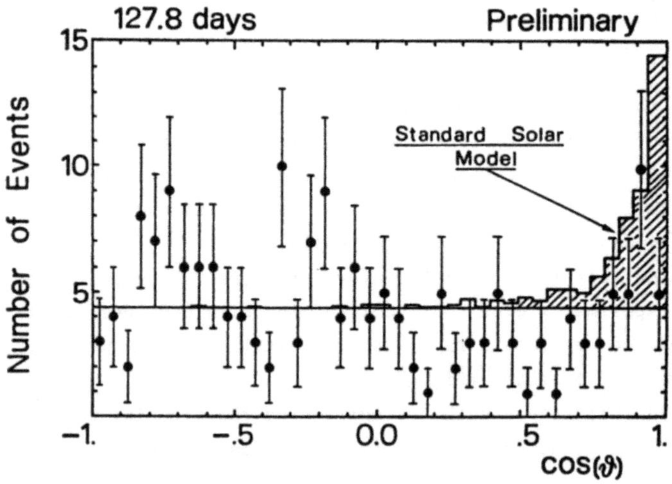

Fig. 9 - Angular distribution for events with E>10.5 MeV.

From this result and assuming that the ν deficit is due to ν oscillation in matter, it is already possible to exclude some of the possible values of (Δm^2, $\sin^2 2\theta$) which could result from the Gallium experiment and from the already existing Clorine measurement [16].

New experiments are in design or in construction stage and some of them have been presented at this same Conference [16, 14]; here I would briefly describe two direct experiments: ICARUS and BOREX. In both detectors it is possible to observe either neutral current events, which can be induced by neutrinos of any flavour and charged current events which can be induced only by electron neutrinos; therefore it is possible, at least in principle, to clarify the origin of the observed deficit of solar neutrinos. Information on energy distribution can be also extracted.

ICARUS is essentially a large liquid argon Time Projection Chamber [18]. According to present design, the volume will be 2600 m^3, corresponding to 3500 tons ($2\ 10^{31}$) nucleons or $5\ 10^{31}$ ^{40}Ar nuclei). The spatial resolution is expected to be a few mm.

Two reactions are important to measure the solar neutrino flux:

$$\nu e^- \rightarrow \nu e^- \qquad \text{(I)}$$

and

$$^{40}Ar\ (\nu_e, e^-)^{40}K^* \quad (4.38\ \text{MeV}) \qquad \text{(II)}$$

according to the SSM 1500 events/(kton year) of type (I) and 1330 events/(kton year) of type (II) with an electron of energy greater than 5 MeV are expected. It is interesting to consider the ratio:

$$R = \frac{\nu e^- \rightarrow \nu e^-}{^{40}Ar(\nu_e\ e^-)^{40}K^*}$$

whose value has to be 1.1. if only ν_e are interacting. If neutrinos fully oscillate, f.i. in ν_μ and ν_τ, R becomes

$$R = \frac{\frac{1}{3}(\nu_e\ e^-) + \frac{1}{3}(\nu_\mu\ e^-) + \frac{1}{3}(\nu_\tau\ e^-)}{^{40}Ar(\nu_e,\ e^-)^{40}K^*} \sim 1.4$$

The two reactions can be distinguished by their angular distribution with respect the direction of the Sun; however it has to be remarked that background could be very different in the two cases.

The construction of such a detector, which will be installed in hall C of the Gran Sasso Laboratory, is a big technical effort and many problems have to be solved.

Major difficulties are related to gas purification and to the read-out system. A first prototype has been constructed at CERN and exposed to a test beam. A scheme of chamber set-up is shown in Fig. 10.

The purification system was based on the use of Oxisorb and azeolites. Signals (Fig. 11) and tracks have been seen with the expected features. As a second step, a larger cryostat is in construction; its volume will be 2 m^3 and it will be possible to drift electrons along 1.5 - 2. m.

In parallel, test on a different purification system, based on the use of getters are in progress in collaboration with SAES-GETTERS; preliminary results give half-lives of ~1 ms (corresponding to a drift length of 2 m with a field of 1 kV/cm).

In the same time, the final design of the cryostat is in progress.

Time scale for the construction will depend on the strategy of the Collaboration; in any way the minimum time to build and put in operation the detector is probably larger than three years.

Obviously, such a detector will give data on a large variety of phenomena; let me simply mention some of them: neutrino bursts from collapsing stars, atmospheric neutrinos, slow monopoles, muons to search for point sources of high energy cosmic rays etc.

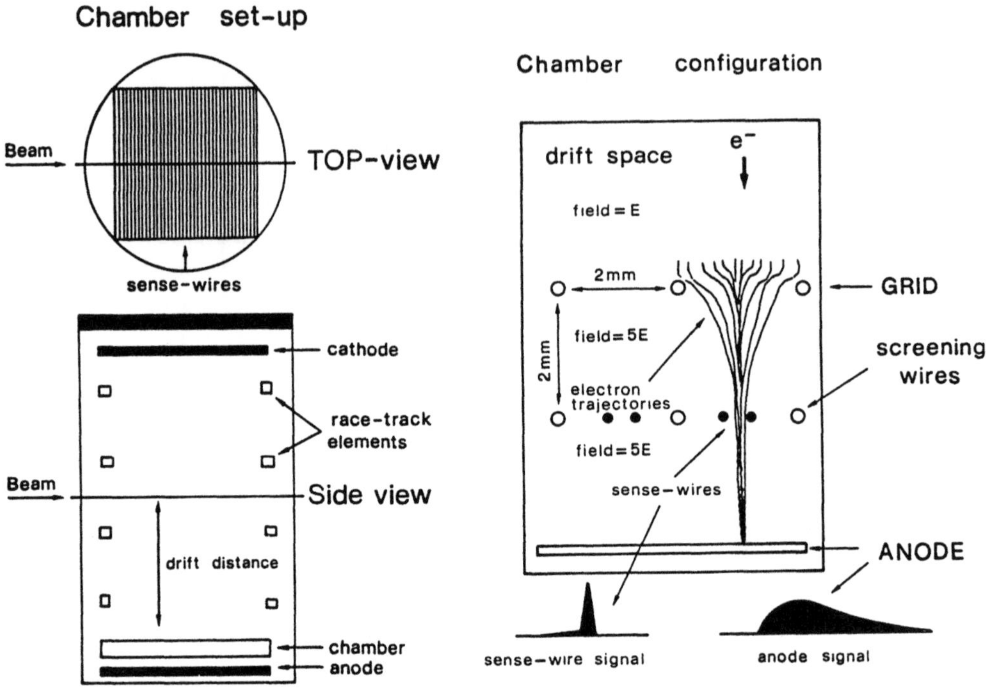

Fig. 10 - a) Chamber set-up of a prototype of ICARUS;
b) Details of the detection system.

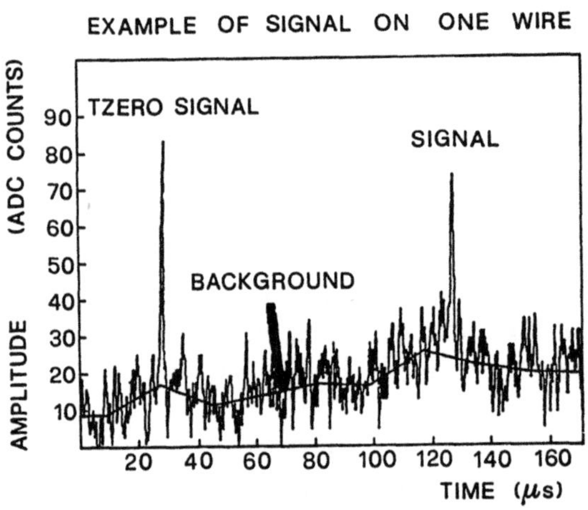

Fig. 11 - Signals in a prototype of ICARUS.

BOREX

R.S. RAGHAVAN, S. PAKVASA and B.A. BROWN [19] propose to use ^{11}B (i.a. 80%) as a target material for solar neutrinos. Fig. 12 shows the nuclear level scheme of ^{11}B and ^{11}C as well as possible transitions induced by neutral current and charged current interactions; obviously purely leptonic interactions will be also present. The following Table reports the expected yields, assuming a flux of 4 10^6 $\nu_e/(cm^2 s)$.

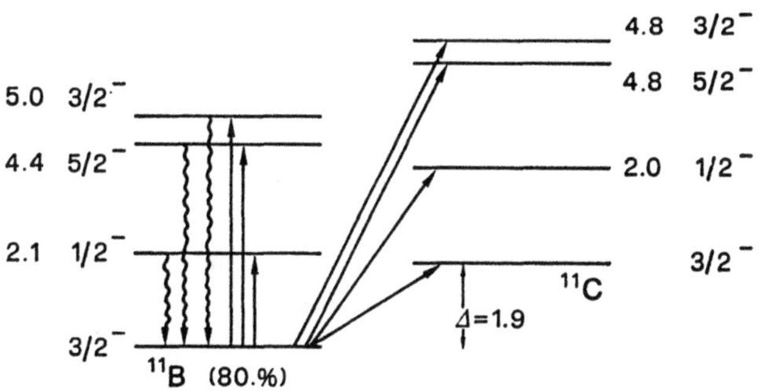

Fig. 12 - Level schemes of ^{11}B on ^{11}C.

TABLE

Nuclear excitations energy level (MeV)	Y*	c.c. interactions energy level (MeV)	Y*^
-	-	g.s.	5760
2.12	415	2.00	1125
4.45	285	4.32	675
5.02	375	4.81	840

* Yield/(kiloton target element) yr

^ Yield for total energy E> 4 MeV.

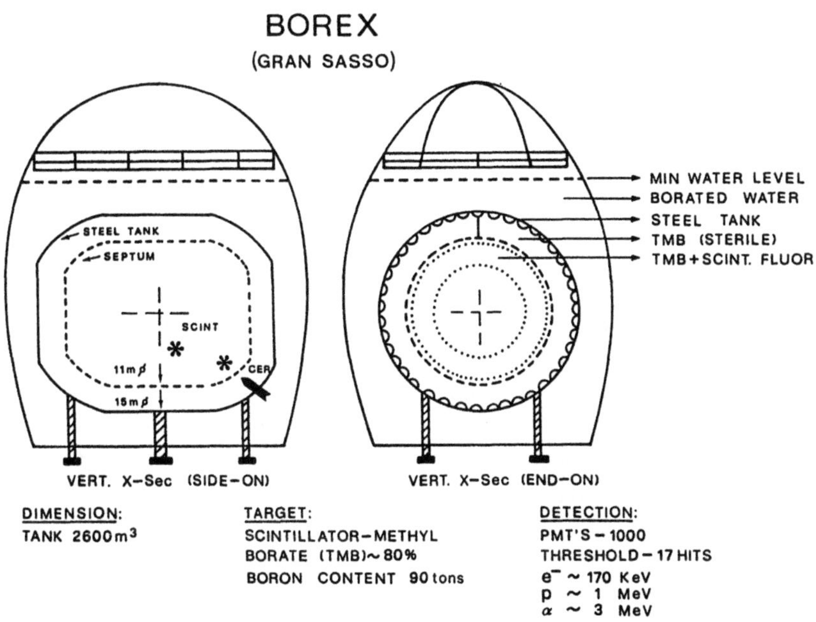

Fig. 13 - A possible scheme of a detector for BOREX.

Therefore, any deviation from the expected ratio of nuclear excitation events to c.c. interactions would indicate that not only electron neutrinos are interacting.

A possible scheme of the detector is reported in Fig. 13.

A large tank, viewed by photomultipliers like in water Cerenkov detectors, and shielded by water against neutrons, contains TMB (trymethilborate) which contains 10% of B. It has to be remarked that the light output could allow the detection of very low energy events, with a good energy resolution. The two kinds of events (nuclear excitations and c.c. interactions) can be distinguished by their energy spectrum, nuclear excitations giving sharp lines while c.c. interactions a continuous spectrum.

A more detailed design of the detector is in progress.

3. Conclusions

Important progresses have been obtained in double beta decay, either experimentally and theoretically; new experiments and new techniques are very promising to improve present values and limits on neutrinoless, two neutrino and majoron decay mode.

Data from Kamiokande tend to confirm the Davis result; in a few years from now, new data on low energy (pp) neutrinos as well as high energy neutrinos (^{8}B) will be provided if the new detectors discussed at this Conference - or at least some of them - will be constructed.

References

1. For a recent revue, see f.i.: E. Fiorini: The Present and Future of Double Beta Decay, Eighth Workshop on Grand Unification, Syracuse, April 16-18, 1987
2. M.K. Moe: this Conference
   S.R. Elliott, A.A. Hahn and M.K. Moe: Direct Evidence for Two Neutrino Double Beta Decay in ^{82}Se UCI #87-34 preprint
3. J.L. Vuillemier: This Conference
4. F.T. Avignone: ib.
5. A.Yu. Smolnikov: ib.
   V. Novikov: ib.
6. A. Alessandrello, E. Bellotti, C. Cattadori, D. Camin, O. Cremonesi, E. Fiorini, C. Liguori, A. Pullia, L. Rossi, S. Ragazzi, P.P. Sverzellati and L. Zanotti: Nucl. Instr. and Methods B17, 411 (1986)
7. Proceedings of a Workshop on Low Temperature Detectors for Neutrinos and Dark Matter, Ringberg Castle, Tegernsee, May 12-13, 1987 ed. K. Pretzl, N. Schmitz and L. Stodolsky, Springer Verlag Berlin, Heidelberg, New York, London, Paris, Tokyo
8. E. Fiorini and T.O. Niinikoski: Nucl. Instr. and Methods 224, 83 (1984)
9. D.Mc Cammon, S.H. Moseley, J.C. Mather and R.F. Mushotzky: J. Apll. Phys. 56, 1263 (1984)
   D.Mc Cammon, M. Juda, J. Zhang, R.L. Kelley, S.H. Moseley and A.E. Szymkowiak: IEEE Trans. on Nucl. Sci. 33, 236 (1986)

10. N. Coron, G. Dambier, G.J. Focker, P.G. Hansen, G. Jegoudez, B. Jonson, J. Leblanc, J.P. Moalic, H.L. Ravn, H.H. Stroke and O. Tesdard: Nature 314, 928 (1985)
11. H.H. Stroke, G. Artzner, N. Coron, G. Dambiez: IEEE Trans. on Nucl. Sci. 33, 759 (1986)
12. N. Coron, G. Artzner, H.H. Stroke, J. Rich, H. Ravn, P.G. Hansen, B. Jonson, R. Rocchia, G. Dambier, G. Jegoudez, J. Leblanc, J.P. Lepeltier and O. Testard: Thermal Spectroscopy with Cooled Composite Bolometers and Measurement of New Phenomena to be published in Proceedings of VIIth Moriond Workshop Searches for New and Exotic Phenomena - Jan 24-31, 1987
13. J. C. Overlay, H.F. Lefevre, J.G. Nolt, J.V. Radostitz, S. Predleo and P.A.R. Ade: Nucl. Instr. and Methods B10/11, 928 (1985)
14. T.O. Niinikoski, A. Rijllart, A. Alessandrello, E. Fiorini and A. Giuliani: Europhys. Lett. 1, 499 (1986)
A. Alessandrello, D.V. Camin, E. Fiorini and G. Pessina: Thermal Detectors of Particles Using Bolometric Devices to be publ. in Nucl. Instr. and Methods
15. The Kamiokande II Collaboration: Search for ^{8}B Solar Neutrinos at Kamiokande II UT-ICEPP-87-04; UPR-0144E preprint
16. W. Hampel: this Conference
17. D.O. Sinclair: id
18. CERN, Harward, Milano, Padova, Roma, Tokyo, Winsconsin coll.: Searching for New Underground Phenomena with High Resolution Visual Techniques and Magnetic Analysis (ICARUS). A proposal for the Gran Sasso Laboratory INFN/AE-85/7 (1985)
19. R.S. Raghavan, S. Pakvasa and B.A. Brown: Phys. Rev. Lett. 57, 1801 (1986)
R.S. Raghavan: private communication

List of Participants

Alexejev, E.	Baksan Neutrino Lab., Acad. of Sci., USSR
Avignone, F.T.	Dept. of Phys., Univ. of South Carolina,Columbia,USA
Backe, H.	Inst. für Kernphysik, Univ. Mainz, Germany
Barnes, C. A.	California Inst. of Techn.,Pasadena, USA
Bauer, M.	Inst. für Hochenergiephysik, Heidelberg, Germany
Bean, M.	Physik. Inst., Univ. Heidelberg,Germany
Becker, H.-W.	Inst. für Kernphysik, Univ. Münster, Germany
Beer, H.	Inst. für Kernphysik II, KfK Karlsruhe, Germany
Bellotti, E.	Dipart. di Fisica del Universita, Milano, Italy
Bender, E.	MPI für Kernphysik, Heidelberg, Germany
Bergkvist, K.E.	Research Inst. of Physics, Stockholm, Sweden
Bernreuther, W.	Inst. für Theor. Phys., Univ. Heidelberg, Germany
Bertolani, O.	Inst. für Theor. Phys., Univ. Heidelberg, Germany
Blümer, H.	Institut für Physik, Univ. Mainz, Germany
Blum, J.	MPI für Kernphysik, Heidelberg, Germany
Boer, F.W.N. de	NIKHEF, Amsterdam, Netherlands
Börner, G.	MPI für Astrophysik, Garching, Germany
Bonn, J.	Inst. für Physik, Univ. Mainz, Germany
Borzumati, F.	SISSA, (ISAS), Trieste, Italy
Bottino, A.	Dipart. Fisica Teorica, Univ. di Torino, Italy
Brüll, A.	Univ. Freiburg, Germany
Bruß, D.	Boxgraben 144, Aachen, Germany
Carioli, M.	MPI für Kernphysik, Germany
Carnoy, A.-S.	Inst. de Phys. Nucl., Univ. Cath. de Louvain, Belgium
Cavaignac, J.-F.	Inst. des Sciences Nucl., Grenoble, France
Cousinou, M. C.	C.P.P.M., Fac. d. Sciences, Marseille, France
Csikor, F.	Dept. of Atomic Phys., Eoetvoes Univ., Budapest, Hungary
Decker, R. N.	Inst. für Theor. Kernphysik, Univ. Karlsruhe, Germany
Depommier, P.	Nucl. Phys. Lab., Univ. of Montreal, Quebec,Canada
Deutsch, J.	Inst. de Phys. Nucl., Univ. Cath. de Louvain, Belgium
Dincklage, R. von	IMP der ETHZ, c/o SIN, Villigen, Switzerland
Döhner, J.	Phys. Inst., Univ. Heidelberg, Germany
Doll, P.	Inst. für Kernphysik I, KfK Karlsruhe, Germany
Dostrovsky, I.	Weizmann Institute of Science, Rehovot, Israel
Dragon, N.	Inst. für Theor. Physik, Univ. Hannover, Germany
Engelfried, J.	Physik. Inst., Univ. Heidelberg, Germany
Engler, J.	IK1, KfK , Karlsruhe, Germany
Faessler, A.	Inst. für Theor. Physik, Univ. Tübingen, Germany
Faissner, H.	III. Physik. Inst., TH Aachen, Germany
Feilitzsch, F. von	Physik Dept., TU München, Garching, Germany
Fetscher, W.	Inst. f.Mittelenergiephys.,ETH Zürich,Villigen, Switzerland
Finlay, R. W.	Dept. of Physics, Ohio University, Athens, Ohio,USA
Freedman, S.	Argonne National Laboratory, Argonne, USA
Galster, S.	KFK-NEP, Karlsruhe, Germany
Geiges, R.	Inst. für Hochenergiephysik, Univ. Heidelberg, Germany
Gelmini, G.	ICTP, Trieste, Italy
Geweniger, Ch.	Inst. für Hochenergiephysik, Univ. Heidelberg, Germany
Giorginis, G.	Inst. für Kernphysik, IK1/KfK, Karlsruhe, Germany
Hampel, W.	MPI für Kernphysik, Heidelberg, Germany
Hartwig, D.	GSI, Darmstadt, Germany

Hawelka, P.	III. Physik. Institut, TH Aachen, Germany
Hein, K.	Inst. für Theor. Hochenergiephysik, Heidelberg, Germany
Hermanni, A.	Inst. für Physik, Univ. Mainz, Germany
Heusser, G.	MPI für Kernphysik, Heidelberg, Germany
Heusser, E.	MPI für Kernphysik, Heidelberg, Germany
Hillebrandt, W.	MPI für Physik und Astrophysik, Garching, Germany
Hubert, Ph.	Centre d'Etudes Nucl., Bordeaux-Gradignan, France
Irvine, J.M.	The Univ. of Manchester, England
Janka, H. T.	MPI für Physik und Astrophysik, Garching, Germany
Kabuß, E.	MPI für Kernphysik, Heidelberg, Germany
Kalbitzer, S.	MPI für Kernphysik, Heidelberg, Germany
Kerret, H. de	Lab. de Phys.,College de France, Paris, France
Kirsten, T.	MPI für Kernphysik, Heidelberg, Germany
Klages, H. O.	KfK Karlsruhe IK1 , Karlsruhe, Germany
Klapdor, H. V.	MPI für Kernphysik, Heidelberg, Germany
Kleinknecht, K.	Inst. für Physik, Univ. Mainz, Germany
Klemt, E.	Physik. Inst., Univ. Heidelberg, Germany
Knapp, J.	KfK IK1, Karlsruhe, Germany
Knoepfle, K.-T.	MPI für Kernphysik, Heidelberg, Germany
Koang, D.-H.	I.S.N.'53, Grenoble Cedex, France
Königsmann, K.	DESY F31, Hamburg, Germany
Koepke, R.	BMFT, Bonn, Germany
Körner, J.	Inst. für Physik, Univ. Mainz, Germany
Kretschmer, W.	Phys. Inst., Univ. Erlangen-Nürnberg, Germany
Kündig, W.	Physik-Institut, Univ. Zürich, Switzerland
Kuhlmann, E.	Inst. für Exp.-Physik I, Ruhr-Univ. Bochum, Germany
Kurihara, Y.	MPI für Kernphysik, Heidelberg, Germany
Langacker, P.	Dept. of Phys., Univ. of Pennsylvania, Philadelphia,USA
Ljungfelt, S.	Phys. Inst., Univ. Heidelberg, Germany
Löw, U.	Inst. für Theor. Physik, Univ. Heidelberg, Germany
Lubimow, V.	Inst. of Theor. and Exp. Physics, Moscow, USSR
Luck, W.	MPI für Kernphysik, Heidelberg, Germany
Malvetti, M.	Univ. Karlsruhe, Germany
Maschuw, R.	IKI, KFZ Karlsruhe, Germany
Mennrath, P.	Centre d'Etudes Nucl., Bordeaux-Gradignan,France
Metsch, B.	Inst. für Theor. Physik, Bonn, Germany
Metzinger, J.	MPI für Kernphysik, Heidelberg, Germany
Meyer, H.	Univ. Wuppertal, Germany
Missimer, J.	SIN, Villigen, Switzerland
Mizuno, Y.	EP-Division Cern, Genf, Switzerland
Moe, M. K.	Phys. Dept., Univ. of California, Irvine, USA
Mohapatra, R. N.	Dept. of Phys. and Astrophys., Univ. of Maryland, USA
Morales, A.	Dept. de Fisica Nucl.,Univ. de Zaragoza, Spain
Morales, J.	Dept. de Fisica Nucl.,Univ. de Zaragoza, Spain
Müller, B.	Inst. für Theor. Physik, Univ. Frankfurt, Germany
Müller, B.	MPI für Kernphysik, Heidelberg, Germany
Muto, K.	Dept. of Physics, Tokyo Inst. of Techn., Tokyo, Japan
Nachtmann, O.	Inst. für Theor. Physik, Univ. Heidelberg, Germany
Naviliat C., O.	Inst. de Phys., Univ. Cath. de Louvain, Begium
Nötzold, D.	MPI für Physik und Astrophysik, München, Germany
Novikov, V.	Baksan Neutrino Lab., Acad. of Sci., USSR
Oberauer, L.	Phys. Dept. E15, TU München, Garching, Germany
Osipowicz, A.	Physik. Inst., Univ. Mainz, Germany
Otter, G.	III. Physik. Inst. B, RWTH Aachen, Germany
Paul, S.	CERN, Geneva, Switzerland
Pauli, H.C.	MPI für Kernphysik, Heidelberg, Germany
Picard, A.	Inst. für Physik, Univ. Mainz, Germany
Piepke, A.	MPI für Kernphysik, Heidelberg, Germany
Pirner, H. J.	Inst. für Theor. Physik, Univ. Heidelberg, Germany
Polzine, K.	Kurfürstenanlage 61, Heidelberg, Germany
Povh, B.	MPI für Kernphysik, Heidelberg, Germany
Prieels, R.	Inst. de Phys. Nucl., Univ. Cath. de Louvain, Belgium
Ragazzi, S.	INFN, Univ. Milano, Italy
Rein, D.	III. Physik. Inst., RWTH Aachen, Germany
Renk, B.	Univ. Mainz, Germany

Rieseberg, H.	Phys. Inst., Univ. Heidelberg, Germany
Rith, K.	MPI für Kernphysik Heidelberg, Germany
Roehn, St.	Inst. für Physik, Univ. Mainz, Germany
Roos, M.	Dept. of High Energy Phys., Univ.of Helsinki, Finland
Rowley, K.	Dept. of Phys., Brookhaven Nat. Lab., Upton, USA
Saavedra, O.	Istituto di Cosmo-Geofisica del CNR, Torino, Italy
Sauerland, Th.	Inst. für Experimentalphysik I, Univ. Bochum, Germany
Schaden, M.	Inst. für Theor. Physik I, Univ. Regensburg, Germany
Schael, St.	Inst. für Hochenergiephysik, Univ. Heidelberg, Germany
Schmidt, M.G.	Inst. für Theor. Phys., Univ. Heidelberg, Germany
Schmidt-Rohr, U.	MPI für Kernphysik, Heidelberg, Germany
Schneider, J.	MPI für Kernphysik, Heidelberg, Germany
Scholtz, Ch.	EP-Division, C E R N, Geneva, Switzerland
Schubert, K.R.	Inst. für Hochenergiephysik, Univ. Heidelberg, Germany
Schütte, D.	Inst. für Theor. Kernphysik, Bonn, Germany
Shibata, T.-A.	CERN, EP-Division, Genf, Switzerland
Sinclair, D.	Dept. of Theor. Physics, Univ. of Oxford, England
Smirnov, A.Yu.	Inst. of Nucl. Res., Acad. of Sci., Moscow, USSR
Smolnikov, A.A.	Inst. of Nucl. Res., Acad. of Sci., Moscow, USSR
Soloshenko, V.	Inst. of Theor. and Experim. Phys., Moscow, USSR
Spitzer, J.	Phys. Inst., Univ. Heidelberg, Germany
Stamopoulos, C.	MPI für Kernphysik, Heidelberg, Germany
Steeg, F.	Inst. für Physik, Univ. Mainz, Germany
Steyaert, J.	Inst. de Phys. Nucl., Univ. Cath. de Louvain, Belgium
Suhonen, J.	Inst. für Theor. Physik, Tübingen, Germany
Taiget, Th.	Inst. für Theor. Physik, Tübingen, Germany
Tasiaux,	Inst. de Phys. Nucl. Univ. Cath. de Louvain, Belgium
Tittel, K.	Inst. für Hochenergiephysik, Univ. Heidelberg, Germany
Tomoda, T.	C I N, Villigen, Switzerland
Valle, J.W.F.	Dept. de Fisica Teorica, Univ. de Barcelona, Spain
Vignaud, D.	DPhPE, Saclay, Gif-Sur-Yvette, France
Vuilleumier, J. L.	Inst. de Phys., Univ. de Neuchatel, Switzerland
Weneser, J.	Dept. of Physics, Brookhaven Nat. Lab., Upton, USA
Wichers, V.	KVI Kernfysisch Versneller Inst., Groningen, Netherlands
Wiedner, Ch.A.	MPI für Kernphysik, Heidelberg, Germany
Wiedner, U.	Inst. für Experim. Kernphysik, Karlsruhe, Germany
Wink, R.	MPI für Kernphysik, Heidelberg, Germany
Winter, K.	CERN, Geneva, Switzerland
Wirbel, M.	Inst. für Physik, Univ. Dortmund T III, Germany
Wolschin, G.	Spektrum der Wissenschaft, Heidelberg, Germany
Wuthrick,	College de France Lab. de Physique, Paris, France
Zacek, V.	CERN, EP-Division, Geneve, Switzerland
Zeitnitz, B.	IEKP, KfK Karlsruhe, Germany
Zetsche, F.	MPI für Kernphysik, Heidelberg, Germany
Zirnbauer, M.	Caltech, Pasadena CA, USA
Zuber, K.	Am Fürstenweiher 14, Heidelberg, Germany
Zugan, D.	Dept. of Phys., Chin. Acad. of Sci.,Bejing, China

Index of Contributors

MIX
Papier aus verantwortungsvollen Quellen
Paper from responsible sources
FSC® C105338

If you have any concerns about our products,
you can contact us on
ProductSafety@springernature.com

In case Publisher is established outside the EU,
the EU authorized representative is:
Springer Nature Customer Service Center GmbH
Europaplatz 3, 69115 Heidelberg, Germany

Printed by Libri Plureos GmbH
in Hamburg, Germany